REGULATORY CHEMICALS OF HEALTH AND ENVIRONMENTAL CONCERN

REGULATORY CHEMICALS OF HEALTH AND ENVIRONMENTAL CONCERN

William H. Lederer, Ph.D., J.D.

VNR VAN NOSTRAND REINHOLD COMPANY
——— New York

Copyright © 1985 by Van Nostrand Reinhold Company Inc.

Library of Congress Catalog Card Number: 84-13016
ISBN: 0-442-26018-0

All rights reserved. No part of this work covered by the copyright hereon may be reproduced or used in any form or by any means—graphic, electronic, or mechanical, including photocopying, recording, taping, or information storage and retrieval systems—without permission of the publisher.

Manufactured in the United States of America

Published by Van Nostrand Reinhold Company Inc.
135 West 50th Street
New York, New York 10020

Van Nostrand Reinhold Company Limited
Molly Millars Lane
Wokingham, Berkshire RG11 2PY, England

Van Nostrand Reinhold
480 Latrobe Street
Melbourne, Victoria 3000, Australia

Macmillan of Canada
Division of Gage Publishing Limited
164 Commander Boulevard
Agincourt, Ontario M1S 3C7, Canada

15 14 13 12 11 10 9 8 7 6 5 4 3 2 1

Library of Congress Cataloging in Publication Data

Lederer, William H.
  Regulatory chemicals of health and environmental concern.

  1. Poisons. 2. Pollutants. 3. Hazardous substances.
4. Chemicals. 5. Toxiocology. 6. Environmental health.
I. Title.
RA1216.L34   1985   615.9   84-13016
ISBN 0-442-26018-0

To my wife, Carol, who makes it all possible.

In Memory of
WILLIAM HAROLD LEDERER
July 20, 1943–March 20, 1984

PREFACE

This book was written to provide information on regulations, standards, and relevant information affecting chemicals. The regulations and the listings, which are the primary data for this book, are scattered in diverse sources including the *Federal Register,* Code of Federal Regulations, voluntary standards, and consent decrees. This book brings together these diverse sources in one systematized alphabetical compilation. The book is titled "regulatory" rather than regulation since it includes chemicals that are not regulated in the strict sense of the term in that an excursion from the standard would not be a violation of the law. Some of the listings of chemicals are not from regulations per se but arise from voluntary standards, for example American Conference of Governmental Industrial Hygienists Threshold Limit Values, NIOSH Criteria Documents, NIOSH Current Intelligence Bulletins, and National Fire Protection Association hazardous chemicals. Other lists represent proposals for chemicals of importance that have not been enacted into law, such as the OSHA Candidate List of carcinogens. In addition the list of the National Cancer Institute (now the National Toxicology Program) animal carcinogens and the International Agency for Research on Cancer (IARC) human and animal carcinogens are included. In all a total of over 18 different lists have been compiled in developing the information on the chemicals found in this book.

The book is organized with the health information on each chemical first followed by environmental information. The order of data is organized as followed:

1. OSHA permissible exposure levels or standards
2. ACGIH threshold limit values
3. NIOSH Criteria Documents
4. NIOSH Current Intelligence Bulletins
5. OSHA Candidate List of Potential Occupational Carcinogens
6. Carcinogens listed by either NCI or IARC
7. FIFRA pesticide information
8. National Fire Protection Association hazardous chemicals ratings
9. TSCA 4(e)—Interagency Testing Committee recommendations
10. TSCA 8(a)—Chemical Information Rule
11. TSCA 8(d)—Unpublished Health and Safety Studies Reporting
12. TSCA Chemical Hazard Information Profiles (CHIPs) available
13. CWA 304(a)(1)—Water Quality Criteria
14. CWA 307(a)(1)—Priority Pollutants
15. CWA 311(b)(2)(A)—Hazardous Materials Discharge Reportable Quantities
16. National Drinking Water Standards
17. Clean Air Act—Hazardous Air Pollutants and Traditional Pollutants (Sections 111 and 112)
18. Resource Conservation and Recovery Act hazardous waste numbers (P and U lists), 40 CFR 261

It should be noted that a regulatory notation preceded by an asterisks (*) indicates that the regulation represents a legal requirement. Not included in this compilation are Food and Drug Administration (FDA) regulated chemicals or Department of Transporation (DOT) information. The book is complete until November 1983, including the Twelfth Interagency Testing Committee (ITC) list under TSCA 4(e).

This book is intended for toxicologists, occupational physicians, health professionals, industrial hygienists, environmental scientists, engineers, attorneys, plant managers, safety specialists, and regulatory personnel. It is hoped that it will provide a single reference source with a concise description of the regulations and recommendations for chemicals of health and environmental concern.

Every attempt was made to ensure the accuracy of the information, but errors are inevitable in compilations of this magnitude. It is recommended that this book be used as a guide, and if data are required for legal purposes, the original source, which is referenced, be consulted.

WILLIAM H. LEDERER

ACKNOWLEDGMENT

Numerous individuals contributed to this book in organizing, typing, and proofreading. I should specifically like to thank Muhammad Qadeer, Chris Rasmussen, and Georg Keleti for their excellent assistance.

GLOSSARY

ACGIH	American Conference of Governmental Industrial Hygienists.
ACGIH-TLV	*See* TLV.
Action level	The level at which certain provisions of the proposed OSHA standards must be initiated, such as periodic measurements of employee exposure, training of employees, and medical surveillance (if appropriate for the particular substance).
ANSI	American National Standards Institute.
*****	An asterisk before a regulation means it is legally required as opposed to a voluntary standard or a proposal.
BAT	Best available technology.
CAA	Clean Air Act.
Ceiling	(ACGIH definition) Threshold limit value-ceiling—the concentration that should not be exceeded even instantaneously. For some substances, for example irritant gases, only one category, the TLV-ceiling, may be relevant. For other substances, either two or three categories may be relevant, depending upon their physiologic action. It is important to observe that if any one of these three TLVs is exceeded, a potential hazard from that substance is presumed to exist. The ACGIH committee holds to the opinion that limits based on physical irritation should be considered no less binding than those based on physical impairment. There is increasing evidence that physical irritation may initiate, promote, or accelerate physical impairment through interaction with other chemical or biologic agents.
CFR	Code of Federal Regulations.
CHIP	Chemical Hazard Information Profile—documents prepared by the EPA on the hazards of specific chemicals.
CWA	Clean Water Act.
	40 CFR Part 120, Section 304(a)(1)—Water Quality Criteria: Pursuant of section 304(a)(1) of the Clean Water Act, 33 U.S.C. 1314(a)(1), EPA is required to periodically review and publish criteria for water quality accurately reflecting the latest scientific knowledge: (A) on the kind and extent of all identifiable effects on health and welfare including, but not limited to, plankton, fish, shellfish, wildlife, plant life, shorelines, beaches, esthetics, and recreation which may be expected from the presence of pollutants in any body of water, including groundwater, (B) on the concentration and dispersal of pollutants, or their byproducts, through biological, physical, and chemical processes, and (C) on the effects of pollutants on biological community diversity, productivity, and stability, including information on the factors affecting rates of eutrophication and rates of organic and inorganic sedimentation for varying types of receiving waters. Section 304(a)(1) criteria contain two essential types of

information: (1) discussions of available scientific data on the effects of pollutants on public health and welfare, aquatic life, and recreation, and (2) quantitative concentrations or qualitative assessments of the pollutants in water which will generally ensure water quality adequate to support a specified water use. Under section 304(a)(1), these criteria are based solely on data and scientific judgments on the relationship between pollutant concentrations and environmental and human health effects. Criteria values do not reflect considerations of economic or technological feasibility. Section 304(a)(1) criteria are not rules and they have no regulatory impact. Rather, these criteria present scientific data and guidance on the environmental effect of pollutants which can be useful in the derivation of regulatory requirements based on considerations of water quality impacts. Under the Clean Water Act, these regulatory requirements may include the promulgation of water quality-based effluent limitations under section 302, water quality standards under section 303, or toxic pollutant effluent standards under section 307. States are encouraged to begin to modify or, if necessary, develop new programs necessary to support the implementation of regulatory controls for toxic pollutants. As appropriate, states may incorporate criteria for toxic pollutants, based on this guidance, into their water quality standards.

Section 307(a)(1), Priority Pollutants: The 1976 consent decree between National Resources Defense Council and the U.S. Environmental Protection Agency (*Natural Resources Defense Council* v. *Train* 8 ERC 2120), identified 65 toxic pollutant classes to be given priority. Under the Federal Clean Water Act the Environmental Protection Agency was assigned the responsibility to develop water quality criteria and both BAT and categorical pretreatment standards for these pollutant classes. Subsequently, the basic terms of the consent decree were incorporated into the 1977 amendment to the Federal Clean Water Act. In partial fulfillment of this requirement, EPA (with guidance from a nationwide committee of experts) identified a list of 129 (3 have since been deleted) compounds and elements from the 65 pollutant categories. The list of elements and compounds (the "priority pollutant list") includes substances with known toxic effects on human and aquatic life and those known or suspected to be carcinogens, mutagens, or teratogens. The original list which has since been modified included 13 heavy metals, asbestos, cyanides, and 114 organic compounds. The organic compounds are further categorized into the following broad groups: volatile compounds (31); acid compounds (11); base/neutral compounds (46); pesticides (25); and dioxin (1). EPA has conducted an extensive screening and verification sampling program to determine the types and levels of "priority pollutants" present in wastewaters from those industries which are known or suspected users or manufacturers of these pollutants (i.e., the primary industries). EPA was to have developed BAT and categorical pretreatment standards for primary industries to address priority pollutants. This effort was to have been completed in December 1979, but has been severely delayed due to a variety of problems (including challenges from many industry groups). In addition to developing BAT and pretreatment standards and in order to set effluent standards that will ensure water quality sufficient to protect specified uses, EPA has developed recommended water quality criteria for the priority pollutants. These recommended criteria were published in the *Federal Register* of November 28, 1980 (Vol. 45, No. 231, pp. 79318–79379). Because there are no BAT regulations or mandatory water quality criteria for dealing with priority pollutants, this strategy has been developed to assist permit writers in issuing and reissuing NPDES permits to primary industries. For some types of industries BAT limitations may eventually be developed for certain of these priority pollutants. These BAT limits will need to be considered in addition to the water quality-related limitations that are developed using this strategy.

40 CFR Part 117, Section 311(b)(2)(A) (See also RQ*):* This regulation establishes reportable quantities for substances designated as hazardous under section 311(b)(2)(A) of the Clean Air Act and sets forth requirements for notification in the event of such discharges. This regulation should be read in conjuction with 40 CFR Part 116: Designation of Hazardous Substances, as well as "Memorandum of Understanding Between the Environmental Protection Agency and the United States Coast Guard Concerning the Assessment of Civil Penalties for Discharges of Oil and Designated Hazardous Substances Under Section 311 of the Clean Water Act"; Proposed Regulations to Delete Lime from the List of 299 Hazardous Substances; and Request for Additional Comments on Proposed Application Requirements for National Pollutant Discharge Elimination System (NPDES) Permits and on Proposed Rules for Regulating Indicator Parameters.

FIFRA	Federal Insecticide, Fungicide and Rodenticide Act (administered by the EPA).
FR	*Federal Register.*
IARC	International Agency for Research on Cancer.
ITC	Interagency Testing Committee. *See* TSCA, 40 CFR Part 712, Section 4(e).
m^3	Cubic meters of air.
m	*Meta.*
μg	Micrograms.
mg	Milligrams.
ml	Milliliters.
mppcf	Millions of particles per cubic foot.
NCI	National Cancer Institute (now part of the National Toxicology Program).
NFPA	National Fire Protection Association (hazardous chemicals).

Health:
4—A few whiffs of the gas or vapor could cause death; or the gas, vapor, or the liquid could be fatal on penetrating the fire fighters' normal full protective clothing which is designed for resistance to heat. For most chemicals having a health 4 rating, the normal full protective clothing available to the average fire department will not

provide adequate protection against skin contact with these materials. Only special protective clothing designed to protect against the specific hazard should be worn.

3—Materials extremely hazardous to health, but areas may be entered with extreme care. Full protective clothing, including self-contained breathing apparatus, rubber gloves, boots, and bands around legs, arms, and waist should be provided. No skin surface should be exposed.

2—Materials hazardous to health, but areas may be entered freely with self-contained breathing apparatus.

1—Materials only slightly hazardous to health. It may be desirable to wear self-contained breathing apparatus.

0—Materials which on exposure under fire conditions would offer no health hazard beyond that of ordinary combustible material.

Flammability:

4—Very flammable gases, very volatile flammable liquids, and materials that in the form of dusts or mists readily form explosive mixtures when dispersed in air. Shut off flow of gas or liquid and keep cooling water streams on exposed tanks or containers. Use water spray carefully in the vincinity of dusts so as not to create dust clouds.

3—Liquids that can be ignited under almost all normal temperature conditions. Water may be ineffective on these liquids because of their low flash points. Solids that form coarse dusts, solids in shredded or fibrous form that create flash fires, solids that burn rapidly, usually because they contain their own oxygen, and any material that ignites spontaneously at normal temperatures in air.

2—Liquids that must be moderately heated before ignition will occur and solids that readily give off flammable vapors. Water spray may be used to extinguish the fire because the material can be cooled to below its flash point.

1—Materials that must be preheated before ignition can occur. Water may cause frothing of liquids with this flammability rating number if it gets below the surface of the liquid and turns to steam. However, water spray gently applied to the surface will cause a frothing that will extinguish the fire. Most combustible solids have a flammability rating of 1.

0—Materials that will not burn.

Reactivity:

4—Materials that in themselves are readily capable of detonation or of explosive decomposition or explosive reaction at normal temperatures and pressures. Includes materials that are sensitive to mechanical or localized thermal shock. If a chemical with this hazard rating is in an advanced or massive fire, the area should be evacuated.

3—Materials that in themselves are capable of detonation or of explosive decomposition or of explosive reaction but that require a strong initiating source or that must be heated under confinement before initiation. Includes materials that are sensitive to thermal or mechanical shock at elevated temperatures and pressures or that react explosively with water without requiring heat or confinement. Fire fighting should be done from an explosive-resistant location.

2—Materials that in themselves are normally unstable and readily undergo violent chemical change but do not detonate. Includes materials that can undergo chemical change with rapid release of energy at normal temperatures and pressures or that can undergo violent chemical change at elevated temperatures and pressures. Also includes those materials that may react violently with water or that may form potentially explosive mixtures with water. In advanced or massive fires, fire fighting should be done from a protected location.
1—Materials that in themselves are normally stable but that may become unstable at elevated temperatures and pressures or that may react with water with some release of energy but not violently. Caution must be used in approaching the fire and applying water.
0—Materials that are normally stable even under fire exposure conditions and that are not reactive with water. Normal fire fighting procedures may be used.

NIOSH National Institute for Occupational Safety and Health.
Criteria Documents: The Occupational Safety and Health Act of 1970 emphasizes the need for standards to protect the health and provide for the safety of workers occupationally exposed to an ever-increasing number of potential hazards. The National Institute for Occupational Safety and Health (NIOSH) evaluates all available research data and criteria and recommmends standards for occupational exposure. The Secretary of Labor will weigh these recommendations along with other considerations, such as feasibility and means of implemenation, in promulgating regulatory standards. NIOSH will periodically review the recommended standards to ensure protection of workers and will make successive reports as new research and epidemiologic studies are completed and as sampling and analytical methods are developed. The Criteria Documents are developed in accordance with sections 6(b)(7) and 20(a)(3) of the Occupational Safety and Health Act of 1970. The recommended standards within the documents have no legal status.

no. Number.
NPDES National pollution discharge elimination system.
NTIS National Technical Information Service.
Nuisance dusts Excessive concentrations of nuisance dusts in the workroom air may seriously reduce visibility, may cause unpleasant deposits in the eyes, ears, and nasal passages (Portland Cement dust), or cause injury to the skin or mucous membranes by chemical or mechanical action per se or by the rigorous skin cleansing procedures necessary for their removal. A threshold limit of 10 mg/m^3, or 30 mppcf, of total dust < 1% quartz, or 5 mg/m^3 respirable dust is recommended for substances in these categories, for which no specific threshold limits have been assigned. This limit, for a normal workday, does not apply to brief exposures at higher concentrations. Neither does it apply to those substances which may cause physiologic impairment at lower concentrations but for which a threshold limit has not yet been adopted.

o *Ortho.*
OSHA Occupational Safety and Health Act of 1970.
p *Para.*

PEL	Permissible exposure limit.
PEL(Ceiling)	(OSHA definition) An employee's exposure to any material that has a ceiling value (e.g., boron trifluoride) shall at no time exceed the ceiling value.
PEL(TWA)	(OSHA definition) An employee's exposure to any material in any 8-hour work shift of a 40-hour workweek shall not exceed the 8-hour time-weighted average limit given for that material.
ppb	Parts per billion.
ppm	Parts per million.
RCRA	Resources Conservation and Recovery Act, 40 CFR 261.
RPAR	Rebuttable presumption against registration (of a pesticide).
RQ	The reportable quantity that may be harmful as set forth in 40 CFR 117.3, the discharge of which is a violation of section 311(b)(3) of the CWA and requires notice as set forth in 117.21. Any person in charge of a vessel or an onshore or an offshore facility shall, as soon as he has knowledge of any discharge of a designated hazardous substance from such vessel or facility in quantities equal to or exceeding in any 24-hour period the reportable quantity determined by this part, immediately notify the appropriate agency of the United States Government of this discharge. Notice shall be given in accordance with such procedures as the Secretary of Transportation has set forth in 33CFR 153.203. Theis provision applies to all discharges not specifically excluded or reserved by another section of these regulations.
sec	Secondary.
Skin	Listed substances followed by the designation "skin" refer to the potential contribution to the overall exposure by the cutaneous route including mucous membranes and eye, either by air borne, or more particularly, by direct contact with the substance. Vehicles can alter skin absorption. This attention-calling designation is intended to suggest appropriate measures for the prevention or cutaneous absorption so that the threshold limit is not invalidated.
STEL	See TLV-STEL.
Syn	Synonym.
tert (or *t*)	Tertiary.
TLV	Threshold limit value. Threshold limit values refer to airborne concentrations of substances and represent conditions under which it is believed that nearly all workers may be repeatedly exposed day after day without adverse effect. Because of wide variation in individual susceptibility, however, a small percentage of workers may experience discomfort from some substances at concentrations at or below the threshold limit; a smaller percentage may be affected more seriously by aggravation of a preexisting condition or by development of an occupational illness.
TLV-STEL	Threshold Limit Value-Short Term Exposure Limit. The concentration to which workers can be exposed continously for a short period of time without suffering from (1) irritation, (2) chronic or irreversible tissue change, or (3) narcoses of sufficient degree to increase the likelihood of accidental injury, impair self-rescue, or materially

reduce work efficiency, and provided that the daily TLV-TWA also is not exceeded. It is not a separate independent exposure limit; rather it supplements the time-weighted average (TWA) limit where there are recognized acute effects from a substance whose toxic effects are primarily of a chronic nature. STELs are recommended only where toxic effects have been reported from high short-term exposures in either humans or animals. A STEL is defined as a 15-minute time-weighted average exposure that should not be exceeded at any time during a workday even if the 8-hour time-weighted average is within the TLV. Exposures at the STEL should not be longer than 15 minutes and should not be repeated more than four times per day. There should be at least 60 minutes between successive exposures at the STEL. An averaging period other than 15 minutes may be recommended when this is warranted by observed biological effects.

TSCA Toxic Substances Control Act of 1976.

40 CFR Part 712, Section 4(e): Section 4(e) of TSCA established an Interagency Testing Committee (ITC) to make recommendations to the administrator of EPA of chemical substances to be given priority consideration in proposing test rules under section 4(a). Section 4(e) directs the committee to revise its list of recommendations at least every 6 months as it determines to be necessary. The ITC also may designate chemicals for priority consideration by EPA within 12 months of the date of designation. The total number of designated chemicals on the list may not exceed 50 at any one time. For each designation chemical EPA must, within 12 months, either initiate rulemaking or publish in the *Federal Register* its reasons for not so doing. Section 4(a) of TSCA authorizes the adminstrator of EPA to promulgate regulations requiring testing of chemical substances in order to develop data relevant to determining the risks that such chemical substances may present to health and the environment.

40 CFR Part 712, Section 8(a): The Preliminary Assessment Information rule requires chemical manufacturers (including certain producers and importers) to submit information on chemicals. The information sought from manufacturers includes data on the quantities of chemicals manufactured, the amounts directed to certain classes of uses, and the potential exposures and environmental releases associated with the manufacturer's own and his immediate customer's processing of the chemicals. The information collected under this rule is to answer a critical need for basic data that can be used in setting priorities for testing chemicals and for assessing risks associated with chemicals.

40 CFR Part 716, Section 8(d): This rule requires the submission of unpublished health and safety studies on specifically listed chemicals by chemical manufacturers, processors, and others in possession of such studies. Under this rule, EPA will acquire unpublished health and safety studies on specified chemicals from manufacturers and processors of the chemicals. The agency will use the studies to support its investigations of the risks posed by chemicals and, in

	particular, to support its decisions whether to require industry to test chemicals under section 4 of TSCA. The addition of chemicals to the rule will occur by notice of amendment in the *Federal Register*.
TWA	The time-weighted average concentration for a normal 8-hour workday and a 40-hour workweek, to which nearly all workers may be repeatedly exposed, day after day, without adverse effect.
U.S.C.	United States Code.

A

ACENAPHTHENE. [208-96-8]

CWA 304(a)(1), 45 FR 79318, November 28, 1980, Water Quality Criteria

Freshwater Aquatic Life

The available data for acenaphthene indicate that acute toxicity to freshwater aquatic life occurs at concentrations as low as 1700 µg/liter and would occur at lower concentrations among species that are more sensitive than those tested. No data are available concerning the chronic toxicity of acenaphthene to sensitive freshwater aquatic animals, but toxicity to freshwater algae occurs at concentrations as low as 520 µg/liter.

Saltwater Aquatic Life

The available data for acanaphthene indicate that acute and chronic toxicity to saltwater aquatic life occur at concentrations as low as 970 and 710 µg/liter, respectively, and would occur at lower concentrations among species that are more sensitive than those tested. Toxicity to algae occurs at concentrations as low as 500 µg/liter.

Human Health

Sufficient data are not available for acenaphthene to derive a level that would protect against the potential toxicity of this compound. Using available organoleptic data, for controlling undesirable taste and odor quality of ambient water, the estimated level is 20 µg/liter. It should be recognized that organoleptic data as a basis for establishing a water quality criterion have limitations and have no demonstrated relationship to potential adverse human health effects.

CWA 307(a)(1), Priority Pollutant

ACENAPHTHENE, 5-NITRO-. Syn: 1,2-Dihydro-5-nitroacenaphthylene. [602-87-9]

OSHA Candidate List of Potential Occupational Carcinogens, 45 FR 53672, August 12, 1980

ACENAPHTHYLENE. [208-96-8] See Acenaphthene.

ACEPHATE (ORTHENE®). [30560-19-1]

FIFRA—Registration Standard Development—Data Evaluation and Development of Regulatory Position (insecticide)

ACETALDEHYDE. [75-07-0]

*OSHA 29 CFR 1910.1000 Table Z-1
PEL(TWA): 200 ppm, 360 mg/m^3

ACGIH-TLV
TWA: 100 ppm, 180 mg/m^3
STEL: 150 ppm, 270 mg/m^3

NFPA—hazardous chemical
health—2, flammability—4, reactivity—2

TSCA-CHIP available

*CWA 311(b)(2)(A), 40 CFR 116, 117
Discharge RQ = 1000 pounds (454 kilograms)

Chemical proposed to be assessed as a Hazardous Air Pollutant by a House of Representatives bill to amend the CAA

*RCRA 40 CRF 261
EPA hazardous waste no. U001 (hazardous substance, ignitable waste)

ACETALDEHYDE, TRICHLORO-.

*RCRA 40 CFR 261
EPA hazardous waste no. U034

ACETAMIDE. Syn: Methanecarboxamide. [60-35-5]

OSHA Candidate List of Potential Occupational Carcinogens, 45 FR 53672, August 12, 1980

IARC—carcinogenic in animals

ACETAMIDE, N-(4-ETHOXYPHENYL)-. [62-44-2]

*RCRA 40 CFR 261
EPA hazardous waste no. U187

ACETAMIDE, N-9H-FLUOREN-2-YL-. [53-96-3]

*RCRA 40 CFR 261
EPA hazardous waste no. U005

ACETANILIDE, 4'-PHENYL-. Syn: 4-acetylaminobiphenyl. [4075-79-0]

OSHA Candidate List of Potential Occupational Carcinogens, 45 FR 53672, August 12, 1980

ACETIC ACID. [64-19-7]

*OSHA 29 CFR 1910.1000 Table Z-1
PEL(TWA): 10 ppm, 25 mg/m^3
ACGIH-TLV
TWA: 10 ppm, 25 mg/m^3
STEL: 15 ppm, 37 mg/m^3

NFPA—hazardous chemical
health—2, flammability—2, reactivity—1

*CWA 311(b)(2)(A), 40 CFR 116, 117
Discharge RQ = 1000 pounds (454 kilograms)

ACETIC ACID, BROMO-, ETHYL ESTER. Syn: Antol. [105-36-2]

OSHA Candidate List of Potential Occupational Carcinogens, 45 FR 53672, August 12, 1980

ACETIC ACID, ETHYL ESTER. [141-78-6]

*RCRA 40 CFR 261
EPA hazardous waste no. U112 (ignitable waste)

ACETIC ACID, LEAD SALT. [301-04-2]

*RCRA 40 CFR 261
EPA hazardous waste no. U144 (hazardous substance)

ACETIC ACID, THALLIUM SALT. [563-68-8]

*RCRA 40 CFR 261
EPA hazardous waste no. U214 (ignitable waste)

ACETIC ANHYDRIDE. [108-24-7]

*OSHA 29 CFR 1910.1000 Table Z-1
PEL(TWA): 5 ppm, 20 mg/m^3
ACGIH-TLV
TWA(ceiling): 5 ppm, 20 mg/m^3

NFPA—hazardous chemical health—2, flammability—2, reactivity—1, W water may be hazardous in fire fighting

*CWA 311(b)(2)(A), 40 CFR 116, 117
Discharge RQ = 1000 pounds (454 kilograms)

ACETOMETHOXAN. See m-Dioxan-4-ol, 2,6-dimethyl-, acetate. [828-00-2]

ACETONE. [67-64-1]

*OSHA 29 CFR 1910.1000 Table Z-1
PEL(TWA): 1000 ppm, 2400 mg/m^3
ACGIH-TLV (1982 Addition)
TWA: 750 ppm, 1780 mg/m^3
STEL: 1000 ppm, 2375 mg/m^3

See Ketones, NIOSH Criteria Document.

NFPA—hazardous chemical (flammable chemical)
health—1, flammability—3, reactivity—0

*RCRA 40 CFR 261
EPA hazardous waste no. U002 (ignitable waste)

ACETONE CYANOHYDRIN. See 2-Hydroxy-2-methylpropanenitrile. [75-86-5]

See Nitriles, NIOSH Criteria Document.

*CWA 311(b)(2)(A), 40 CFR 116, 117
Discharge RQ = 10 pounds (4.54 kilograms)

ACETONITRILE. [75-05-8]

*OSHA 29 CFR 1910.1000 Table Z-1
PEL(TWA): 40 ppm, 70 mg/m^3
ACGIH-TLV

Skin

TWA: 40 ppm, 70 mg/m^3
STEL: 60 ppm, 105 mg/m^3
See Nitriles, NIOSH Criteria Document.
NFPA—hazardous chemical
health—2, flammability—3, reactivity—0
TSCA 4(e), ITC
Fourth Report of the Interagency Testing Committee to the administrator, Environmental Protection Agency, April 1979; 44 FR 31866, June 1, 1979, responded to by EPA administrator 47 FR 58020, December 29, 1982
*TSCA 8(a), 40 CFR 712, 47 FR 26992, June 22, 1982, Chemical Information Rule
*TSCA 8(d), Unpublished Health and Safety Studies Reporting, 40 CFR 716, 47 FR 387, September 2, 1982
TSCA-CHIP available
*RCRA 40 CFR 261
EPA hazardous waste no. U003 (ignitable waste, toxic waste)

p-ACETOPHENETIDE, 3'-NITRO-.

Syn: 3-Nitro-p-acetophenetide.
[1777-84-0]
OSHA Candidate List of Potential Occupational Carcinogens, 45 FR 53672, August 12, 1980
NCI—may be carcinogenic in animals

p-ACETOPHENETIDIDE, 3'-AMINO-.

Syn: 3-Amino-4-ethoxyacetanilide.
[17026-81-2]
OSHA Candidate List of Potential Occupational Carcinogens, 45 FR 53672, August 12, 1980

ACETOPHENONE. [98-86-2]

NFPA—hazardous chemical (flammable chemical)
health—1, flammability—2, reactivity—0
*RCRA 40 CFR 261
EPA hazardous waste no. U004

4-ACETYLAMINOBIPHENYL. See

Acetanilide, 4'-phenyl-. [4075-79-0]

2-ACETYLAMINOFLUORENE.

[53-96-3]
*OSHA Standard 29 CFR 1910.1014
Cancer-Suspect Agent
Shall not apply to solid or liquid mixtures containing less than 1.0% by weight or volume of 2-acetylaminofluorene.
OSHA Candidate List of Potential Occupational Carcinogens (EPA Carcinogen Assessment Group List), 45 FR 53672, August 12, 1980
*RCRA 40 CFR 261
EPA hazardous waste no. U005

ACETYL BROMIDE. [506-96-7]

*CWA 311(b)(2)(A), 40 CFR 116, 117
Discharge RQ = 5000 pounds (2270 kilograms)

ACETYL CHLORIDE. [75-36-5]

NFPA—hazardous chemical
health—3, flammability—3, reactivity—2,
W—water may be hazardous in fire fighting
*CWA 311(b)(2)(A), 40 CFR 116, 117
Discharge RQ = 5000 pounds (2270 kilograms)
*RCRA 40 CFR 261
EPA hazardous waste no. U006 (hazardous substance, corrosive waste, reactive waste, toxic waste)

ACETYLENE. [74-86-2]

ACGIH—TLV
TWA: Asphyxiant, gas or vapors, when present in high concentrations in air, act primarily as simple asphyxiant without other significant physiologic effects.

NIOSH Criteria Document (Pub. No. 76-195), NTIS Stock No. PB 267068, July 1, 1976

Acetylene is also known as ethine, ethyne, and narcylene. "Occupational exposure to acetylene" is defined as exposure to airborne acetylene at concentrations greater than the environmental limit. Exposure at lower environmental concentrations will not require adherence to all sections of the standard with the exception of Sections 3, 4(a), 5, 6, 7, and 8. See the Criteria Document for details. If exposure to other chemicals also occurs, as would be the case with various commercial grades of acetylene (which, depending on the source, may contain phosphine, arsine, hydrogen sulfide, methyl acetylene, or other contaminants), provisions of any applicable standard for these chemicals shall also be followed. Occupational exposure to airborne acetylene shall be controlled so that no employees will be exposed to acetylene at a concentration in excess of 2500 ppm (2662 mg/m^3 of air). The standard is designed to protect the health and safety of workers for up to a 10-hour work shift, 40-hour workweek, over a working lifetime.

NFPA—hazardous chemical
health—1, flammability—4, reactivity—3
When dissolved in acetone in closed cylinder:
health—1, flammability—4, reactivity—2

ACETYLENE DICHLORIDE. See 1,2-Dichloroethylene. [540-59-0]

ACETYLENE TETRABROMIDE. See 1,1,2,2-Tetrabromoethane. [79-27-6]

ACETYL PEROXIDE. See Diacetyl peroxide. [110-22-5]

ACETYLSALICYLIC ACID (ASPIRIN). [50-78-2]
AGCIH-TLV
TWA: 5 mg/m^3

ACIFLUORFEN.
FIFRA—Data Call In, 45 FR 75488, November 14, 1980 Status: letter being drafted

ACRIDINE. See Coal tar pitch volatiles. [260-94-6]

ACROLEIN. [107-02-8]
*OSHA 29 CFR 1910.1000 Table Z-1
PEL(TWA): 0.1 ppm, 0.25 mg/m^3
ACGIH-TLV
TWA: 0.1 ppm, 0.25 mg/m^3
STEL: 0.3 ppm, 0.8 mg/m^3
NFPA—hazardous chemical
health—3, flammability—3, reactivity—2
TSCA-CHIP available
CWA 304(a)(1), 45 FR 79318, November 28, 1980, Water Quality Criteria

Freshwater Aquatic Life

The available data for acrolein indicate that acute and chronic toxicity to freshwater aquatic life occurs at concentrations as low as 68 and 21 µg/liters, respectively, and would occur at lower concentrations among species that are more sensitive than those tested.

Saltwater Aquatic Life

The available data for acrolein indicate that acute toxicity to saltwater aquatic life occurs at concentrations as low as 55 µg/liter and would occur at lower concentrations among species that are more sensitive than those tested. No data are available concerning the chronic toxicity of acrolein to sensitive saltwater aquatic life.

Human Health

For the protection of human health from the toxic properties of acrolein ingested through water and contaminated aquatic organisms, the ambient water criterion is determined to be 320 µg/liter. For the protection of human health from the toxic properties of acrolein ingested through contaminated aquatic organisms alone, the ambient water criterion is determined to be 780 µg/liter.

*CWA 311(b)(2)(A), 40 CFR 116, 117
Discharge RQ = 1 pound (0.454 kilogram)
CAA 112
Chemical proposed to be assessed as a Hazardous Air Pollutant by a House of Representatives bill to amend the CAA
*RCRA 40 CFR 261

EPA hazardous waste no. P003 (hazardous substance)

ACROLEIN DIMER. [100-73-2]
NFPA—hazardous chemical
health—1, flammability—2, reactivity—1

ACRONYCINE. [7008-42-6]
NCI—carcinogenic in animals

ACRYLAMIDE. [79-06-1]
*OSHA 29 CFR 1910.1000 Table Z-1

Skin

PEL(TWA): 0.3 mg/m^3
ACGIH—TLV

SKIN

TWA: 0.3 mg/m^3
STEL: 0.6 mg/m^3
NIOSH CRITERIA DOCUMENT (Pub. No. 77-112), NTIS Publication No. PB 273871, October 21, 1976

Synonyms for acrylamide include propenamide, acrylic amide, and akrylamid. "Action level" is defined as a time-weighted average (TWA) concentration of one-half the environmental limit. "Occupational exposure to acrylamide," because of systemic effects and dermal irritation produced by contact of acrylamide with the skin, is defined as work in an area where acrylamide is stored, produced, processed, or otherwise used, except as an unintentional contaminant in other materials at a concentration of less than 1% by weight. If an employee is occupationally exposed to airborne concentrations of acrylamide in excess of the action level, then all sections of the recommended standard shall be complied with; if the employee is occupationally exposed at or below the action level, then all sections of the recommended standard shall be complied with except Section 8.

The employer shall control workplace concentrations of acrylamide so that no employee is exposed at a concentration greater than 0.3 mg/m^3 of air determined as a TWA concentration for up to a 10-hour work shift, 40-hour workweek. The standard is designed to protect the health and safety of employees for up to a 10-hour work shift, 40-hour workweek, over a working lifetime.

TSCA 4(e), ITC
Second Report of the TSCA Interagency Testing Committee to the administrator, Environmental Protection Agency, April 1978; 43 FR 16684, April 19, 1978, responded to by the EPA administrator 44 FR 28095, May 14, 1979, 45 FR 48510, July 18, 1980; 48 FR 724, January 6, 1983

*TSCA 8(d) Unpublished Health and Safety Studies Reporting, 40 CFR 716, 47 FR 387, September 2, 1982

*RCRA 40 CFR 261
EPA hazardous waste no. U007

ACRYLIC ACID. [79-10-7]
ACGIH-TLV
TWA: 10 ppm, 30 mg/m^3
NFPA—hazardous chemical
health—3, flammability—2, reactivity—2

TSCA-CHIP available

*RCRA 40 CFR 261
EPA hazardous waste no. U008 (ignitable waste)

ACRYLONITRILE. [107-13-1]
*OSHA Standard 29 CFR 1910.1045
Cancer Hazard

Permissible exposure limits. (1) *Inhalation.* (i) *Time-weighted average limit (TWA).* The employer shall assure that no employee is exposed to an airborne concentration of acrylonitrile in excess of 2 parts acrylonitrile per million parts of air (2 ppm) as an 8-hour time-weighted average. (ii) *Ceiling limit.* The employer shall assure that no employee is exposed to an airborne concentration of acrylonitrile in excess of 10 ppm as averaged over any 15-minute period during the workday. (2) *Dermal and eye exposure.* The employer shall assure that no employee is exposed to skin contact or eye contact with liquid AN. Does not apply to exposures that result solely from the processing, use, and handling of the following materials: (i) ABS resins, SAN resins, nitrile barrier resins, solid nitrile elastomers, and acrylic and modacrylic fibers, when these listed materials are in the form of finished polymers, and products fabricated from such finished polymers; (ii) materials made from and/or containing AN for which objective data is reasonably relied upon to demonstrate that the material is not capable of releasing AN in airborne concentrations in excess of 1 ppm as an 8-hour time-weighted average, under the expected conditions of processing, use, and handling which will cause the greatest possible release; and (iii) solid materials made from and/or containing AN which will not be heated above 170°F during handling, use, or processing.

ACGIH-TLV

Skin

TWA: 2 ppm, 4.5 mg/m^3
Human carcinogen—recognized to have carcinogenic or cocarcinogenic potential
Intended changes for 1982:
(1982 Revision or Addition)
TWA: 2 ppm, 4.5 mg/m^3
Industrial substance suspect of carcinogenic potential for humans—which is suspect of inducing cancer, based on either (1) limited epidemiologic evidence, exclusive of clinical reports of single cases, or (2) demonstration of carcinogenesis in one or more animal species by appropriate methods. Worker exposure by all routes should be carefully controlled to levels consistent with the animal and human experience data.

NIOSH Criteria Document (Pub. No. 78-116), NTIS Stock No. PB 81-225617, September 29, 1977

Synonyms for acrylonitrile include acrylon, carbacryl, cyanoethylene, fumigrain, 2-propenenitrile, VCN, ventox, and vinyl cyanide. "Occupational exposure to acrylonitrile" refers to any workplace situation in which acrylonitrile is manufactured, polymerized, used, handled, or stored. All sections of the standard shall apply where there is occupational exposure to acrylonitrile. Acrylonitrile shall be controlled in the workplace so that the concentration of airborne acrylonitrile, sampled and analyzed according to the procedures in Appendix I of the Criteria Document is not greater than 4 ppm (approximately 8.7 mg/m^3) of breathing zone air. The standard is designed to protect the health and provide for the safety of employees for up to a 10-hour work shift, 40-hour workweek over a working lifetime.

NIOSH Current Intelligence Bulletin no. 18—cancer-related bulletin

OSHA Candidate List of Potential Occupational Carcinogens (EPA Carcinogen Assessment Group List), 45 FR 53672, August 12, 1980

NFPA—hazardous chemical
health—4, flammability—3, reactivity—2

CWA 304(a)(1), 45 FR 79318, November 28, 1980, Water Quality Criteria

Freshwater Aquatic Life

The available data for acrylonitrile indicate that acute toxicity to freshwater aquatic life occurs at concentrations as low as 7550 µg/liter and would occur at lower concentrations among species that are more sensitive than those tested. No definitive data are available concerning the chronic toxicity of acrylonitrile to sensitive freshwater aquatic life, but mortality occurs at concentrations as low as 2600 µg/liter with a fish species exposed for 30 days.

Saltwater Aquatic Life

Only one saltwater species has been tested with acrylonitrile and no statement can be made concerning acute or chronic toxicity.

Human Health

For the maximum protection of human health from the potential carcinogenic effects due to exposure of acrylonitrile through ingestion of contaminated water and contaminated aquatic organisms, the ambient water concentrations should be zero based on the nonthreshold assumption for this chemical. However, zero level may not be attainable at the present time. Therefore, the levels that may result in incremental increase of cancer risk over the lifetime are estimated at 10^{-5},

10^{-6}, and 10^{-7}. The corresponding criteria are 0.58 µg/liter, 0.058µg/liter, and 0.006 µg/liter, respectively. If the above estimates are made for consumption of aquatic organisms only, excluding consumption of water, the levels are 6.5 µg/liter, 0.65 µg/liter, and 0.065 µg/liter, respectively. Other concentrations representing different risk levels may be calculated by use of the guidelines. The risk estimate range is presented for information purposes and does not represent an agency judgment on an "acceptable" risk level.

CWA 307(a)(1), Priority Pollutant

*CWA 311(b)(2)(A), 40 CFR 116, 117
Discharge RQ = 100 pounds (45.4 kilograms)

*RCRA 40 CFR 261
EPA hazardous waste no. U009 (hazardous substance)

40 CFR 162.11, FIFRA-RPAR
Current Status: Final Action Voluntary Cancellation, 43 FR 26310 (6/19/78)
Criteria Possibly Met or Exceeded Oncogenicity, Teratogenicity, and Neurotoxicity

CCA 112
Chemical proposed to be assessed as a Hazardous Air Pollutant by a House of Representatives bill to amend the CAA

ACTINOLITE. See Asbestos. [13768-00-8]

*TSCA 8(d) Unpublished Health and Safety Studies Reporting, 40 CFR 716, 47 FR 387, September 2, 1982

ACTINOMYCINS.

IARC—carcinogenic in animals

ADIPATE ESTER PLASTICIZERS.
[103-23-1, 105-97-5, 105-99-7, 106-19-4, 110-29-2, 110-32-7, 123-79-5, 141-04-8, 141-17-3, 141-28-6, 151-32-6, 627-93-0, 849-99-0, 1330-86-5, 7790-07-0, 10022-60-3, 22707-35-3, 25101-03-5, 27178-16-1]
TSCA-CHIP available

ADIPIC ACID. [124-04-9]
NFPA—hazardous chemical (flammable chemical)

health—1, flammability—1, reactivity—0
*CWA 311(b)(2)(A), 40 CFR 116, 117
Discharge RQ = 5000 pounds (2270 kilograms)

ADIPONITRILE. See Nitriles,
NIOSH Criteria document. [111-69-3]

NFPA—hazardous chemical
health—4, flammability—2, reactivity—0

AFLATOXINS.
OSHA Candidate List of Potential Occupational Carcinogens (EPA Carcinogen Assessment Group List), 45 FR 53672, August 12, 1980

IARC—carcinogenic in humans

ALACHLOR. [15972-60-8]
FIFRA—Registration Standard Development—Data Collection (herbicide)
FIFRA—Data Call In, 45 FR 75488, November 14, 1982
Status: letter issued 1/5/82

ALANINE, 3-[p-BIS(2-CHLOROETHYL)AMINO]PHENYL-,L-.
[148-82-3]
*RCRA 40 CFR 261
EPA hazardous waste no. U150

ALDICARB. [116-06-3]
FIFRA—Data Call In, 45 FR 75408, November 14, 1980
Status: letter issued 9/21/81

ALDRIN. [309-00-2]
The assigned common name for an insecticidal product containing 95% or more of 1,2,3,4,10,10-hexachloro-1,4,4a,5,8,8a-hexahydro-1,4,5,8-endoexodimethanonaphthalene
*OSHA 29 CFR 1910.1000 Table Z-1

Skin
PEL(TWA): 0.25 mg/m^3
ACGIH-TLV

Skin

TWA: 0.25 mg/m^3
STEL: 0.75 mg/m^3

OSHA Candidate List of Potential Occupational Carcinogens (EPA Carcinogen Assessment Group List), 45 FR 53672, August 12, 1980

(Aldrin/Dieldrin)
40 CFR 162.11, FIFRA-RPAR, Notice of Intent to Cancel/Suspend Issued Current Status: 39 FR 37216 (10/18/74) (cancellation of most uses)

NCI—carcinogenic in animals

CWA 304(a)(1), 45 FR 79318, November 28, 1980, Water Quality Criteria

Freshwater Aquatic Life

For freshwater aquatic life the concentration of aldrin should not exceed 3.0 µg/liter at any time. No data are available concerning the chronic toxicity of aldrin to sensitive freshwater aquatic life.

Saltwater Aquatic Life

For saltwater aquatic life the concentration of aldrin should not exceed 1.3 µg/liter at any time. No data are available concerning the chronic toxicity of aldrin to sensitive saltwater aquatic life.

Human Health

For the maximum protection of human health from the potential carcinogenic effects due to exposure of aldrin through ingestion of contaminated water and contaminated aquatic organisms, the ambient water concentration should be zero based on the nonthreshold assumption for the chemical. However, zero level may not be attainable at the present time. Therefore, the levels that may result in incremental increase of cancer risk over the lifetime are estimated at 10^{-5}, 10^{-6}, and 10^{-7}. The corresponding criteria are 0.74 ng/liter, 0.074 ng/liter, and 0.0074 ng/liter, respectively. If the above estimates are made for consumption of aquatic organisms only, excluding consumption of water, the levels are 0.79 ng/liter, 0.079 ng/liter, and 0.0079 ng/liter, respectively. Other concentrations representing different risk levels may be calculated by use of the guidelines. The risk estimate range is presented for information purposes and does not represent an agency judgment on an "acceptable" risk level.

CWA 307(a)(1), Priority Pollutant

*CWA 311(b)(2)(A), 40 CFR 116, 117
Discharge RQ = 1 pound (0.454 kilogram)

NFPA—hazardous chemical
when solution: health—3, flammability—1, reactivity—0
when dry: health—2, flammability—0, reactivity—0

ALKANES (C_5-C_8).

NIOSH Criteria Document (Pub. No. 77-151), NTIS Stock No. PB 273817, March 29, 1977

These criteria and the recommended standard apply to exposure of workers to alkanes which are aliphatic hydrocarbons with the empirical formula $C_{(n)}H_{(2n+2)}$ where $n = 5, 6, 7,$ or 8. These alkanes are hereinafter referred to as pentane, hexane, heptane, and octane, respectively. The prefix "n-" will be used to refer to the straight-chain isomeric form of an alkane, for example n-pentane. An alkane without the "n-" prefix is a mixture of isomeric forms, unless otherwise designated.

"Action level" is defined as an airborne time-weighted average (TWA) concentration of 200 mg/m^3 of air of these alkanes for up to a 10-hour work shift in a 40-hour workweek. "Occupational exposure" to alkanes is defined as exposure above the action level. Exposure at lower concentrations will not require adherence to all sections of the standard except Sections 3a, 4a, 4b, 5a, 6, and 7a. See the Criteria Document for details.

Occupational exposure to airborne C_5-C_8 alkanes shall be controlled so that no employee is exposed at concentrations greater than 350 mg/m^3 as a TWA concentration for up to a 10-hour work shift in a 40-hour workweek. This concentration is equivalent to about 120 parts of pentane per million parts of air (ppm), 100 ppm of hexane, 85 ppm of heptane, or 75 ppm of octane.

If an employee is exposed to a mixture of C_5–C_8 alkanes, total alkane exposure shall not be greater than 350 mg/m^3. In addition, no employee shall be exposed to individual C_5–C_8 alkanes or mixtures of these alkanes at ceiling concentrations greater than 1800 mg/m^3 as determined over a sampling time of 15 minutes. This concentration is equivalent to about 610 ppm pentane, 510 ppm hexane, 440 ppm heptane, or 385 ppm octane. The standard is designed to protect the health and safety of employees for up to a 10-hour work shift in a 40-hour workweek over a working lifetime.

ALKYLALUMINUMS (20% or less by weight in hydrocarbon solution).

NFPA—hazardous chemical
health—3, flammability—3, reactivity—3, W—water may be hazardous in fire fighting

ALKYL EPOXIDES (including all noncyclic hydrocarbons with one or more epoxy functional groups). *See* individual alkyl epoxides.

R_1 = H or alkyl
R_2 = H or alkyl
R_3 = H or alkyl
R_4 = H or alkyl

Groups R_1–R_4 may contain one or more epoxide functions

TSCA 4(e), ITC
Initial Report to the administrator, Environmental Protection Agency, TSCA Interagency Testing Committee, October 1, 1977; 45 FR 55026, October 12, 1977, responded to by the EPA administrator, 43 FR 50134, October 26, 1978

*TSCA 8(d) Unpublished Health and Safety Studies Reporting, 40 CFR 716, 47 FR 387, September 2, 1982

ALKYL PHTHALATES. All alkyl esters of 1,2-benzene dicarboxylic acid (orthophthalic acid). *See* individual alkyl phthalates.

[84-61-7, 84-64-0, 84-66-2, 84-69-5, 84-72-0, 84-74-2, 84-75-3, 84-76-4, 84-78-6, 85-68-7, 85-69-8, 85-70-1, 85-71-2, 89-13-4, 89-19-0, 117-81-7, 117-82-8, 117-83-9, 117-84-0, 119-06-2, 119-07-3, 131-11-3, 131-15-7, 131-17-9, 146-50-9, 3648-20-2, 5334-09-8, 25724-58-7, 26761-40-0, 27215-22-1, 27554-26-3, 28553-12-0, 61702-81-6, 61886-60-0]

R_1 = alkyl
R_2 = alkyl

*TSCA 8(d) Unpublished Health and Safety Studies Reporting, 40 CFR 716, 47 FR 387, September 2, 1982

TSCA-CHIP available

TSCA 4(e), ITC
This category consists of all high-production (e.g., 10 million pounds per year greater) alkyl esters of 1,2-benzene dicarboxylic acid (orthophthalic acid). Initial Report to the administrator, Environmental Protection Agency, TSCA Interagency Testing Committee, October 1, 1977; 42 FR 55026, October 12, 1977, responded to by the EPA administrator 43 FR 50134, October 26, 1978, EPA responded to the committee's recommendations for testing 46 FR 53775, October 30, 1981.

ALKYLTIN COMPOUNDS. *See* Dibutyltin dilaurate, dimethyltin bis(isooctylmercaptoacetate), Dibutyltin bis(isooctylmercaptoacetate), Dibutyltin bis(isooctylmaleate), Dibutyltin bis(laurylmercaptide), Monobutyltin tris(isooctylmercaptoacetate).

*TSCA 8(d) Unpublished Health and Safety Studies Reporting, 40 CFR 716, 40 FR 54624, December 3, 1982

TSCA 4(e), ITC
The general formula for compounds that comprise this category is: $R_n SnY_{4-n}$.

Where:

R represents an alkyl group containing one to eight carbon atoms covalently bonded to the tin atom

n represents the number of alkyl groups covalently bonded to the tin atom; n can have a value between 1 and 4.

Y represents a singly charged anion or anionic organic group bonded to the tin atom.

Sn is the chemical symbol for the element tin.

The alkyl groups of commercially important alkyltin compounds are methyl, ethyl, n-butyl, and n-octyl groups. These compounds are listed below.

CAS No.	Chemical Name
[25852-70-4]	Butyltin-tris(isooctylmercaptoacetate)
[54849-38-6]	Methyltin-tris(isooctylmercaptoacetate)
[1185-81-5]	Dibutyltin-bis(laurylmercaptide)
[29575-02-8]	Dibutyltin-bis(isooctylmaleate)
[26401-97-8]	Di(n-octyl)tin-S,S'-bis(isooctylmercaptoacetate)
[16091-18-2]	Di(n-octyl)tin maleate polymers
[77-58-7]	Dibutyltin dilaurate
[2781-10-4]	Dibutyltin-bis(2-ethyl hexoate)
[56-35-9]	Bis(tributyltin) oxide
[1983-10-4]	Tributyltin fluoride
[1118-46-3]	Monobutyltin trichloride
[993-16-8]	Monomethyltin trichloride
[683-18-1]	Dibutyltin dichloride
[3349-36-8]	Dibutyltin dibutoxide
[7324-75-6]	Dibutyltin-bis(laurylmaleate)
[4371-77-5]	Dibutyltin dinonylate
[1067-55-6]	Dibutyltin dimethoxide
[818-08-6]	Dibutyltin oxide
[753-73-1]	Dimethyltin dichloride
[2273-45-2]	Dimethyltin oxide
[870-08-6]	Dioctyltin oxide
[3542-36-7]	Dioctyltin-dichloride
[26636-01-1]	Dimethyltin-bis(isooctylmercaptoacetate)
[51287-84-4]	Dimethyltin-bis(dodecylmercaptide)
[4342-30-7]	Tributyltin(2-hydroxypropylmaleate)
[1461-22-9]	Tributyltin chloride
[1067-97-6]	Tributyltin hydroxide
[2155-70-6]	Tributyltin (2-methyl-2-propenoate)
[994-32-1]	Trimethyltin hydroxide
[1066-45-1]	Trimethyltin chloride
[1461-25-2]	Tetrabutyltin
[597-64-8]	Tetraethyltin
[3590-84-9]	Tetractyltin

Seventh Report of the Interagency Testing Committee to the Administrator, Environmental Protection Agency, October 1980; 45 FR 78432, November 25, 1980

Removed by the Interagency Testing Committee from the TSCA 4(e) Priority List for reconsideration, Ninth Report of the Interagency Testing Committee to the administrator, Environmental Protection Agency, October 30, 1981; 47 FR 5456, February 5, 1982

ALLYL ALCOHOL. [107-18-6]

*OSHA 29 CFR 1910.1000 Table Z-1

Skin

PEL(TWA): 2 ppm, 5 mg/m^3

ACGIH-TLV

Skin

TWA: 2ppm, 5 mg/m^3
STEL: 4ppm, 10 mg/m^3

NFPA—hazardous chemical
health—3, flammability—3, reactivity—0

*CWA 311(b)(2)(A) 40 CFR 116, 117
Discharge RQ = 100 pounds (454 kilograms)

ALLYLAMINE. [107-11-9]

NFPA—hazardous chemical
health—3, flammability—3, reactivity—1

ALLYL BROMIDE. [106-95-6]

NFPA—hazardous chemical
health—3, flammability—3, reactivity—1

ALLYL CHLORIDE. [107-05-1]

*OSHA 29 CFR 1901.1000 Table Z-1
PEL(TWA): 1ppm, 3mg/m^3

ACGIH-TLV
TWA: 1 ppm, 3 mg/m^3
STEL: 2 ppm, 6 mg/m^3

NIOSH Criteria Document (Pub. No. 76-204), NTIS Stock No. PB 267071, September 21, 1976

Allyl chloride is the common synonym for the compound 3-chloropropene, also referred to as 3-chloro-1-propene. The recommendations in this chapter apply to all places of employment where allyl chloride is manufactured, used, stored, or handled and where employees may be exposed by dermal or eye contact, inhalation, or ingestion. "Overexposure to allyl chloride vapor" is defined as known or suspected exposure above the time-weighted average (TWA) environmental level or ceiling limit. If exposure to other chemicals also occurs, for example from contamination of epichlorohydrin with allyl chloride, provisions of any applicable standards for the other chemicals also shall apply. The "action level" is defined as half the recommended TWA environmental limit. When environmental concentrations are at or below the action level, adherence to Section 8(a) and (b) is not required.

Exposure to allyl chloride vapor shall be controlled so that employees are not exposed at a concentration greater than 1.0 part per million parts of air (ppm) by volume (approximately 3.1 mg/m^3 of air) determined as a TWA concentration for up to a 10-hour workday in a 40-hour workweek, or at a ceiling concentration of 3.0 ppm (9.4 mg/m^3) for any 15-minute sampling period.

The standard is designed to protect the health and safety of employees for up to a 10-hour workday in a 40-hour workweek over a working lifetime.

NCI—may be carcinogenic in animals

NFPA—hazardous chemical
health—3, flammability—3, reactivity—1

TSCA-CHIP available

*CWA 311(b)(2)(A) 40 CFR 116, 117

Discharge RQ = 100 pounds (454 kilograms)

CAA 112
Chemical proposed to be assessed as a Hazardous Air Pollutant by a House of Representatives bill to amend the CAA

ALLYL CHLOROCARBONATE.
[2937-50-0]

NFPA—hazardous chemical
health—3, flammability—3, reactivity—1

ALLYL GLYCIDYL ETHER (AGE).
[106-92-3]

*OSHA 29 CFR 1910.1000 Table Z-1
PEL(Ceiling): 10 ppm, 45 mg/m^3

ACGIH-TLV

Skin

TWA: 5 ppm, 22 mg/m^3
STEL: 10 ppm, 44 mg/m^3

See Glycidyl ethers, NIOSH Criteria Document

TSCA 4(e), ITC
Third Report of the TSCA Interagency Testing Committee to the administrator, Environmental Protection Agency, October 1978; 43 FR 50630, October 30, 1978

ALLYL ISOTHIOCYANATE. [57-06-7]

NCI—carcinogenic in animals

ALLYL ISOVALERATE. [2835-39-4]

NCI—carcinogenic in animals

ALLYL PROPYL DISULFIDE.
[2179-59-1]

*OSHA 29 CFR 1910.1000 Table Z-1
PEL(TWA): 2 ppm, 12 mg/m^3

ACGIH-TLV
TWA: 2 ppm, 12 mg/m^3
STEL: 3 ppm, 18 mg/m^3

α-ALUMINA (PLASTER OF PARIS). [1344-28-1]

ACGIH-TLV
TWA: Nuisance particulate, 30 mppcf or 10 mg/m^3 of total dust <1% quartz, or 5 mg/m^3 respirable dust
STEL: 20 mg/m^3

ALUMINUM, ALKYLS (not otherwise classified). [7429-90-5]

ACGIH-TLV
TWA: 2 mg/m^3

ALUMINUM AND ALUMINUM COMPOUNDS. [97-93-8, 100-99-2, 139-12-8, 142-03-0, 555-31-7, 555-35-1, 555-75-9, 688-37-9, 1318-16-7, 1327-36-2, 1327-41-9, 1344-28-1, 7429-90-5, 7446-70-0, 7784-18-1, 7784-21-6, 7784-25-0, 7784-26-1, 7784-30-7, 10043-01-3, 10043-67-1, 10102-71-3, 11121-16-7, 12005-16-2, 12005-48-0, 12068-56-3, 12656-43-8, 13473-90-0, 13771-22-7, 15477-33-5, 16853-85-3, 18917-91-4, 21645-51-2]

TSCA-CHIP available

ALUMINUM CHLORIDE (ANHYDROUS). [7446-70-0]

NFPA—hazardous chemical
health—3, flammability—0, reactivity—2, W—water may be hazardous in fire fighting.

ALUMINUM (DUST or POWDER). [7429-90-5]

NFPA—hazardous chemical
health—0, flammability—1, reactivity—1

ALUMINUM, METAL and OXIDE. [7429-90-5]

ACGIH-TLV
TWA: 10 mg/m^3
STEL: 20 mg/m^3

ALUMINUM PHOSPHIDE. [20859-73-8]

FIFRA—Registration Standard—Issued October 1981 (insecticide)

ALUMINUM, PYRO POWDERS. [7429-90-5]

ACGIH-TLV
TWA: 5 mg/m^3

ALUMINUM, SOLUBLE SALTS. [7429-90-5]

ACGIH-TLV
TWA: 2 mg/m^3

ALUMINUM SULFATE. [10043-01-3]
*CWA 311(b)(2)(A), 40 CFR 116, 117
Discharge RQ = 5000 pounds (2270 kilograms)

ALUMINUM, WELDING FUMES. [7429-90-5]

ACGIH-TLV
TWA: 5 mg/m^3

AMIBEN. See Benzoic acid, 3-amino-2,5-dichloro-. [133-90-4]

2-AMINO-9, 10-ANTHRACENEDIONE. See Anthraquinone, 2-amino-. [117-79-3]

2-AMINOANTHRAQUINONE. See Anthraquinone, 2-amino-. [117-79-3]

p-AMINOAZOBENZENE. [60-09-3]
IARC—carcinogenic in animals

o-AMINOAZOTOLUENE. [97-56-3]
IARC—carcinogenic in animals

4-AMINOBIPHENYL. See 4-aminodiphenyl.

1-AMINO-3-CHLORO-6-METHYLBENZENE. See o-Toluidine, 5-chloro-. [95-79-4]

7-AMINO-2,2-DIMETHYL-2,3-DIHYDROBENZOFURAN. See Carbofuran intermediates. [68298-46-4]

TSCA 4(e), ITC
Eleventh Report of the Interagency Testing Committee to the administrator, Environmental Protection Agency, November 1982; 47 FR 54626, December 3, 1982, recommended but not designated for response within 12 months

TSCA 8(a), 40 CFR 712, 48 FR 22697, May 19, 1983, Chemical Information Rule
TSCA 8(d), Unpublished Health and Safety Studies Reporting, 40 CRF 716, 48 FR 28483, June 22, 1983

4-AMINODIPHENYL. [92-67-1]
*OSHA Standard 29 CFR 1910.1011
Cancer-Suspect Agent
Shall not apply to solid or liquid mixtures containing less than 0.1% by weight or volume of 4-aminodiphenyl

ACGIH-TLV

Skin

TWA: Human carcinogen—no exposure or contact by any route—respiratory, skin, or oral, as detected by the most sensitive methods—shall by permitted. The worker should be properly equipped to insure virtually no contact with the carcinogen.

STEL: Human carcinogen—no exposure or contact by any route—respiratory, skin, or oral as detected by the most sensitive methods—shall be permitted. The worker should be properly equipped to insure virtually no contact with the carcinogen.

OSHA Candidate List of Potential Occupational Carcinogens (EPA Carcinogen Assessment Group List), 45 FR 53762, August 12, 1980

IARC—carcinogenic in humans

2-AMINOETHANOL. See Ethanolamine. [141-43-5]

3-AMINO-4-ETHOXYACETANILIDE. See p-Acetophenetidide, 3'-amino-. [17026-81-2]

NCI—carcinogenic in animals

3-AMINO-N-ETHYLCARBAZOLE. See Carbazole, 3-amino-9-ethyl-. [132-32-1]

6-AMINO-1,1A,2,8,8A,8B,-HEXAHYDRO-8-(HYDROXYMETHYL)-8A-METHOXY-5-METHYLCARBAMATE AZIRINO (2',3':3,4) PYRROLO (1,2-A)INDOLE-4,7-DIONE(ESTER). Syn: Mitomycin C. [50-07-7]

OSHA Candidate List of Potential Occupational Carcinogens (EPA Carcinogen Assessment Group List), 45 FR 53672, August 12, 1980

IARC—carcinogenic in animals
*RCRA 40 CFR 261
EPA hazardous waste no. U010

AMINOMESITYLENE. See Aniline, 2,4,6-trimethyl- [88-05-1]

1-AMINO-2-METHYLANTHRAQUINONE. [82-28-0]
NCI—carcinogenic in animals

1-AMINO-2-METHYLBENZENE. See o-Toluidine. [95-53-4]

2-AMINO-4-NITROANISOLE. See o-Anisidine, 5-nitro-. [99-59-2]

2-AMINO-5-(5-NITRO-2-FURYL)-1,3,4-THIADIAZOLE. [712-68-5]
IARC—carcinogenic in animals

4-AMINO-2-NITROPHENOL. [119-34-6]
NCI—carcinogenic in animals

2-AMINO-5-NITROTHIAZOLE. [121-66-4]
NCI—carcinogenic in animals

2-AMINOPYRIDINE. [504-29-0]
*OSHA 29 CFR 1910.1000 Table Z-1
PEL(TWA): 0.5 ppm, 2 mg/m^3
CCGIH—TLV
TWA: 0.5 ppm, 2 mg/m^3
STEL: 2 ppm, 4 mg/m^3

4—AMINOPYRIDINE (AVITROL®). [504-24-5]
FIFRA—Registration Standard—Issued September 1980, NTIS Stock No. 540/Rs 81-003 80 (bird chemosterilants, toxicants, and repellents)

3-AMINO-1,2,4-TRIAZOLE. See Amitrol. [61-82-5]

11-AMINOUNDECANOIC ACID. [2432-99-7]
NCI—carcinogenic in animals
TSCA-CHIP available

AMITRAZ (BAAM). [33089-61-1]

FIFRA—Registration Standard Development—Data Evaluation and Development of Regulatory Position (insecticide)

40 CFR 162.11, FIFRA-RPAR Completed Current Status: PD 1 published, 42 FR 18299 (4/6/77); comment period closed 7/18/77: NTIS# PB80 212046. PD 2/3 completed and Notice of Determination published, 44 FR 2678 (1/12/79); comment period closed 2/12/79: NTIS# PB80 211436. PD 4 published, 44 FR 32736 (6/7/79); 44 FR 59938 (10/7/79) (correction): NTIS# PB80 211428.
— conditional registration approved for pears but not for apples.
— registrants will conduct an oncogenicity study and collect additional benefits data.
— tolerance for pears approved 12/6/79.

Criteria Possibly Met or Exceeded: Oncogenicity

AMITROL(E). [61-82-5]

ACGIH-TLV
TWA: Industrial substance suspect of carcinogenic potential for man—chemical substance or substance associated with industrial processes, which is suspect of inducing cancer, based on either (1) limited epidemiologic evidence, exclusive of clinical reports of single cases, or (2) demonstration of carcinogenesis in one or more animal species by appropriate methods. Worker exposure by all routes should be carefully controlled to levels consistent with the animal and human experience data.

OSHA Candidate List of Potential Occupational Carcinogens (EPA Carcinogen Assesment Group List), 45 FR 53672, August 12, 1980

IARC—carcinogenic in animals

*RCRA 40 CFR 261
EPA hazardous waste no. U011

AMMONIA. [7664-41-7]

*OSHA 29 CFR 1910.1000 Table Z-1
PEL(TWA): 50 ppm, 35 mg/m^3

ACGIH-TLV
TWA: 25 ppm, 18 mg/m^3
STEL: 35 ppm, 27 mg/m^3

NIOSH Criteria Document (Pub. No. 74-136), NTIS Stock No. PB 246699, July 15, 1974.
"Ammonia" is defined as gaseous or liquified anhydrous ammonia and aqueous solutions thereof (aqua ammonia, ammonium hydroxide). "Strong aqua ammonia" is defined as aqueous solutions containing more than 10% ammonia. "Weak aqua ammonia" is defined as solutions of 10% or less. Occupational exposure shall be controlled so that no worker is exposed to ammonia at greater than a ceiling concentration fo 50 ppm as determined by a 5-minute sampling period. The standard is designed to protect the health and safety of workers for a 40-hour workweek over a working lifetime.

*CWA 311(b)(2)(A), 40 CFR 116, 117
Discharge RQ = 100 pounds (45.4 kilograms)

NFPA—hazardous chemical
when gas: health—2, flammability—1, reactivity—0
when liquified: health—3, flammability—1, reactivity—0

AMMONIUM ACETATE. [631-61-8]

*CWA 311(b)(2)(A), 40 CFR 116, 117
Discharge RQ = 5000 pounds (2270 kilograms)

AMMONIUM BENZOATE. [1863-63-4]

*CWA 311(b)(2)(A), 40 CFR 116, 117
Discharge RQ = 5000 pounds (2270 kilograms)

AMMONIUM BICARBONATE. [1066-33-7]

*CWA 311(b)(2)(A), 40 CFR 116, 117
Discharge RQ = 5000 pounds (2270 kilograms)

AMMONIUM BICHROMATE. See

Ammonium dichromate. [7789-09-5]

AMMONIUM BIFLUORIDE.
[1341-49-7]
*CWA 311(b)(2)(A), 40 CFR 116, 117
Discharge RQ = 5000 pounds (2270 kilograms)

AMMONIUM BISULFITE. [10192-30-0]
*CWA 311(b)(2)(A), 40 CFR 116, 117
Discharge RQ = 5000 pounds (2270 kilograms)

AMMONIUM BROMIDE. [12124-97-9]
NFPA—hazardous chemical
when nonfire: health—1, flammability—0, reactivity—0
when fire: health—2, flammability—0, reactivity—0

AMMONIUM CARBAMATE.
[1111-78-0]
*CWA 311(b)(2)(A), 40 CFR 116, 117
Discharge RQ = 5000 pounds (2270 kilograms)

AMMONIUM CARBONATE.
[506-87-6]
*CWA 311(b)(2)(A), 40 CFR 116, 117
Discharge RQ = 5000 pounds (2270 kilograms)

AMMONIUM CHLORIDE. [12125-02-9]
ACGIH-TLV (fume)
TWA: 10 mg/m^3
STEL: 20 mg/m^3
NFPA—hazardous chemical
when nonfire: health—1, flammability—0, reactivity—0
when fire: health—2, flammability—0, reactivity—0
*CWA 311(b)(2)(A), 40 CFR 116, 117
Discharge RQ = 5000 pounds (2270 kilograms)

AMMONIUM CHROMATE. [7788-98-9]
See Chromium (VI), NIOSH Criteria Document.
*CWA 311(b)(2)(A), 40 CFR 116, 117
Discharge RQ = 1000 pounds (454 kilograms)

AMMONIUM CITRATE DIBASIC.
[3012-65-5]
*CWA 311(b)(2)(A), 40 CFR 116, 117
Discharge RQ = 5000 pounds (2270 kilograms)

AMMONIUM DICHROMATE.
[7789-09-5]
See Chromium (VI), NIOSH Criteria Document.
NFPA—hazardous chemical
health—1, flammability—1, reactivity—1, oxy—oxidizing chemical
*CWA 311(b)(2)(A), 40 CFR 116, 117
Discharge RQ = 1000 pounds (454 kilograms)

AMMONIUM FLUOBORATE.
[13826-83-0]
*CWA 311(b)(2)(A), 40 CFR 116, 117
Discharge RQ = 5000 pounds (2270 kilograms)

AMMONIUM FLUORIDE. [12125-01-8]
NFPA—hazardous chemical
health—3, flammability—0, reactivity—0
*CWA 311(b)(2)(A), 40 CRF 116, 117
Discharge RQ = 5000 pounds (2270 kilograms)

AMMONIUM HYDROXIDE.
[1136-21-6]
*CWA 311(b)(2)(A), 40 CFR 116, 117
Discharge RQ = 1000 pounds (454 kilograms)

AMMONIUM NITRATE. [6484-52-2]
NFPA—hazardous chemical
when nonfire: health—0, flammability—0, reactivity—3, oxy—oxidizing chemical
when fire: health—2, flammability—0, reactivity—3, oxy—oxidizing chemical

AMMONIUM OXALATE. [6009-70-7, 5972-73-6, 1458-49-2]
*CWA 311(b)(2)(A), 40 CFR 116, 117
Discharge RQ = 5000 pounds (2270 kilograms)

AMMONIUM PERCHLORATE.
[7790-98-9]

NFPA—hazardous chemical
when nonfire: health—0, flammability—0, reactivity—4, oxy—oxidizing chemical
when fire: health—2, flammability—0, reactivity—4, oxy—oxidizing chemical

AMMONIUM PERMANGANATE.
[13446-10-1]

NFPA—hazardous chemical
when nonfire: health—1, flammability—0, reactivity—3, oxy-oxidizing chemical
when fire: health—2, flammability—0, reactivity—3, oxy—oxidizing chemical

AMMONIUM SILICOFLUORIDE.
[16919-19-0]

*CWA 311(b)(2) (A), 40 CFR 116, 117
Discharge RQ = 1000 pounds (454 kilograms)

AMMONIUM SULFAMATE.
[7773-06-0]

FIFRA—Registration Standard—Issued June 1981 (herbicide)
*OSHA 29 CRF 1910.1000 TAble Z-1
PEL(TWA): 15 mg/m³
ACGIH—TLV
TWA: 10 mg/m³
STEL: 20 mg/m³
*CWA 311(b)(2) (A), 40 CFR 116, 117
Discharge RQ = 5000 pounds (2270 kilograms)

AMMONIUM SULFATE. [7783-20-2]

NFPA—hazardous chemical
when nonfire: health—0, flammability—0, reactivity—0
when fire: health—3, flammability—0, reactivity—0

AMMONIUM SULFIDE. [12135-76-1]

*CWA 311(b)(2) (A), 40 CFR 116, 117
Discharge RQ = 5000 pounds (2270 kilograms)

AMMONIUM SULFITE. [10196-04-0, 10192-30-0]

*CWA 311(b)(2) (A), 40 CFR 116, 117
Discharge RQ = 5000 pounds (2270 kilograms)

AMMONIUM TARTRATE. [3164-29-2, 14307-43-8]

*CWA 311(b)(2) (A), 40 CFR 116, 117
Discharge RQ = 5000 pounds (2270 kilograms)

AMMONIUM THIOCYANATE.
[1762-95-4]

*CWA 311(b)(2) (A), 40 CFR 116, 117
Discharge RQ = 5000 pounds (2270 kilograms)

AMMONIUM THIOSULFATE.
[7783-18-8]

*CWA 311(b)(2) (A), 40 CFR 116, 117
Discharge RQ = 5000 pounds (2270 kilograms)

AMOSITE (CUMMINGONITE-GRUNERITE). See Asbestos. [12172-73-5]

*TSCA 8(d) Unpublished Health and Safety Studies Reporting, 40 CFR 116, 47 FR 387, September 2, 1982

n-AMYL ACETATE. [628-63-7]

*OSHA 29 CFR 1910.1000 Table Z-1
PEL(TWA): 100 ppm, 525 mg/m³
ACGIH-TLV
TWA: 100 ppm, 530 mg/m³
STEL: 150 ppm, 800 mg/m³
*CWA 311(b)(2) (A), 40 CFR 116, 117
Discharge RQ = 1,000 pounds (454 kilograms)

sec-AMYL ACETATE. [626-38-0]

*OSHA 29 CFR 1910.1000 Table Z-1
PEL(TWA); 125 ppm, 650 mg/m³
ACGIH-TLV
TWA: 125 ppm, 670 mg/m³
STEL: 150 ppm, 800 mg/m³

AMYLAMINE. See Pentylamine.

AMYL MERCAPTANS. See Pentanethiols.

AMYL NITRATE. [1002-16-0]
NFPA—hazardous chemical
when nonfire: health—1, flammability—2, reactivity—0, oxy—oxidizing chemical
when fire: health—2, flammability—2, reactivity—0, oxy—oxidizing chemical

ANILINE. Syn: Benzenamine. [62-53-3, 142-04-1]
*OSHA 29 CFR 1910.1000 Table Z-1

Skin

PEL(TWA): 5 ppm, 19 mg/m^3
ACGIH-TLV

Skin

TWA: 2 ppm, 10 mg/m^3
STEL: 5 ppm, 20 mg/m^3
OSHA Candidate List of Potential Occupational Carcinogens, 45 FR 53762, August 12, 1980
NFPA—hazardous chemical
health—3, flammability—2, reactivity—0
TSCA 4(e), ITC
Fourth Report of the Interagency Testing Committee to the administrator, Environmental Protection Agency, April 1979; 44 FR 31866, June 1, 1979
*TSCA 8(d) Unpublished Health and Safety Studies Reporting, 40 CFR 716, 47 FR 387, September 2, 1982
TSCA—CHIP available
*CWA 311(b)(2) (A), 40 CFR 116, 117
Discharge RQ = 1000 pounds (454 kilograms)
*RCRA 40 CFR 261
EPA hazardous waste no. U012 (hazardous substance, ignitable waste, toxic waste)

ANILINE and CHLORO-, BROMO-, and/or NITRO- ANILINES.

CAS No.	Chemical
[142-04-1]	Aniline
[95-51-2]	o-Chloroaniline
[108-42-9]	m-Chloroaniline
[106-47-8]	p-Chloroaniline
[95-76-1]	3,4-Dichloroaniline
[634-93-5]	2,4,6-Trichloroaniline
[106-40-1]	p-Bromoaniline
[88-74-4]	o-Nitroaniline
[99-09-2]	m-Nitroaniline
[100-01-6]	p-Nitroaniline
[97-02-9]	2,4-Dinitroaniline
[121-87-9]	2-Chloro-4-nitroaniline
[89-63-4]	4-Chloro-2-nitroaniline
[99-30-9]	2,6-Dichloro-4-nitroaniline
[3531-19-9]	2-Chloro-4,6-dinitroaniline
[827-94-1]	2,6-Dibromo-4-nitroaniline
[1817-73-8]	2-Bromo-4,6-dinitroaniline
[99-29-6]	2-Bromo-6-chloro-4-nitroaniline

TSCA 4(e), ITC
This category includes aniline and aniline substituted in one or more positions with a chloro-, bromo-, or nitro- group, or any combination of one or more of these substituent groups. Excluded from the category are anilines substituted in one or more positions with a gorup other than a chloro-, bromo-, and/or nitro- group irrespective as to whether a chloro-, bromo-, and/or nitro- group also is present. This category includes, but is not limited to, those chemicals listed.
Fourth Report of the Interagency Testing Committee to the administrator, Environmental Protection Agency, April 1979; 44 FR 31866, June 1, 1979
*TSCA 8(a), 40 CFR 712, 47 FR 26992, June 22, 1982, Chemical Information Rule (for all anilines or benzenamines except 2-chloro-4,6-dinitroaniline, 2-bromo-6-chloro-4-nitroaniline, and 4-chloro-3-nitroaniline)
*TSCA 8(d) Unpublished Health and Safety Studies Reporting, 40 CFR 716, 47 FR 387, September 2, 1982

ANILINE, N,N-DIMETHYL-p NITROSO-. Syn: 4-Nitrosodimethylaniline. [138-89-6]
OSHA Candidate List of Potential Occupational Carcinogens, 45 FR 53672, August 12, 1980

ANILINE, N,N-DIMETHYL-p-(m-TOLYLAZO)-. Syn: N,N-Dimethyl-4-((3-methylphenyl)azo)benzenamine. [55-80-1]

OSHA Candiate List of Potential Occupational Carcinogens, 45 FR 53672, August 12, 1980

ANILINE HYDROCHLORIDE. Syn: Benzenamine hydrochloride [142-04-1]

OSHA Candidate List of Potential Occupational Carcinogens, 45 FR 53672, August 12, 1980

*TSCA 8(a), 40 CFR 712, 47 FR 26992, June 22, 1982, Chemical Information Rule

NCI—carcinogenic in animals

ANILINE, 4,4'-(IMIDO-CARBONYL) BIS(N,N-DIMETHYL)-. See C.I.Solvent Yellow 34. [492-80-8]

ANILINE, 4,4'-METHYLENEBIS (N,N-DIMETHYL). Syn: Michler's Base. [106-61-1]

OSHA Candidate List of Potential Occupational Carcinogens, 45 FR 53672, August 12, 1980

ANILINE, N-METHYL-N-NITROSO-. Syn: N-methyl-N-nitrosobenzenamine. [614-00-6]

OSHA Candidate List of Potential Occupational Carcinogens, 45 FR 53672, August 12, 1980

ANILINE, p-(PHENYLAZO)-. See C.I.Solvent Yellow 1. [60-09-3]

ANILINE, 4,4'-SULFONYLDI-. Syn: Dapsone.[80-08-0]

OSHA Candidate List of Potential Occupational Carcinogens, 45 FR 53672, August 12, 1980

NCI—carcinogenic in animals

ANILINE, 4,4'-THIODI-. Syn: p,p'-Diaminodiphenyl sulfide. [139-65-1]

OSHA Candidate List of Potential Occupational Carcinogens, 45 FR 53672, August 12, 1980

ANILINE, 2,4,6-TRICHLORO-. See Aniline and Chloro-, bromo-, and/or nitroanilines. [634-93-5]

OSHA Candidate List of Potential Occupational Carcinogens, 45 FR 53672, August 12, 1980

ANILINE, 2,4,6-TRIMETHYL-. Syn: Aminomesitylene. [88-05-1]

OSHA Candiate List of Potential Occupational Carcinogens, 45 FR 53672, August 12, 1980

ANISIDINE (o,p-ISOMERS). [29191-52-4]

*OSHA 29 CRF 1910.1000 Table Z-1

Skin

PEL(TWA): 0.5 mg/m^3

ACGIH-TLV

Skin

TWA: 0.1 ppm, 0.5 mg/m^3

o-ANISIDINE HYDROCHLORIDE. [134-29-2]

NCI—carcinogenic in animals

o-ANISIDINE, 5-METHYL-. Syn: p-Cresidine. [120-71-8]

OSHA Candidate List of Potential Occupational Carcinogens, 45 FR 53672, August 12, 1980

NCI—carcinogenic in animals

o-ANISIDINE, 5-NITRO-. Syn: 2-Amino-4-nitroanisole. [99-59-2]

OSHA Candidate List of Potential Occupational Carcinogens, 45 FR 53672, August 12, 1980

NCI—carcinogenic in animals

ANTHOPHYLLITE. See Asbestos. [17068-78-9]

*TSCA 8(d) Unpublished Health and Safety Studies Reporting, 40 CFR 716, 47 FR 387, September 2, 1982

ANTHRACENE. See Coal tar pitch volatiles. [120-12-7]

CWA 307(a)(1), 40 CFR 125, Priority Pollutant

ANTHRAQUINONE, 2-AMINO-. Syn: 2-Amino-9,10-anthracenedione. [117-79-3]

OSHA Candidate List of Potential Occupational Carcinogens, 45 FR 53672, August 12, 1980

NCI—carcinogenic in animals

TSCA-CHIP available

ANTHRAQUINONE, 1-AMINO-2-METHYL-. See C.I.Disperse Orange 11. [32-28-0]

ANTIMONY and compounds, as Sb. [7440-36-0]

OSHA 29 CFR 1910.1000 Table Z-1
PEL(TWA): 0.5 mg/m^3 (as Sb)
ACGIH-TLV
TWA: 0.5 mg/m^3

NIOSH Criteria Document (Pub. No. 78-216), NTIS Stock No. PB 81 226060, September 28, 1978

"Antimony" refers to elemental antimony and all antimony compounds except the gas stibine (SbH3). The "action level" is defined as one-half the appropriate recommended time-weighted average (TWA) concentration limit. "Occupational exposure to antimony" is defined as exposure to antimony at a concentration greater than one-half the recommended environmental limit. Exposures to airborne antimony concentrations equal to or less than one-half the TWA workplace environmental limit, as determined in accordance with Section 8(a), will require adherence to all sections of the recommended standard except Sections 2(b), 4(c), 8(b), and the monitoring provisions of 8(c). (See the Criteria Document for details.)

If exposure to other chemicals occurs (e.g., from contamination of antimony with arsenic or free silica), provisions of any applicable standards for the other chemicals shall also be followed.

Exposure to antimony shall be controlled so that employees are not exposed to antimony at a concentration greater than 0.5 mg/m^3 of air determined as a time-weighted average concentration limit for up to a 10-hour work shift in a 40-hour workweek. The recommended standard is designed to protect the health and provide for the safety of employees for up to a 10-hour work shift, 40-hour workweek, over a working lifetime.

TSCA 4(e), ITC
Fourth Report of the Interagency Testing Committee to the administrator, Environmental Protection Agency, April 1979; 44 FR 31866, June 1, 1979; 48 FR 716, January 6, 1983

*TSCA 8(a), 40 CFR 712, 47 FR 26992, June 22, 1982, Chemical Information Rule

*TSCA 8(d) Unpublished Health and Safety Studies Reporting, 40 CFR 716, 47 FR 387, September 2, 1982

CWA 304(a)(1), 45 FR 79318, November 28, 1980, Water Quality Criteria

Freshwater Aquatic Life

The available data for antimony indicate that acute and chronic toxicity to freshwater aquatic life occur at concentrations as low as 9000 and 1600 µg/liter, respectively, and would occur at lower concentrations among species that are more sensitive than those tested. Toxicity to algae occurs at concentrations as low as 610 µg/liter.

Saltwater Aquatic Life

No saltwater organisms have been adequately tested with antimony, and no statement can be made concerning acute or chronic toxicity.

Human Health

For the protection of human health from the toxic properties of antimony ingested through water and contaminated aquatic organisms, the ambient water criterion is determined to be 146 µg/liter.

For the protection of human health from the toxic properties of antimony ingested through contaminated aquatic organisms alone, the ambient water criterion is determined to be 45,000 µg/liter.

CWA 307(a)(1), Priority Pollutant

ANTIMONY OXIDE. [1309-64-4]

*TSCA 8(a), 40 CFR 712, 47 FR 26992, June 22, 1982, Chemical Information Rule

ANTIMONY PENTACHLORIDE.
[7647-18-9]

NFPA—hazardous chemical
health—3, flammability—0, reactivity—1

*CWA 311(b)(2)(A), 40 CFR 116, 117
Discharge RQ = 1000 pounds (454 kilograms)

ANTIMONY PENTAFLUORIDE.
[7783-70-2]

NFPA—hazardous chemical
health—3, flammability—0, reactivity—1

ANTIMONY PENTASULFIDE. See Antimony sulfide. [1345-04-6]

ANTIMONY POTASSIUM TARTRATE.
[28300-74-5]

*CWA 311(b)(2) (A), 40 CFR 116, 117
Discharge RQ = 1000 pounds (454 kilograms)

ANTIMONY RED. See Antimony pentasulfide. [1345-04-5]

ANTIMONY SULFIDE. [1345-04-6]

NFPA—hazardous chemical when nonfire: health—0, flammability—1, reactivity—1 when fire: health—3, flammability—1, reactivity—1

TSCA 4(e), ITC
Fourth Report of the Interagency Testing Committee to the administrator, Environmental Protection Agency, April 1979; 44 FR 31866, June 1, 1979, 48 FR 716, January 6, 1983

*TSCA 8(d) Unpublished Health and Safety Studies Reporting, 40 CFR 716, 47 FR 387, September 2, 1982

*TSCA 8(a), 40 CFR 712, 47 FR 26992, June 22, 1982, Chemical Information Rule

ANTIMONY TRIBROMIDE.
[7789-61-9]

*CWA 311(b)(2)(A), 40 CFR 116, 117
Discharge RQ = 1000 pounds (454 kilograms)

ANTIMONY TRICHLORIDE.
[10025-91-9]

*CWA 311(b)(2)(A), 40 CFR 116, 117
Discharge RQ = 1000 pounds (454 kilograms)

ANTIMONY TRIFLUORIDE.
[7783-56-4]

*CWA 311(b)(2)(A), 40 CFR 116, 117
Discharge RQ = 1000 pounds (454 kilograms)

ANTIMONY TRIOXIDE. [1327-33-9]

*TSCA 8(d) Unpublished Health and Safety Studies Reporting, 40 CFR 716, 47 FR 387, September 2, 1982

TSCA 4(e), ITC
Fourth Report of the Interagency Testing Committee to the administrator, Environmental Protection Agency, April 1979; 44 FR 31886, June 1, 1979; 48 FR 716, January 6, 1983

TSCA-CHIP available

*CWA 311(b)(2)(A), 40 CFR 116, 117
Discharge RQ = 5000 pounds (2270 kilograms)

ANTIMONY TRIOXIDE, handling and use, as Sb. [1327-33-9]

ACGIH-TLV
TWA 0.5 mg/m^3

ANTIMONY TRIOXIDE, production.
[1327-33-9]

ACGIH-TLV
TWA: Industrial substance suspect of carcinogenic potential for humans—which is suspect of inducing cancer, based on either (1) limited epidemiologic evidence, exclusive of clinical reports of single cases, or (2) demonstration of carcinogenesis in one or more animal species by appropriate methods. Worker exposure by all routes should be carefully controlled to levels consistent with the animal and human experience data.

ANTOL. See Acetic acid, bromo-, ethylester. [105-36-2]

ANTU. Syn: Alpha-napthylthiourea.
[86-88-4]

*OSHA 29 CFR 1910.1000 Table Z-1
PEL(TWA): 0.3 mg/m^3

ACGIH-TLV
TWA: 0.3 mg/m^3
STEL: 0.9 mg/m^3

*RCRA 40 CFR 261
EPA hazardous waste no. P072

ARAMITE. [140-57-8]

OSHA Candidate List of Potential Occupational Carcinogens (EPA Carcinogen Assessment Group List), 45 FR 53672, August 12, 1980

IARC—carcinogenic in animals

ARGON. [7440-37-1]

ACGIH-TLV
TWA: Asphyxiants, gas or vapors, when present in high concentrations in air, act primarily as simple asphyxiants without other significant physiologic effects.

AROCLOR 1016.*

CWA 307(a)(1), Priority Pollutant

AROCLOR 1221.* [11104-28-2]

CWA 307(a)(1), Priority Pollutant

AROCLOR 1232.* [11141-16-5]

CWA 307(a)(1), Priority Pollutant

AROCLOR. 1248.* [12672-29-6]

CWA 307(a)(1), Priority Pollutant

AROCLOR 1254.* [11097-69-1]

CWA 307(a)(1), Priority Pollutant

AROCLOR 1260.* [11096-82-5]

CWA 307(a)(1), Priority Pollutant
One of a series of polychlorinated polyphenols.

ARSENIC, INORGANIC. [7440-38-2]

*OSHA Standard 29 CFR 1910.1018
Cancer hazard

*One of a series of polychlorinated polyphenols.

Permissible Exposure Limit

The employer shall assure that no employee is exposed to inorganic arsenic at concentrations greater than 10 μg/m^3, averaged over any 8-hour period.

"Action level" means a concentration of inorganic arsenic of 5 μg/m^3, averaged over any 8-hour period.

Does not apply to employee exposures in agriculture or resulting from pesticide application, the treatment of wood with preservatives, or the utilization of arsenically preserved wood.

"Inorganic arsenic" means copper acetoarsenite and all inorganic compounds containing arsenic except arsine, measured as arsenic (As).

NIOSH Criteria Document revised (Pub. No. 75-149), NTIS Stock No. PB 246701, 1975

"Arsenic" is defined as elemental arsenic and all of its inorganic compounds. "Exposure to arsenic" is defined as exposure at or above 0.002 mg (2.0 μg) As/m^3. Arsine and other arsenical gases should be controlled to the same concentration as other forms of inorganic arsenic. Suitable sampling and analytical methods for arsenical gases are not yet available but are being developed.

Inorganic arsenic shall be controlled so that no worker is exposed to a concentration of arsenic in excess of 0.002 mg (2.0 μg) As/m^3 as determined by a 15-minute sampling period. The standard is designed to protect the health and safety of workers for a 40-hour week over a working lifetime. Compliance with all sections of the standard will prevent noncarcinogenic adverse effects of exposure to inorganic arsenic in the workplace air and by skin exposure, and should at the minimum materially reduce the risk of arsenic-induced cancer.

Respiratory Protection:
NIOSH Current Intelligence Bulletin No. 14

ARSENIC and soluble compounds as As.
[7440-38-2]

ACGIH-TLV
TWA: 0.2 mg/m^3

ARSENIC compounds.

IARC—carcinogenic in humans

ARSENIC ACID [7778-39-4]

OSHA Candidate List of Potential Occupational Carcinogens (EPA Carcinogen Assessment Group List), 45 FR 53672, August 12, 1980

ARSENICALS (INORGANIC).

Arsenic acid
Arsenic pentoxide
Arsenic trioxide
Sodium arsenate
Sodium pyroarsenate
Lead arsenate (basic and standard)
Calcium arsenate
sodium arsenite
copper acetoarsenite

40 CFR 162.11 FIFRA-RPAR issued
Current Status: PD 1 published, 43 FR 42867 (10/18/78); comment period closed 2/12/79.
PD 2/3 (wood uses only) completed and Notice of Determination published in 46 FR 13020 (2/19/81); comment period closed 5/20/81.
PD 4 being developed.
Criteria Possibly Met or Exceeded: Oncogenicity, Mutagenicity, and Teratogenicity
CWA 304(a)(1), 45 FR 79318, November 28, 1980, Water Quality Critieria

Freshwater Aquatic Life

For freshwater aquatic life the concentration of total recoverable trivalent inorganic arsenic should not exceed 440 µg/liter at any time. Short-term effects on embryos and larvae of aquatic vertebrate species have been shown to occur at concentrations as low as 40 µg/liter.

Saltwater Aquatic Life

The available data for total recoverable trivalent inorganic arsenic indicate that acute toxicity to saltwater aquatic life occurs at concentrations as low as 508 µg/liter and would occur at lower concentrations among species that are more sensitive than those tested. No data are available concerning the chronic toxicity of trivalent inorganic arsenic to sensitive saltwater aquatic life.

Human Health

For the maximum protection of human health from the potential carcinogenic effects due to exposure of arsenic through ingestion of contaminated water and contaminated aquatic organisms, the ambient water concentration should be zero based on the nonthreshold assumption for this chemical. However, zero level may not be attainable at the present time. Therefore, the levels that may result in incremental increase of cancer risk over the lifetime are estimated at 10^{-5}, 10^{-6}, and 10^{-7}. The corresponding criteria are 22 ng/liter, 2.2 ng/liter, and 0.22 ng/liter, respectively. If the above estimates are made for consumption of aquatic organisms only, excluding consumption of water, the levels are 175 ng/liter, 17.5 ng/liter, and 1.75 ng/liter, respectively. Other concentrations representing different risk levels may be calculated by use of the guidelines. The risk estimate range is presented for information purposes and does not represent an agency judgment on an "acceptable" risk level.

CWA 307(a)(1), Priority Pollutant

*National Interim Primary Drinking Water Regulations, 40 CFR 141; 40 FR 59565, December 24, 1975; amended by 41 FR 28402, July 9, 1976; 44 FR 68641, November 29, 1979; corrected by 45 FR 15542, March 11, 1980; 45 FR 57342, August 27, 1980; 47 FR 18998, March 3, 1982; corrected by 47 FR 10998, March 12, 1982
Maximum contaminant level—0.05 mg/liter

CAA 122, 45 FR 37886 June 5, 1980, Hazardous Air Pollutant (proposed)

ARSENIC CHLORIDE. [7784-34-1]

NFPA—hazardous chemical
health—3, flammability—0, reactivity—0
*CWA 311(b)(2)(A), 40 CFR 116, 117
Discharge RQ = 5000 pounds (2270 kilograms)

ARSENIC DISULFIDE. [1303-32-8]

*CWA 311(b)(2)(A), 40 CFR 116, 117
Discharge RQ = 5000 pounds (2270 kilograms)

ARSENIC(III) OXIDE. See Arsenic trioxide. [1327-53-3]

ARSENIC(V) OXIDE. See Arsenic pentoxide. [1303-28-2]

ARSENIC PENTOXIDE. [1303-28-2]
OSHA Candidate List of Potential Occupational Carcinogens (EPA Carcinogen Assessment Group List), 45 FR 53672, August 12, 1980
*CWA 311(b)(2)(A), 40 CFR 116, 117
Discharge RQ = 5000 pounds (2270 kilograms)
*RCRA 40 CFR 261
EPA hazardous waste no. P011 (hazardous substance)

ARSENIC TRICHLORIDE. [7784-34-1]

ARSENIC TRIOXIDE. [1327-53-3]
ACGIH-TLV (production)
TWA: Industrial substance suspect of carcinogenic potential for humans—which is suspect of inducing cancer, based on either (1) limited epidemiologic evidence, exclusive of clinical reports of single cases, or (2) demonstration of carcinogenesis in one or more animal species by appropriate methods. Worker exposure by all routes should be carefully controlled to levels consistent with the animal and human experience data.
OSHA Candidate List of Potential Occupational Carcinogens (EPA Carcinogen Assessment Group List), 45 FR 53672, August 12, 1980
40 CFR 162.11, FIFRA-RPAR
Current Status: Final Action Voluntary Cancellation, 42 FR 45944 (9/13/77)
Criteria Possibly Met or Exceeded: Oncogenicity, Mutagenicity, and Teratogenicity
*CWA 311(b)(2)(A), 40 CFR 116, 117
Discharge RQ = 5000 pounds (2270 kilograms)
*RCRA 40 CFR 261
EPA hazardous waste no. P012 (hazardous substance)

ARSENIC TRISULFIDE. [1303-33-9]
NFPA—hazardous chemical
when nonfire: health—2, flammability—1, reactivity—0
when fire: health—3, flammability—1, reactivity—0
*CWA 311(b)(2)(A), 40 CFR 116, 117
Discharge RQ = 5000 pounds (2270 kilograms)

ARSENIOUS CHLORIDE. See Arsenic chloride. [7784-34-1]

ARSENOUS CHLORIDE. See Arsenic chloride. [7784-34-1]

ARSINE. [7784-42-1]
*OSHA 29 CFR 1910.1000 Table Z-1
PEL(TWA): 0.05 ppm, 0.2 mg/m^3
ACGIH-TLV
TWA: 0.05 ppm, 0.2 mg/m^3
See Arsenic, inorganic, NIOSH Criteria Document.

Poisoning in the Workplace, Arsine (Arsenic Hydride): NIOSH Current Intelligence Bulletin No. 32

ARSINE, DIETHYL-.
*RCRA 40 CFR 261
EPA hazardous waste no. P038

ARYL PHOSPHATES. This category consists of phosphate esters of phenol or of alkyl-substituted phenols. Tri-aryl and mixed alkyl and aryl esters are included, but tri-alkyl esters are excluded.
See individual aryl phosphates.

$$O=P\begin{matrix}OR_1\\OR_2\\OR_3\end{matrix}$$

R_1 = phenyl, either unsubstituted or substituted with one or more alkyl or aryl groups

R_2 = alkyl; or phenyl, either unsubstituted or substituted with one ore more alkyl or aryl groups

R_3 = alkyl; or phenyl, either unsubstituted or substituted with one or more alkyl or aryl groups

*TSCA 8(d) Unpublished Health and Safety Studies Reporting, 40 CFR 716, 47 FR 387, September 2, 1982

TSCA 4(e), ITC
Second Report of the TSCA Interagency Testing Committee to the administrator, Environmental Protection Agency, April 1978; 43 FR 16684, April 19, 1978, responded to by the EPA administrator 44 FR 28095, May 14, 1979

ASBESTOS. [1332-21-4]
*OSHA Standard 29 CFR 1910.1001
Permissible exposure to airborne concentrations of asbestos fibers: The 8-hour time-weighted average airborne concentrations of asbestos fibers to which any employee may be exposed shall not exceed five fibers, longer than 5 μm, per cubic centimeter of air.

ACGIH-TLV
TWA: Human carcinogens—recognized to have carcinogenic or cocarcinogenic potential
STEL: Human carcinogens—recognized to have carcinogenic or cocarcinogenic potential

NIOSH Criteria Document, revised (Pub. No. 77-169), NTIS Stock No. PB 273965, December 1976.

The forms of asbestos are chrysotile, crocidolite, amosite, and anthophyllite. Available studies provide conclusive evidence that exposure to asbestos fibers causes cancer and asbestosis in humans. Excessive cancer risks have been demonstrated at all fiber concentrations studied to date. Evaluation of all available human data provides no evidence for a threshold or for a "safe" level of asbestos exposure. In view of this, the standard should be set at the lowest level detectable by available analytical techniques, an approach consistent with NIOSH's most recent recommendations for other carcinogens (i.e., arsenic and vinyl chloride). Such a standard should also prevent the development of asbestosis.

Since phase contrast microscopy is the only generally available and practical analytical technique at the present time, this level is defined as 100,000 fibers >5 μm in length/m^3 (0.1 fiber/cc), on an 8-hour TWA basis with peak concentrations not exceeding 500,000 fibers >5 μm in length/m^3 (0.5 fiber/cc) based on a 15-minute sample period. Sampling and analytical techniques should be performed as specified by NIOSH publication USPHS/NIOSH Membrane Filter Method for Evaluating Airborne Asbestos Fibers—T.R. 84 (1976).

This recommended standard of 100,000 fibers >5 μm in length/m^3 is intended to (1) protect against the noncarcinogenic effects of asbestos, (2) materially reduce the risk of asbestos-induced cancer (only a ban can assure protection against carcinogenic effects of asbestos) and (3) be measured by techniques that are valid, reproducible, and available to industry and official agencies.

This standard was not designed for the population-at-large, and any extrapolation beyond general occupational exposures is not warranted. The standard was designed only for the processing, manufacturing, and use of asbestos and asbestos-containing products as applicable under the Occupational Safety and Health act of 1970.

IARC—carcinogenic in humans

*TSCA 8(a)—Asbestos Reporting Requirements, 40 CFR 763, 47 FR 33198, July 30, 1982

*TSCA 8(d) Unpublished Health and Safety Studies Reporting, 40 CFR 716, 47 FR 387, September 2, 1982

CWA 304(a)(1), 45 FR 79318, November 28, 1980, Water Quality Criteria

Freshwater Aquatic Life

No freshwater organisms have been tested with any asbestiform mineral and no statement can be made concerning acute or chronic toxicity.

Saltwater Aquatic Life

No saltwater organisms have been tested with any asbestiform mineral and no statement can be made concerning acute or chronic toxicity.

Human Health

For the maximum protection of human health from the potential carcinogenic effects due to exposure of asbestos through ingestion of contaminated water and contaminated aquatic organisms, the ambient water concentration should be zero based on the nonthreshold assumption for this chemical. However, zero level may not be attainable at the present time. Therefore, the levels that may result in incremental increase of cancer risk over the lifetime are estimated at 10^{-5}, 10^{-6}, and 10^{-7}. The corresponding criteria are 300,000 fibers/liters, 30,000 fibers/liters and 3000 fibers/liter, respectively. Other concentrations representing different risk levels may be calculated by use of the guidelines. The risk estimate range is presented for information purposes and does not represent an agency judgment on an "acceptable" risk level.

CWA 307(a)(1), (fibrous), Priority Pollutant

*CAA 112, 40 CFR 61, Hazardous Air Pollutant

ASBESTOS, AMOSITE. [12172-73-5]

ACHIG-TLV

0.5 fiber >5 μm/cc: Human carcinogen—recognized to have carcinogenic or cocarcinogenic potential

ASBESTOS—ASBESTIFORM Varieties of: CHRYSOLTILE (SERPENTINE); CROCIDOLITE (RIEBECKITE); AMOSITE (CUMMINGTONITE-GRUNERITE); ANTHOPHYLLITE; TREMOLITE; and ACTINOLITE. See individual asbestos types. See below for ACGIH-TLV information.

*TSCA 8(d) Unpublished Health and Safety Studies Reporting, 40 CFR 176, 47 FR 387, September 2, 1982

ASBESTOS—Asbestos exposure during servicing of motor vehicle brake and clutch assemblies.

NIOSH Current Intelligence Bulletin no. 5—cancer-related bulletin

ASBESTOS, CHRYSOTILE.
[12001-29-5]

ACGIH-TLV

2 fibers > 5 μm/cc; Human carcinogen—recognized to have carcinogenic or cocarcinogenic potential

ASBESTOS, CROCIDOLITE.
[12001-28-4]

ACGIH-TLV

0.2 fiber > 5 μm/cc: Human carcinogen—recognized to have carcinogenic or cocarcinogenic potential

ASBESTOS, other forms.

ACHIG-TLV

2 fibers > 5 μm/cc: Human carcinogen—recognized to have carcinogenic or cocarcinogenic potential

ASPHALT (PETROLEUM) FUMES.
[8052-42-4]

ACGIH-TLV

TWA: 5 mg/m^3

STEL: 10 mg/m^3

NIOSH Criteria Document (Pub. No. 78-106), NTIS Stock No. PB 277333, September 27, 1977

"Asphalt fumes" are defined as the cloud of small particles created by condensation from the gaseous state after volatilization of asphalt. "Occupational exposure to asphalt fumes" is defined as exposure in the workplace at a concentration of one-half or more of the recommended occupational exposure limit. If exposure to other chemicals also occurs, as is the case when asphalt is mixed with a solvent, emulsified, or used concurrently with other materials such as tar or pitch, provisions of any applicable standard for the other chemicals shall also be followed.

Occupational exposure to asphalt fumes shall be controlled so that employees are not exposed to the airborne particulates at a concentration greater than 5 mg/m^3, determined during any 15-minute period. The standard is designed to protect the health and provide for the safety of employees for up to a 10-hour work shift, 40-hour workweek, over a working lifetime.

OSHA Candidate List of Potential Occupational Carcinogens, 45 FR 53672, August 12, 1980

NFPA—hazardous chemical (flammable chemical)
health—0, flammability—1, reactivity—0

ASPON® [3244-90-4]
FIFRA—Registration Standard—Issued September 1980, NTIS# PB81 112484 (insecticide)

ATRAZINE. [1912-24-9]
ACGIH-TLV
TWA: 10 mg/m^3
Intended changes for 1982:
TWA: 5 mg/m^3
FIFRA—Registration Standard Development—Data Collection (herbicide)
FIFRA—Data Call In, 45 FR 75488, November 14, 1980
Status: letter being drafted

AURAMINE. [2465-27-2, 492-80-8]
OSHA Candidate List of Potential Occupational Carcinogens (EPA Carcinogen Assessment Group List), 45 FR 53672, August 12, 1980
IARC—carcinogenic in humans
TSCA-CHIP available
*RCRA 40 CFR 261
EPA hazardous waste no. U014

AUROTHIOGLUCOSE. [12192-57-3]
IARC—carcinogenic in animals

5-AZACYTIDINE. [320-67-2]
NCI—carcinogenic in animals

1-AZANAPHTHALENE. See quinoline. [91-22-5]

AZASERINE. [115-02-6]
OSHA Candidate List of Potential Occupational Carcinogens
(EPA Carcinogen Assessment Group List), 45 FR 53672 August 12, 1980
IARC—carcinogenic in animals
*RCRA 40 CFR 261
EPA hazardous waste no. U015

AZIDE explosive hazard. [26628-22-8]
NIOSH Current Intelligence Bulletin No. 13

AZINPHOS-METHYL. [86-50-0]
*OSHA 29 CFR 1910.1000 Table Z-1

Skin
PEL(TWA): 0.2 mg/m^3
ACGIH-TLV

Skin
TWA: 0.2 mg/m^3
STEL: 0.6 mg/m^3
NCI—may be carcinogenic in animals

AZIRIDINE. See Ethyleneimine. [151-56-4]
IARC—carcinogenic in animals
*RCRA 40 CFR 261
EPA hazardous waste no. P054

2-(1-AZIRIDINYL) ETHANOL. [1072-52-2]
IARC—carcinogenic in animals

AZIRIDYL BENZOGUINONE. [800-24-8]
IARC—carcinogenic in animals

AZIRINO (2',3':3,4)PYRROLO (1,2-a) INDOLE-4,7-DIONE, 6-AMINO-8-[((AMINOCARBONYL)OXY)METHYL]-1,1a,2,8,8a,8b-HEXAHYDRO-8a-METHOXY-5-METHYL-2.
*RCRA 40 CFR 261
EPA hazardous waste no. U010

AZOBENZENE. Syn: Benzeneazobenzene. [103-33-3]
OSHA Candidate List of Potential Occupational Carcinogens, 45 FR 53672, August 12, 1980
IARC—carcinogenic in animals
NCI—carcinogenic in animals
TSCA-CHIP available

B

BARIUM (SOLUBLE COMPOUNDS).
[7440-39-3]
*National Interim Primary Drinking Water Regulations, 40 CFR 141; 40 FR 59565, December 24, 1975; amended by 41 FR 28402, July 9, 1976; 44 FR 68641, November 29, 1979; corrected by 45 FR 15542, March 11, 1980; 45 FR 57342, August 27, 1980; 47 FR 18998, March 3, 1982; corrected by 47 FR 10998, March 12, 1982
Maximum contaminant level—1.0 mg/liter
*OSHA 29 CFR 1910.1000 Table Z-1
PEL (TWA) : 0.5 mg/m^3
ACGIH-TLV
TWA: 0.5 mg/m^3

BARIUM CHLORATE. [13477-00-4]
NFPA—hazardous chemical
when nonfire: health—0, flammability—0, reactivity—2
when fire: health—1, flammability—0, reactivity—2, oxy—oxidizing chemical

BARIUM CYANIDE. [542-62-1]
*CWA 311 (b)(2)(A), 40 CFR 116, 117
Discharge RQ = 10 pounds (4.54 kilograms)
*RCRA 40 CFR 261
EPA hazardous waste no. P013 (hazardous substance)

BARIUM NITRATE. [10022-31-8]
NFPA—hazardous chemical
when nonfire: health—0, flammability—0, reactivity—0, oxy—oxidizing chemical
when fire: health—1, flammability—0, reactivity—0, oxy—oxidizing chemical

BARIUM PEROXIDE. [1324-29-6]
NFPA—hazardous chemical
health—1, flammability—0, reactivity—0, oxy—oxidizing chemical

BARIUM SUPEROXIDE. *See* Barium peroxide. [1304-29-6]

BENOMYL. [17084-35-2]
ACGIH-TLV
TWA: 0.8 ppm, 10mg/m^3
STEL: 1.3 ppm, 15 mg/m^3
40 CFR 162.11, FIFRA-RPAR issued
Current Status: PD 1 published, 42 FR 61788 (12/6/77).
PD $^{2}/_{3}$ completed and Notice of Determination published, 44 FR 51166 (8/30/79); comment period closed 9/28/79. Risk/benefit assessment is ongoing.
Criteria Possibly Met or Exceeded: Reduction in nontarget species (rebutted), Mutagenicity, Teratogenicity, Reproductive Effects, and Hazard to Wildlife

BENTAZON. [25057-89-0]
FIFRA—Data Call Inc 45 FR 75488, November 14, 1980
Status: letter being drafted

BENZ[j]ACEANTHRYLENE, 1,2-DIHYDRO-3-METHYL-. [56-49-5]
*RCRA 40 CFR 261
EPA hazardous waste no. U157

BENZ(a)ACRIDINE.
IARC—carcinogenic in animals

BENZ(c)ACRIDINE. [255-51-4]
OSHA Candidate List of Potential Occupational Carcinogens (EPA Carcinogen Assessment Group List), 45 FR 53672, August 12, 1980
IARC—carcinogenic in animals
*RCRA 40 CFR 261
EPA hazardous waste no. U016

3,4-BENZACRIDINE. *See* Benz(c)acridine. [255-51-4]

BENZAL CHLORIDE. [98-87-3]
TSCA-CHIP available
*RCRA 40 CFR 261
EPA hazardous waste no. U107

BENZALDEHYDE. [100-52-7]
NFPA—hazardous chemical
health—2, flammability—2, reactivity—0

BENZ(a)ANTHRACENE. [56-55-3]
OSHA Candidate List of Potential Occupational Carcinogens (EPA Carcinogen Assessment Group List), 45 FR 53672, August 12, 1980
*RCRA 49 CFR 172.101
EPA hazardous waste no. U018

1,2-BENZANTHRACENE. *See* Benz(a)anthracene. [56-55-3]

1,2-BENZANTHRACENE, 7,12-DIMETHYL-. [57-97-6]
*RCRA 40 CFR 261
EPA hazardous waste no. U094

BENZENAMINE. *See* Aniline. [62-53-3]

BENZENAMINE, 4-BROMO-. *See* Aniline and chloro-, bromo-, and/or nitro- anilines. [106-40-1]

BENZENAMINE, 2-BROMO-4,6-DINITRO-. *See* Aniline and chloro-, bromo-, and/or nitro- anilines. [1817-73-8]

BENZENAMINE, 4,4'-CARBONIMISOYLBIS(N,N-DIMETHYL—. [492-80-8]
*RCRA 49 CFR 172.101
EPA hazardous waste no. U104

BENZENAMINE, 2-CHLORO-. *See* Aniline and chloro-, bromo-, and/or nitro- anilines. [95-51-2]

BENZENAMINE, 3-CHLORO-. *See* Aniline and chloro-, bromo-, and/or nitro- anilines. [108-42-9]

BENZENAMINE, 4-CHLORO-. *See* p-Chloroaniline. [106-47-8]

BENZENAMINE, 4-CHLORO-2,6-DINITRO-. *See* Aniline and chloro-, bromo-, and/or nitro- anilines. [5388-62-5]

BENZENAMINE, N-(2-CHLOROETHYL)-N-ETHYL-. [92-49-9]
*TSCA 8(a), 40 CFR 712, 47 FR 26992, June 22, 1982, Chemical Information Rule

BENZENAMINE, 3-CHLORO-, HYDORCHLORIDE. [141-85-5]
*TSCA 8(a), 40 CFR 712, 47 FR 26992, June 22, 1982, Chemical Information Rule

BENZENAMINE, 4-CHLORO-2-METHYL-. [95-69-2]
*RCRA 40 CFR 261
EPA hazardous waste no. U049

BENZENAMINE, 2-CHLORO-4-NITRO-. *See* Aniline and chloro-, bromo-, and/or nitro- anilines. [121-87-9]

BENZENAMINE, 2-CHLORO-5-NITRO-. *See* Aniline and chloro-, bromo-, and/or nitro- anilines. [6283-25-6]

BENZENAMINE, 4-CHLORO-2-NITRO-. *See* Aniline and chloro-, bromo-, and/or nitro- anilines. [89-63-4]

BENZENAMINE, 2,6-DIBROMO-4-NITRO-. *See* Aniline and chloro-, bromo-, and/or nitro- anilines. [827-94-1]

BENZENAMINE, 2,3-DI-CHLORO-. *See* Aniline and chloro-, bromo-, and/or nitro- anilines. [608-27-5]

BENZENAMINE, 2,4-DICHLORO-. *See* Aniline and chloro-, bromo-, and/or nitro- anilines. [554-00-7]

BENZENAMINE, 2,5-DICHLORO-. *See* Aniline and chloro-, bromo-, and/or nitro- anilines. [95-82-9]

BENZENAMINE, 3,4-DICHLORO-. *See* 3,4-Dichloroaniline. [95-76-1]

BENZENAMINE, 3,5-DICHLORO-.
See Aniline and chloro-, bromo-, and/or nitro- anilines. [626-43-7]

BENZENAMINE, 2,6-DICHLORO-4-NITRO-. See Aniline and chloro-, bromo-, and/or nitro- anilines. [99-30-9]

BENZENAMINE, N,N-DIMETHYL-4-PHENYLAZO-. [60-11-7]
*RCRA 40 CFR 261
EPA hazardous waste no. U093

BENZENAMINE, 2,4-DINITRO-.
See 2,4 Dinitroaniline. [97-02-9]

BENZENAMINE, HYDROCHLORIDE.
See Aniline hydrochloride. [142-04-1]

BENZENAMINE, 4,4'-METHYLENEBIS-. [101-77-9]
*TSCA 8(a), 40 CFR 712, 47 FR 26992, June 22, 1982, Chemical Information Rule

BENZENAMINE, 4,4'-METHYLENEBIS(2-CHLORO-. [101-14-4]
*RCRA 40 CFR 261
EPA hazardous waste no. U158

BENZENAMINE, 2-METHYL-, HYDROCHLORIDE. [636-21-5]
*RCRA 40 CFR 261
EPA hazardous waste no. U222

BENZENAMINE, 2-METHYL-5-NITRO-. [99-55-8]
*RCRA 40 CFR 261
EPA hazardous waste no. U181

BENZENAMINE, 2-NITRO-. See Aniline and chloro-, bromo-, and/or nitro- anilines. [88-74-4]

BENZENAMINE, 3-NITRO-. See Aniline and chloro-, bromo-, and/or nitro- anilines. [99-09-2]

BENZENAMINE, 4-NITRO-.
See p-Nitroaniline [100-01-6]

BENZENAMINE, 2,4,6-TRIBROMO-.
See Aniline and chloro-, bromo-, and/or nitro- anilines. [147-82-0]

BENZENAMINE, 2,4,6-TRICHLORO-.
See Aniline, 2,4,6-trichloro-. [634-93-5]

BENZENE. [71-43-2]
*OSHA Standard 29 CFR 1910.1028—Cancer Hazard (Standard rescinded by Supreme Court)
Permissible Exposure Limits—(1) *Inhalation*—(i) *Time-weighted average limit (TWA).* The employer shall assure that no employee is exposed to an airborne concentration of benzene in excess of 1 part benzene per million parts of air (1 ppm) as an 8-hour time-weighted average.
(ii) *Ceiling limit.* The employer shall assure that no employee is exposed to an airborne concentration of benzene in excess of 5 ppm, as averaged over any 15-minute period.
(2) *Dermal and eye exposure limit.* The employer shall assure that no employee is exposed to eye contact with liquid benzene; or to skin contact with liquid benzene, unless the employer can establish that the skin contact is an isolated instance.
"Action level" means an airborne concentration of benzene of 0.5 ppm, averaged over an 8-hour workday.
Shall not apply to:
i. storage, transportation, distribution, dispensing, sale, or use as fuel of gasoline, motor fuels, or other fuels subsequent to discharge from bulk terminals; or
ii. storage, transportation, distribution, or sale of benzene in intact containers sealed in such a manner as to contain benzene vapors or liquid;
iii. work operations where the only exposure to benzene is from liquid mixtures containing 0.5% (0.1% after June 27, 1981) or less of benzene by volume, or the vapors released from such liquids.
*OSHA 1910.1000 Table Z-2
10 ppm   8-hour TWA
25 ppm   Acceptable ceiling concentrations

50 ppm/10 minutes   Acceptable maximum peak

ACGIH-TLV
TWA: 10 ppm, 30 mg/m^3
STEL: 25 ppm, 75 mg/m^3

ACGIH-TLV
TWA: 10 ppm, 30 mg/m^3
STEL: 25 ppm, 75 mg/m^3

Industrial substance suspect of carcinogenic potential for humans—which is suspect of inducing cancer, based on either (1) limited epidemiologic evidence, exclusive of clinical reports of single cases, or (2) demonstration of carcinogenesis in one or more animal species by appropriate methods. Worker exposure by all routes should be carefully controlled to levels consistent with the animal and human experience data.

NIOSH Criteria Document, revised, no NIOSH publication number is assigned. Available only from NIOSH. August 25, 1976

NIOSH recommends that occupational exposure be controlled so that no worker will be exposed to benzene in excess of 1 ppm (3.2 mg/m^3) in air as determined by an air sample collected at 1 liter/minute for 2 hours.

NIOSH considers the accumulated evidence from clinical as well as from epidemiologic data to be conclusive at this time that benzene is leukemogenic. Because it causes progressive, malignant disease of the blood-forming organs, NIOSH recommends that, for regulatory purposes, benzene be considered carcinogenic in humans. In view of this conclusion and since it is not possible at this time to establish an exposure level at which benzene may be regarded to be without danger, NIOSH recommends that exposure to benzene be kept as low as possible. The use of benzene as a solvent or diluent in open operations should be prohibited. Furthermore, product substitution should be a paramount consideration. Wherever benzene is identifies or its presence suspected, especially with concurrent indications of alterations in the blood or the hematopoietic system, it should be replaced with less harmful substitutes wherever feasible.

OSHA Candidate List of Potential Occupational Carcinogens (EPA Carcinogen Assessment Group List), 45 FR 53672, August 12, 1980

IARC—carcinogenic in humans

40 CFR 162.11, FIFRA-RPAR, Final Action Voluntary Cancellation
Current Status: Voluntary cancellations being processed Criteria Possibly Met or Exceeded: Oncogenicity, Mutagenicity, and Blood Disorders

NFPA—hazardous chemical
health—2, flammability—3, reactivity—0

CWA 304(a)(1), 45 FR 79318, November 28, 1980, Water Quality Criteria

Freshwater Aquatic Life

The available data for benzene indicate that acute toxicity to freshwater aquatic life occurs at concentrations as low as 5300 µg/liter and would occur at lower concentrations among species that are more sensitive than those tested. No data are available concerning the chronic toxicity of benzene to sensitive freshwater aquatic life.

Saltwater Aquatic Life

The available data for benzene indicate that acute toxicity to saltwater aquatic life occurs at concentrations as low as 5100 µg/liter and would occur at lower concentrations among species that are more sensitive than those tested. No definitive data are available concerning the chronic toxicity of benzene to sensitive saltwater aquatic life, but adverse effects occur at concentrations as low as 700 µg/liter with a fish species exposed for 168 days.

Human Health

For the maximum protection of human health from the potential carcinogenic effects due to exposure of benzene through ingestion of contaminated water and contaminated aquatic organisms, the ambient water concentrations should be zero based on the nonthreshold assumption for this chemical. However, zero level may not be attainable at the present time. Therefore, the levels that may result in incremental increase of cancer risk over the lifetime are estimated at 10^{-5}, 10^{-6}, and 10^{-7}. The corresponding criteria are 6.6 µg/liter, respectively. If the above estimates are made for consumption of aquatic organisms only, excluding consumption of water, the levels are 400 µg/liter, 40.0 µg/liter, and 4.0 µg/liter, respectively. Other concentrations representing different risk levels may be calculated by use of the guidelines. The risk estimate range is presented for information purposes and does not represent an agency judgment on an "acceptable" risk level.

CWA 307(a)(1), Priority Pollutant
*CWA 311(b)(2)(A), 40 CFR 116, 117
Discharge RQ = 1000 pounds (454 kilograms)
*CAA 112, Hazardous Air Pollutant (proposed)
*RCRA 40 CFR 261
EPA hazardous waste no. U019 (hazardous substance ignitable waste, toxic waste)

BENZENEACETIC ACID, .ALPHA.-(HYDROXYMETHYL)-, 9-METHYL-3-OXA-9-AZATRICYCLO(3.3.1.02.4) NON-7-YL ESTER. [7(S)-(1.ALPHA, 2.BETA,4.BETA.5.ALPHA,7.BETA.)]-, compd. with METHYL NITRATE (1:1). [6106-46-3]

*TSCA 8(a), 40 CFR 712, 47 FR 26992, June 22, 1982, Chemical Information Rule

BENZENEACETIC ACID, .ALPHA.-(HYDROXYMETHYL)-, -9-METHYL-3-OXA-9-AZATRICYCLO(3.3.1.02.4) NON-7-YL ESTER, HYDROBROMIDE. [7-(S)-(1.ALPHA.,2.BETA.,4.BETA., 5.ALPHA.,7.BETA.)]-. [114-49-8]

*TSCA 8(a), 40 CFR 712, 47 FR 26992, June 22, 1982, Chemical Information Rule

BENZENEACETIC ACID, 4-CHLORO-ALPHA-(4-CHLOROPHENYL)-ALPHAHYDROXY-, ETHYL ESTER.

*RCRA 40 CFR 261
EPA hazardous waste no. U038

BENZENAMINE. See Aniline. [62-53-3]

BENZENEAZOBENZENE. See Azobenzene. [103-33-3]

BENZENE, 1-BROMO-4-PHENOXY-.
*RCRA 40 CFR 261
EPA hazardous waste no. U030

BENAENECARBONAL. See Benzaldehyde. [100-52-7]

BENZENE, CHLORO-. [108-90-7]
*TSCA 8(a), 40 CFR 712, 47 FR 26992, June 22, 1982, Chemical Information Rule
*RCRA 40 CFR 261
EPA hazardous waste no. U037 (hazardous substance)

BENZENE, 1-CHLORO-2-METHYL.
[95-41-8]
*TSCA 8(a), 40 CFR 712, 47 FR 26992, June 22, 1982, Chemical Information Rule

BENZENE, CHLOROMETHYL-.
[100-44-7]
*TSCA 8(a), 40 CFR 712, 47 FR 26992, June 22, 1982, Chemical Information Rule
*RCRA 40 CFR 261
EPA hazardous waste no. P028 (hazardous substance)

1,2-BENZENEDIAMINE. [95-54-5]
*TSCA 8(a), 40 CFR 712, 47 FR 26992, June 22, 1982, Chemical Information Rule

1,3-BENZENEDIAMINE. [108-45-2]
*TSCA 8(a), 40 CFR 712, 47 FR 26992, June 22, 1982, Chemical Information Rule

1,4-BENZENEDIAMINE. [106-50-3]
*TSCA 8(a), 40 CFR 712, 47 FR 26992, June 22, 1982, Chemical Information Rule

1,2-BENZENEDIAMINE, 5-CHLORO-3-NITRO-. See Phenylendiamines.
[42389-30-0]

TSCA 4(e), ITC
Sixth Report of the Interagency Testing Committee to the administrator, Environmental Protection Agency, April 1980; 45 FR 35897, May 28, 1980, responded to by

the EPA administrator, 47 FR 973, January 8, 1982

*TSCA 8(d), Unpublished Health and Safety Studies Reporting, 40 CFR 716, 47 FR 38800, September 2, 1982; 48 FR 13178, March 30, 1983

1,2-BENZENEDIAMINE, 4-CHLOROSULFATE (1:1).
See Phenylenediamines. [68459-98-3]

TSCA 4(e), ITC
Sixth Report of the Interagency Testing Committee to the administrator, Environmental Protection Agency, April 1980; 45 FR 35897, May 28, 1980, responded to by the EPA administrator, 47 FR 973, January 8, 1982

*TSCA 8(d), Unpublished Health and Safety Studies Reporting, 40 CFR 716, 47 FR 38800, September 2, 1982; 48 FR 13178, March 30, 1983

1,2-BENZENEDIAMINE DIHYDROCHLORIDE. See o-Phenylenediamine dihydrochloride. [615-28-1]

1,4-BENZENEDIAMINE DIHYDROCHLORIDE. [624-18-0]

*TSCA 8(a), 40 CFR 712, 47 FR 26992, June 22, 1982, Chemical Information Rule

1,4-BENZENEDIAMINE ETHANEDIOATE. See Phenylenediamines. [62654-17-5]

TSCA 4(e), ITC
Sixth Report of the Interagency Testing Committee to the administrator, Environmental Protection Agency, April 1980; 45 FR 35897, May 28, 1980, responded to by the EPA administrator, 47 FR 973, January 8, 1982

*TSCA 8(d), Unpublished Health and Safety Studies Reporting 40 CFR 716, 47 FR 38800, September 2, 1982; 48 FR 13178, March 30, 1983

1,3-BENZENEDIAMINE, 4-ETHOXY-, DIHYDROCHLORIDE. See Phenylenediamines. [67801-06-3]

TSCA 4(e), ITC
Sixth Report of the Interagency Testing Committee to the administrator, Environmental Protection Agency, April 1980; 45 FR 35897, May 28, 1980, responded to by the EPA administrator, 47 FR 973, January 8, 1982

*TSCA 8(d), Unpublished Health and Safety Studies Reporting, 40 CFR 716, 47 FR 38800, September 2, 1982; 48 FR 13178, March 30, 1983

1,3-BENZENEDIAMINE, 4-ETHOXYSULFATE (1:1). See Phenylenediamines. [60815-98-5]

TSCA 4(e), ITC
Sixth Report of the Interagency Testing Committee to the administrator, Environmental Protection Agency, April 1980; 45 FR 35897, May 28, 1980, responded to by the EPA administrator, 47 FR 973, January 8, 1982

*TSCA 8(d), Unpublished Health and Safety Studies Reporting, 40 CFR 716, 47 FR 38800, September 2, 1982; 48 FR 13178, March 30, 1983

1,3-BENZENEDIAMINE, AR-ETHYL-AR-METHYL. See Phenylenediamines. [68966-84-7]

TSCA 4(e), ITC
Sixth Report of the Interagency Testing Committee to the administrator, Environmental Protection Agency, April 1980; 45 FR 35897, May 28, 1980, responded to by the EPA administrator, 47 FR 973, January 8, 1982

*TSCA 8(d), Unpublished Health and Safety Studies Reporting, 40 CFR 716, 47 FR 38800, September 2, 1982; 48 FR 13178, March 30, 1983

1,3-BENZENEDIAMINE, 4-METHOXY-. [615-05-4]

*TSCA 8(d), 40 CFR 712, 47 FR 26992, June 22, 1982, Chemical Information Rule

1,3-BENZENEDIAMINE, 4-METHOXY-, SULFATE (1:1). See 2,4-Diaminoanisole sulfate. [39156-41-7]

1,2-BENZENEDIAMINE, 3-METHYL-.
[2687-25-4]

*TSCA 8(d), 40 CFR 712, 47 FR 26992, June 22, 1982, Chemical Information Rule

1,2-BENZENEDIAMINE, 4-METHYL-.
[496-72-0]

*TSCA 8(d), 40 CFR 712, 47 FR 26992, June 22, 1982, Chemical Information Rule

1,2-BENZENEDIAMINE, 2-METHYL-.
[823-40-5]

*TSCA 8(d), 40 CFR 712, 47 FR 26992, June 22, 1982, Chemical Information Rule

1,4-BENZENEDIAMINE, 2-METHYL-, DIHYDROCHLORIDE. See Phenylene-diamines. [615-45-2]

TSCA 4(e), ITC
Sixth Report of the Interagency Testing Committee to the administrator, Environmental Protection Agency, April 1980; 45 FR 35897, May 28, 1980, responded to by the EPA administrator, 47 FR 973, January 8, 1982

*TSCA 8(d), Unpublished Health and Safety Studies Reporting, 40 CFR 716, 47 FR 38800, September 2, 1982; 48 FR 13178, March 30, 1983

1,4-BENZENEDIAMINE, 2-METHYL-, SULFATE. [6369-59-1]

TSCA 4(e), ITC
Sixth Report of the Interagency Testing Committee to the administrator, Environmental Protection Agency, April 1980; 45 FR 35897, May 28, 1980, responded to by the EPA administrator, 47 FR 973, January 8, 1982

*TSCA 8(a), 40 CFR 712, 47 FR 26992, June 22, 1982, Chemical Information Rule

*TSCA 8(d), Unpublished Health and Safety Studies Reporting, 40 CFR 716, 47 FR 38800, September 2, 1982; 48 FR 13178, March 30, 1983

1,2-BENZENEDIAMINE, 4-NITRO-.
[99-56-9]

*TSCA 8(a), 40 CFR 712, 47 FR 26992, June 22, 1982, Chemical Information Rule

1,4-BENZENEDIAMINE, 2-NITRO-, SULFATE (1:1). See Phenylene-diamines.
[68239-83-8]

TSCA 4(e), ITC
Sixth Report of the Interagency Testing Committee to the administrator, Environmental Protection Agency, April 1980; 45 FR 35897, May 28, 1980, responded to by the EPA administrator, 47 FR 973, January 8, 1982

*TSCA 8(d), Unpublished Health and Safety Studies Reporting, 40 CFR 716, 47 FR 38800, September 2, 1982; 48 FR 13178, March 30, 1983

1,2-BENZENEDIAMINE, 4-NITRO-, SULFATE (1:1). [68239-82-7]

TSCA 4(e), ITC
Sixth Report of the Interagency Testing Committee to the administrator, Environmental Protection Agency, April 1980; 45 FR 35897, May 28, 1980, responded to by the EPA administrator, 47 FR 973, January 8, 1982

1,2-BENZENEDICARBOXYLIC ACID ANHYDRIDE. [85-44-9]

*RCRA 40 CFR 261
EPA hazardous waste no. U190

1,2—BENZENEDICARBOXYLIC ACID, See Alkyl phthalates. BIS (2-ETHYLHEXYL)ESTER. [117-81-7]

*TSCA 8(a), 40 CFR 712, 47 FR 26992, June 22, 1982, Chemical Information Rule

*TSCA 8(d), Unpublished Health and Safety Studies Reporting, 40 CFR 716, 47 FR 387, September 2, 1982

*RCRA 40 CFR 261
EPA hazardous waste no. U028

1,2-BENZENEDICARBOXYLIC ACID, 2-BUTOXY-2-OXOETHYL BUTYL ESTER. [85-70-1]

*TSCA 8(a), 40 CFR 712, 47 FR 26992, June 22, 1982, Chemical Information Rule

1,2-BENZENEDICARBOXYLIC ACID, BUTYL PHENYLMETHYL ESTER.
[85-68-7]

1,2-BENZENEDICARBOXYLIC ACID, DECYL OCTYL ESTER

*TSCA 8(a), 40 CFR 712, 47 FR 26992, June 22, 1982, Chemical Information Rule

1,2-BENZENEDICARBOXYLIC ACID, DECYL OCTYL ESTER. See Alkyl phthalates. [119-07-3]

*TSCA 8(a), 40 CFR 712, 47 FR 26992, June 22, 1982, chemical Information Rule

*TSCA 8(d), Unpublished Health and Safety Studies Reporting, 40 CFR 716, 47 FR 387, September 2, 1982

1,2-BENZENEDICARBOXYLIC ACID, DIBUTYL ESTER. See Alkyl phthalates. [84-74-2]

*TSCA 8(a), 40 CFR 712, 47 FR 26992, June 22, 1982, Chemical Information Rule

*TSCA 8(d), Unpublished Health and Safety Studies Reporting, 40 CFR 716, 47 FR 387, September 2, 1982

*RCRA 40 CFR 261
EPA hazardous waste no. U069 (hazardous substance)

1,2-BENZENEDICARBOXYLIC ACID, DICYCLOHEXYL ESTER. See Alkyl phthalates. [84-61-7]

*TSCA 8(a), 40 CFR 712, 47 FR 26992, June 22, 1982, Chemical Information Rule

*TSCA 8(d), Unpublished Health and Safety Studies Reporting, 40 CFR 716, 47 FR 387, September 2, 1982

1,2-BENZENEDICARBOXYLIC ACID, DIETHYL ESTER. See Alkyl phthalates. [84-66-2]

*TSCA 8(a), 40 CFR 712, 47 FR 26992, June 22, 1982, Chemical Information Rule

*TSCA 8(d), Unpublished Health and Safety Studies Reporting, 40 CFR 716, 47 FR 387, September 2, 1982

*RCRA 40 CFR 261
EPA hazardous waste no. U088

1,2-BENZENEDICARBOXYLIC ACID, DIISODECYL ESTER. See Alkyl phthalates. [26761-40-0]

*TSCA 8(a), 40 CFR 712, 47 FR 26992, June 22, 1982, Chemical Information Rule

*TSCA 8(d), Unpublished Health and Safety Studies Reporting, 40 CFR 716, 47 FR 387, September 2, 1982

1,2-BENZENEDICARBOXYLIC ACID, DIISOOCTYL ESTER. See Alkyl phthalates. [27554-26-3]

*TSCA 8(d), Unpublished Health and Safety Studies Reporting, 40 CFR 716, 47 FR 387, September 2, 1982

1,2-BENZENEDICARBOXYLIC ACID, DIMETHYL ESTER. See Alkyl phthalates. [131-11-3]

*TSCA 8(a), 40 CFR 712, 47 FR 26992, June 22, 1982, Chemical Information Rule

TSCA 8(d), Unpublished Health and Safety Studies Reporting, 40 CFR 716, 47 FR 387, September 2, 1982

*RCRA 40 CFR 261
EPA hazardous waste no. U102

1,2-BENZENEDICARBOXYLIC ACID, DIOCTYL ESTER. See Alkyl phthalates. [117-84-0]

*TSCA 8(a), 40 CFR 712, 47 FR 26992, June 22, 1982, Chemical Information Rule

*TSCA 8(d), Unpublished Health and Safety Studies Reporting, 40 CFR 716, 47 FR 387, September 2, 1982

*RCRA 40 CFR 261
EPA hazardous waste no. U107

1,2-BENZENEDICARBOXYLIC ACID, DITRIDECYL ESTER. See Alkyl phthalates. [119-06-2]

*TSCA 8(a), 40 CFR 712, 47 FR 26992, June 22, 1982, Chemical Information Rule

*TSCA 8(d), Unpublished Health and Safety Studies Reporting, 40 CRF 716, 47 FR 387, September 2, 1982

BENZENE, 1,2-DICHLORO-. [95-50-1]

*TSCA 8(a), 40 CFR 712, 47 FR 26992, June 22, 1982, Chemical Information Rule

*RCRA 40 CFR 261
EPA hazardous waste no. U070 (hazardous substance)

BENZENE, 1,3-DICHLORO-. [541-73-1]
*TSCA 8(a), 40 CFR 712, 47 FR 26992, June 22, 1982, Chemical Information Rule
*RCRA 40 CFR 261
EPA hazardous waste no. U071 (hazardous substance)

BENZENE, 1,4-DICHLORO-. [106-46-7]
*TSCA 8(a), 40 CFR 712, 47 FR 26992, June 22, 1982, Chemical Information Rule
*RCRA 40 CFR 261
EPA hazardous waste no. U072 (hazardous substance)

BENZENE, DICHLOROMETHYL-. [98-87-3]
*TSCA 8(a), 40 CFR 712, 47 FR 26992, June 22, 1982, Chemical Information Rule
*RCRA 40 CFR 261
EPA hazardous waste no. U017

BENZENE, 1,3-DIISOCYANATOMETHYL-. [1321-38-6]
*RCRA 40 CFR 261
RPA hazardous waste no. U223 (reactive waste, toxic waste)

BENZENE, DIMETHYL-. [1330-20-7]
*TSCA 8(a), 40 CFR 712, 47 FR 26992, June 22, 1982, Chemical Information Rule
*RCRA 40 CFR 261
EPA hazardous waste no. U239 (hazardous substance, ignitable waste, toxic waste)

BENZENE, 1,2-DIMETHYL-. [95-47-6]
*TSCA 8(a), 40 CFR 712, 47 FR 26992, June 22, 1982, Chemical Information Rule

BENZENE, 1,3-DIMETHYL-. [108-38-3]
*TSCA 8(a), 40 CFR 712, 47 FR 26992, June 22, 1982, Chemical Information Rule

BENZENE, 1,4-DIMETHYL-. [106-42-3]
*TSCA 8(a), 40 CFR 712, 47 FR 26992, June 22, 1982, Chemical Information Rule

1,3-BENZENEDIOL. [108-46-3]
*RCRA 40 CFR 261
PEA hazardous waste no. U201 (hazardous substance)

1,4-BENZENEDIOL. [123-31-9]
*TSCA 8(a), 40 CFR 712, 47 FR 26992, June 22, 1982, Chemical Information Rule

1,2-BENZENEDIOL, 4-[1-HYDROXY-2(METHYLAMINO)ETHYL]-.
*RCRA 40 CFR 261
EPA hazardous waste no. P042

BENZENE, 1,1'-[1,2-ETHANEDIYLBIS(OXY)]BIS[2,4,6-TRIBROMO-. [37853-59-1]
*TSCA 8(a), 40 CFR 712, 47 FR 26992, June 22, 1982, Chemical Information Rule

BENZENE, ETHENYL-. [100-42-5]
*TSCA 8(a), 40 CFR 712, 47 FR 26992, June 22, 1982, Chemical Information Rule

BENZENE, ETHYL-. [100-41-4]
*TSCA 8(a), 40 CFR 712, 47 FR 26992, June 22, 1982, Chemical Information Rule

BENZENE, ETHYLMETHYL-. [25550-14-5]
*TSCA 8(a), 40 CFR 712, 47 FR 26992, June 22, 1982, Chemical Information Rule

BENZENE HEXACHLORIDE. See Cyclohexane, 1,2,3,4,5,6-hexachloro-. See below [608-73-1]

alpha-BHC (BENZENE HEXACHLORIDE). [319-84-6]
A commercial mixture of isomers of 1,2,3,4,5,6-hexachlorocyclohexane insecticide
CWA 307(a)(1), 40 CFR 125, Priority Pollutant

beta-BHC (BENZENE HEXACHLORIDE). [319-85-7]
A commercial mixture of isomers of 1,2,3,4,5,6-hexachlorocyclohexane insecticide
CWA 307(a)(1), 40 CFR 125, Priority Pollutant

delta-BHC (BENZENE HEXACHLO-RIDE). [319-86-8]

A commercial mixture of isomers of 1,2,3,4,5,6-hexachlorocyclohexane insecticide

CWA 307(a)(1), 40 CFR 125, Priority Pollutant

gamma-BHC (BENZENE HEXACHLO-RIDE. See Lindane. [58-89-9]

A commercial mixture of isomers of 1,2,3,4,5,6-hexachlorocyclohexane insecticide

CWA 307(a)(1), 40 CFR 125, Priority Pollutant

BENZENE, HEXACHLORO-. See Hexachlorobenzene. [608-73-1]

BENZENE, HEXAHYDRO-. See Cyclohexane. [110-82-7]

BENZENE, HYDROXY-. See Phenol. [108-95-2]

BENZENE, ISOCYANATO-. [103-71-9]

*TSCA 8(a), 40 CFR 712, 47 FR 26992, June 22, 1982, Chemical Information Rule

BENZENE, METHYL-. See Toluene. [108-88-3]

BENZENE. 1-METHYL-2,4-DINITRO-. [121-14-2]

*RCRA 40 CFR 261

EPA hazardous waste no. U105 (hazardous substance)

BENZENE, 1-METHYL-2,6-DINITRO-. [606-20-2]

*RCRA 40 CFR 261

EPA hazardous waste no. U106 (hazardous substance)

BENZENE, 1,2-METHYLENEDIOXY-4-ALLYL-. [94-59-7]

*RCRA 40 CFR 261

EPA hazardous waste no. U203

BENZENE, 1,2-METHYLENEDIOXY-4-PROPENYL-. [120-58-1]

*RCRA 40 CFR 261

EPA hazardous waste no. U141

BENZENE, 1,2-METHYLDIOXY-4-PROPYL-. [94-58-6]

*RCRA 40 CFR 261

EPA hazardous waste no. U090

BENZENE, 1-METHYLETHYL-. See Cumene. [98-82-8]

BENZENE, NITRO-. [98-95-3]

*TSCA 8(a), 40 CFR 712, 47 FR 26992, June 22, 1982, Chemical Information Rule

*RCRA 40 CFR 261

EPA hazardous waste no. U169 (hazardous substance ignitable waste, toxic waste)

1,2-BENZENE, 4-NITRO, SULFATE (1:1). See Phenylenediamines. [68239-82-7]

*TSCA 8(d), Unpublished Health and Safety Studies Reporting, 40 CFR 716, 47 FR 38800, September 2, 1982; 48 FR 13178, March 30, 1983

BENZENE, PENTACHLORO-. [608-93-5]

*TSCA 8(a), 40 CFR 712, 47 FR 26992, June 22, 1982, Chemical Information Rule

*RCRA 40 CFR 261

EPA hazardous waste no. U183

BENZENE, PENTACHLORONITRO-. [82-68-8]

*RCRA 40 CFR 261

EPA hazardous waste no. U185

BENZENESULFONIC ACID CHLORIDE. [98-09-0]

*RCRA 40 CFR 261

EPA hazardous waste no. U020 (corrosive waste, reactive waste)

BENZENESULFONIC ACID, 4-[(4,6-DICHLORO-1,3,5-TRIAZIN-2-YL)AMINO]-. [16110-89-7]

*TSCA 8(a), 40 CFR 712, 47 FR 26992, June 22, 1982, Chemical Information Rule

BENZENESULFONYL CHLORIDE. [98-09-9]

*RCRA 40 CFR 261

EPA hazardous waste no. U020 (corrosive waste, reactive waste)

BENZENE, 1,2,3,4-TETRACHLORO-.
[634-90-2]

*TSCA 8(a), 40 CFR 712, 47 FR 26992, June 22, 1982, Chemical Information Rule

BENZENE, 1,2,4,5-TETRACHLORO-.
[95-94-3]

*TSCA 8(a), 40 CFR 712, 47 FR 26992, June 22, 1982, Chemical Information Rule

*RCRA 40 CFR 261
EPA hazardous waste no. U207

BENZENETHIOL. [108-98-5]

*RCRA 40 CFR 261
EPA hazardous waste no. P014

BENZENE, 1,2,3-TRICHLORO-.
[87-61-6]

*TSCA 8(a), 40 CFR 712, 47 FR 26992, June 22, 1982, Chemical Information Rule

BENZENE, 1,2,4-TRICHLORO-.
[120-82-1]

*TSCA 8(a), 40 CFR 712, 47 FR 26992, June 22, 1982, Chemical Information Rule

BENZENE, 1,3,5-TRICHLORO-.
[108-70-3]

*TSCA 8(a), 40 CFR 712, 47 FR 26992, June 22, 1982, Chemical Information Rule

BENZENE, TRICHLOROMETHYL-.
[98-07-7]

*TSCA 8(a), 40 CFR 712, 47 FR 26992, June 22, 1982, Chemical Information Rule

*RCRA 40 CFR 261
EPA hazardous waste no. U023 (corrosive waste, reactive waste, toxic waste)

BENZENE, 1,2,4-TRIMETHYL-.
[95-63-6]

*TSCA 8(a), 40 CFR 712, 47 FR 26992, June 22, 1982, Chemical Information Rule

BENZENE, 1,3,5-TRINITRO-.
[99-35-4]

*RCRA 40 CFR 261
EPA hazardous waste no. U234 (reactive waste, toxic waste)

BENZIDINE. [92-87-5]

*OSHA Standard 29 CFR 1910. 1010 Cancer-Suspect Agent, shall not apply to solid or liquid mixtures containing less than 0.1% by weight or volume in benzidine

ACGIH-TLV (1982 Addition)

Skin

TWA: Human carcinogen—no exposure or contact by any route—respiratory, skin, or oral, as detected by the most sensitive methods—shall be permitted. The worker should be properly equipped to insure virtually no contact with the carcinogen.

OSHA Candidate List of Potential Occupational Carcinogens (EPA Carcinogen Assessment Group List), 45 FR 53672, August 12, 1980

IARC—carcinogenic in humans

CWA 304(a)(1), 45 FR 79318, November 28, 1980, Water Quality Criteria

Freshwater Aquatic Life

The available data for benzidine indicate that acute toxicity to freshwater aquatic life occurs at concentrations as low as 2500 μg/liter and would occur at lower concentrations among species that are more sensitive than those tested. No data are available concerning the chronic toxicity of benzidine to sensitive freshwater aquatic life.

Saltwater Aquatic Life

No saltwater organisms have been tested with benzidine and no statement can be made concerning acute and chronic toxicity.

Human Health

For the maximum protection of human health from the potential carcinogenic effects due to exposure of benzidine through ingestion of contaminated water and contaminated aquatic organisms, the ambient water concentration should be zero based on the nonthreshold assumption for this chemical. However, zero level may not be attainable at the present time. Therefore, the levels that may result in incremental increase of cancer risk over the lifetime are estimated at 10^{-5}, 10^{-6}, and 10^{-7}. The corresponding criteria are 1.2 ng/liter, 0.12 ng/liter, and 0.01 ng/liter, respectively. If the above estimates are made for consumption of aquatic organisms only, excluding consumption of water, the levels are 5.3 ng/liter, 0.53 ng/liter, and 0.05 ng/liter, respectively. Other concentrations representing different risk levels may be calculated by use of the guidelines. The risk estimate range is presented for information purposes and does not represent an agency judgment on an "acceptable" risk level.

CWA 307(a)(1), Priority Pollutant

*RCRA 40 CFR 261

EPA hazardous waste no. U021

BENZIDINE-BASED DYES.

TSCA 4(e), ITC

Fifth Report of the TSCA Interagency Testing Committee to the administrator, Environmental Protection Agency, November 1979; 44 FR 70664, December 7, 1979, EPA responded to the committee's recommendations for testing 46 FR 55005, November 5, 1981

BENZIDINE DIHYDROCHLORIDE.

Syn: (1,1′-Biphenyl)-4,4′-diamine dihydrochloride. [531-85-1]

OSHA Candidate List of Potential Occupational Carcinogens, 45 FR 53672, August 12, 1980

BENZIDINE, 3,3′-DIMETHOXY, DIHYDROCHLORDIE. See C.I. Disperse Black 6, dihydrochloride. [20325-40-0]

BENZIMIDAZOLE, 5-NITRO-. Syn: 6-Nitro-benzimidazole. [94-52-0]

OSHA Candidate List of Potential Occupational Carcinogens, 45 FR 53672, August 12, 1980

NCI—carcinogenic in animals

1,2-BENZISOTHIAZOLIN-3-ONE, 1,1-DIOXIDE. [81-07-2]

*RCRA 40 CFR 261

EPA hazardous waste no. U202

BENZO (a) ANTHRACENE. [56-55-3]

CWA 307(a)(1), 40 CFR 125, Priority Pollutant

BENZO (b) FLUORANTHENE. [205-99-2]

OSHA Candidate List of Potential Occupational Carcinogens (EPA Carcinogen Assessment Group List), 45 FR 53672, August 12, 1980

IARC—carcinogenic in animals

CWA 307(a)(1), 40 CFR 125, Priority Pollutant

BENZO (f) FLUORANTHENE.

IARC—carcinogenic in animals

BENZO (j) FLUORANTHENE. [205-82-3]

OSHA Candidate List of Potential Occupational Carcinogens (EPA Carcinogen Assessment Group List), 45 FR 53672, August 12, 1980

BENZO (k) FLUORANTHENE.

CWA 307(a)(1), 40 CFR 125, Priority Pollutant

BENZO (j,k) FLUORENE. [206-44-0]

*RCRA 40 CFR 261

EPA hazardous waste no. U120

BENZOIC ACID. [65-85-0]

*CWA 311(b)(2)(A), 40 CFR 116, 117

Discharge RQ = 5000 pounds (2270 kilograms)

BENZOIC ACID, 2-[[2-AMINO-6-[[4′-[(3-CARBOXY-4-HYDROXYPHENYL)AZO]-3,3′-DIMETHOXYL[1,1′-BIPHENYL]-4-YL]AZO]-5-HYDROXY-7-SULFO-1-NAPHTHALENYL]AZO]-5-NITRO-, TRI SODIUM SALT. See Bisazobiphenyl dyes. [6739-62-4]

*TSCA 8(d), Unpublished Health and Safety Studies Reporting, 40 CFR 716, 47 FR 387, September 2, 1982

BENZOIC ACID, 3-AMINO-2,5-DICHLORO-. Syn: Amiben. [133-90-4]
OSHA Candidate List of Potential Occupational Carcinogens, 45 FR 53672, August 12, 1980

BENZOIC ACID, 5-[[4'-[(2-AMINO-8-HYDROXY-6-SULFO-1-NAPHTHALENYL)AZO][1,1'-BIPHENYL]-4-YL]AZO]-2-HYDROXY-, DISODIUM SALT. See Bisazobiphenyl dyes. [2429-84-7]
*TSCA 8(d), Unpublished Health and Safety Studies Reporting, 40 CFR 716, 47 FR 387, September 2, 1982

BENZOIC ACID, 5-[[4'-[(7-AMINO-1-HYDROXY-3-SULFO-2-NAPHTHALENYL)AZO][1,1'-BIPHENYL]-4-YL]AZO]-2-HYDROXY-, DISODIUM SALT. See Bisazobiphenyl dyes. [2429-82-5]
*TSCA 8(d), Unpublished Health and Safety Studies Reporting, 40 CFR 716, 47 FR 387, September 2, 1982

BENZOIC ACID, 5-[[4'-[(1-AMINO-4-SULFO-2-NAPHTHALENYL)AZO][1,1'-BIPHENYL]-4-YL]AZO]-2-HYDROXY-, DISODIUM SALT. See Bisazobiphenyl dyes. [2429-79-0]
*TSCA 8(d), Unpublished Health and Safety Studies Reporting, 40 CFR 716, 47 FR 387, September 2, 1982

BENZOIC ACID, 3-CHLORO-, METHYL ESTER. [2905-65-9]
*TSCA 8(a), 40 CFR 712, 47 FR 26992, June 22, 1982, Chemical Information Rule

BENZOIC ACID, 5-[[4'-[[2,6-DIAMINO-3-[[8-HYDROXY-3,6-DISULFO-7-[(4-SULFO-1-NAPHTHALENYL)AZO]-2-NAPHTHALENYL]AZO]-5-METHYLPHENYL]AZO][1,1'-BIPHENYL]-4-YL]AZO]-2-HYDROXY-, TETRASODIUM SALT. See Bisazobiphenyl dyes. [2429-81-4]

*TSCA 8(d), Unpublished Health and Safety Studies Reporting, 40 CFR 716, 47 FR 387, September 2, 1982

BENZOIC ACID, 5-[[4'-[[2,6-DIAMINO-3-METHYL-5-[(4-SULFOPHENYL)AZO]PHENYL]AZO][1,1'-BIPHENYL]-4-YL]AZO]-2-HYDROXY-, DISODIUM SALT. See Bisazobiphenyl dyes. [2586-58-5]
*TSCA 8(d), Unpublished Health and Safety Studies Reporting, 40 CFR 716, 47 FR 387, September 2, 1982

BENZOIC ACID, 5-[[4'-[[2,4-DIHYDROXY-3-[(4-SULFOPHENYL)AZO]PHENYL]AZO][1,1'-BIPHENYL]-4-YL]AZO]-2-HYDROXY-, DISODIUM SALT. See Bisazobiphenyl dyes. [2893-80-3]
*TSCA 8(d), Unpublished Health and Safety Studies Reporting, 40 CFR 716, 47 FR 387, September 2, 1982

BENZOIC ACID, HYDRAZIDE. Syn: Benzoyl hydrazide. [613-94-5]
OSHA Candidate List of Potential Occupational Carcinogens, 45 FR 53672, August 12, 1980

BENZOIC ALDEHYDE. See Benzaldehyde. [100-52-7]

BENZOL. See Benzene. [71-43-2]

BENZONITRILE. [100-47-0]
*CWA 311(b)(2)(A), 40 CFR 116, 117 Discharge RQ = 1000 pounds (454 kilograms)

BENZO (g,h,i) PERYLENE. [191-24-2]
CWA 307(a)(1), 40 CFR 125, Priority Pollutant

BENZOPHENONE, 4,4'-BIS(DIMETHYLAMINO)-. Syn: Michler's ketone. [90-94-8]
OSHA Candidate List of Potential Occupa-

tional Carcinogens, 45 FR 53672, August 12, 1980

NCI—carcinogenic in animals

BENZO (a) PYRENE (BaP). *See* Coal tar pitch volatiles. [50-32-8]

ACGIH-TLV

TWA: Industrial substance suspect of carcinogenic potential for humans—which is suspect of inducing cancer, based on either (1) limited epidemiologic evidence, inclusive of clinical reports of single cases, or (2) demonstration of carcinogenesis in one or more animal species by appropriate methods. Worker exposure by all routes should be carefully controlled to levels consistent with the animal and human experience data.

OSHA Candidate List of Potential Occupational Carcinogens (EPA Carcinogen Assessment Group List), 45 FR 53672, August 12, 1980

IARC—carcinogenic in animals

CWA 307(a)(1), Priority Pollutant

*RCRA 40 CFR 261

EPA hazardous waste no. U022

3,4-BENZOPYRENE. *See* Benzo (a) pyrene. [50-32-8]

p-**BENZOQUINONE.** *See* Quinone. [106-51-4]

p-**BENZOQUINONE DIOXIME.** Syn: Quinone dioxime. [105-11-3]

OSHA Candidate List of Potential Occupational Carcinogens, 45 FR 53672, August 12, 1980

NCI—carcinogenic in animals

BENZOTRICHLORIDE. [98-07-7]

TSCA-CHIP available

*RCRA 40 CFR 261

EPA hazardous waste no. U023 (corrosive waste, reactive waste, toxic waste)

BENZOTRIFLUORIDE. [98-08-8]

NFPA—hazardous chemical
health—4, flammability—3, reactivity—0

BENZOYL CHLORIDE. *See* Dibenzoyl chloride. [98-88-4]

NFPA—hazardous chemical
health—3, flammability—2, reactivity—1

*TSCA 8(a), 40 CFR 712, 47 FR 26992, June 22, 1982, Chemical Information Rule

TSCA-CHIP available

*CWA 311(b)(2)(A), 40 CFR 116, 117
Discharge RQ = 1000 pounds (454 kilograms)

BENZOYL HYDRAZIDE. *See* Benzoic acid, hydrazide. [613-94-5]

BENZOYL PEROXIDE. *See* Dibenzoyl peroxide. [94-36-0]

*OSHA 29 CFR 1910.1000, Table Z-1
PEL (TWA): 5 mg/m^3

ACGIH-TLV
TWA: 5 mg/m^3

NIOSH Criteria Document (Pub. No. 77-166), NTIS Stock No. PB 273819, May 31, 1977

These criteria and the recommended standard apply to employees exposed to any form of the diacyl organic peroxide (C6H5CO)2O2

Synonyms for benzoyl peroxide include benzoyl superoxide and dibenzoly peroxide.

An "action level" is defined as equal to the environmental limit. Occupational exposure to benzoyl peroxide is defined as any work involving handling, storage, use, or manufacture of benzoyl peroxide at a concentration above the action level. Exposure at lower concentrations will not require adherence to all sections, except for Sections 2(a,c), 3(a), 4(a), 5, 6(b,c,d,e), 7, and 8(a,d). See Criteria Document for details.

Exposure to benzoyl peroxide shall be controlled so that employees are not exposed at a concentration greater than 5 mg/m^3 of air, determined as a time-weighted average (TWA) concentration for up to a 10-hour work shift in a 40-hour workweek. The standard is designed to protect the health and provide for the safety of employees for up to a 10-hour work shift, 40-hour workweek, over a working lifetime. See the Criteria Document for details.

NFPA—hazardous chemical
health—1, flammability—4, reactivity—4, oxy—oxidizing chemical

1,2-BENZPHENANTHRENE.
[218-01-9]

*RCRA 40 CFR 261
EPA hazardous waste no. U050

BENZYL ACETATE. [140-11-4]

NCI—carcinogenic in animals
TSCA-CHIP available

BENZYL BUTYL PHTHALATE.
[85-68-7]

TSCA 4(e), ITC
Seventh Report of the TSCA Interagency Testing Committee to the administrator, Environmental Protection Agency, October 1980; 45 FR 78432, November 25, 1980, EPA responded to the committee's recommendations for testing 46FR 53775, October 30, 1981

*TSCA 8(d), Unpublished Health and Safety Studies Reporting, 40 CFR 716, 47 FR 38800, September 2, 1982; 48 FR 13178, March 30, 1983

BENZYL CHLORIDE. [100-44-7]

*OSHA 29 CFR 1910.1000 Table Z-1
PEL(TWA):
1 ppm, 5 mg/m^3

ACGIH-TLV
TWA: 1 ppm, 5 mg/m^3

NIOSH Criteria Document (Pub. No. 78-182). NTIS Stock No. PB81-226698, August 22, 1979

Synonyms for benzyl chloride include alpha-chlorotoluene, (chloromethyl)benzene, omega-chlorotoluene, and chlorophenylmethane.

"Occupational exposure" to benzyl chloride is defined as work in any area where benzyl chloride is manufactured, processed, stored, handled, or used. Compliance with all sections of the recommended standard is required where there is occupational exposure to benzyl chloride. If benzyl chloride is handled or stored only in sealed, intact containers, for example, during shipment or storage, the recommended standard, except for Sections 3, 5(a), and 6(e), shall not apply. See the Criteria Document for details.

Exposure to benzyl chloride shall be controlled to that no employee is exposed to benzyl chloride at a concentration greater than 5.0 mg/liter determined as a ceiling concentration during any 15-minute sampling period.

The recommended standard is designed to protect the health and provide for the safety of employees for up to a 10-hour work shift, 40-hour workweek, over a working lifetime.

IARC—carcinogenic in animals

NFPA—hazardous chemical
health—2, flammability—2, reactivity—1

TSCA-CHIP available

*CWA 311(b)(2)(A), 40 CFR 116, 117
Discharge RQ = 100 pounds (45.4 kilograms)

CAA 112
Chemical proposed to be assessed as a Hazardous Air Pollutant by a House of Representatives bill to amend the CAA

*RCRA 40 CFR 261
EPA hazardous waste no. U028 (hazardous substance)

BENZYL VIOLET 4B. [1694-09-3]

IARC—carcinogenic in animals

BERYLLIUM and BERYLLIUM COMPOUNDS. [7440-41-7]

*OSHA 1910.1000 Table Z-2, ANSI Z 37.29-1970

2 μg/m^3, 8-hour TWA
5 μg/m^3, acceptable ceiling concentration
25 μg/m^3/30 minutes, acceptable maximum peak

ACGIH-TLV
TWA: 0.002 mg/m^3
Industrial substance suspect of carcinogenic potential for humans—which is suspect of inducing cancer, based on either (1) epidemi-

ologic evidence, exclusive of clinical reports of single cases, or (2) demonstration of carcinogensis in one or more animal species by appropriate methods. Worker exposure by all routes should be carefully controlled to levels consistent with the animal and human experience data.

NIOSH Criteria Document (Pub. No. 72-10268), NTIS Stock No. PB 210806, 1972. (Date is from the publication number of the Criteria Document. No date is given in the summary of the document.)

No worker shall be exposed at the place of employment to a concentration of beryllium more than two micrograms of total airborne particulate beryllium per cubic meter of air (2 μg Be/m^3) determined as a time-weighted average (TWA) exposure for an 8-hour workday, and no peak concentration of beryllium to which workers are exposed shall exceed 25 μg Be/m^3 as determined by a minimum sampling time of 30 minutes.

IARC—carcinogenic in animals

NFPA—hazardous chemical
health—4, flammability—1, reactivity—0

CWA 304(a)(1), 45 FR 79318, November 28, 1980, Water Quality Criteria

Freshwater Aquatic Life

The available data for beryllium indicate that acute and chronic toxicity to freshwater aquatic life occurs at concentrations as low as 103 and 5.3 μg/liters, respectively, and would occur at lower concentrations among species that are more sensitive than those tested. Hardness has a substantial effect on acute toxicity.

Saltwater Aquatic Life

The limited saltwater data base available for beryllium does not permit any statement concerning acute or chronic toxicity.

Human Health

For the maximum protection of human health from the potential carcinogenic effects due to exposure of beryllium through ingestion of contaminated water and contaminated aquatic organisms, the ambient water concentration should be zero based on the nonthreshold assumption for this chemical. However, zero level may not be attainable at the present time. Therefore, the levels that may result in incremental increase of cancer risk over the lifetime are estimated at 10^{-5}, 10^{-6}, and 10^{-7}. The corresponding criteria are 37 ng/liter, 3.7 ng/liter, and 0.37 ng/liter, respectively. If the above estimates are made for consumption of water, the levels are 641 ng/liter, 64.1 ng/liter, and 6.41 ng/liter, respectively. Other concentrations representing different risk levels may be calculated by use of the guidelines. The risk estimate range is presented for information purposes and does not represent an agency judgment on an "acceptable" risk level.

CWA 307(a)(1), Priority Pollutant

*CAA 112, 40 CFR 61, Hazardous Air Pollutant

Emissions to the atmosphere from stationary sources subject to the provisions of this subpart shall not exceed 10 g of beryllium over a 24-hour period, except as provided below.

Rather than meet the requirement of the above paragraph of this section, an owner or operator may request approval from the administrator to meet an ambient concentration limit on beryllium in the vicinity of the stationary source of 0.01 μg/m^3, averaged over a 30-day period.

CAA 112
Chemical proposed to be assessed as a Hazardous Air Pollutant by a House of Representatives bill to amend the CAA

*RCRA 40 CFR 261
EPA hazardous waste no. P015

BERYLLIUM CHLORIDE. [7787-47-5]
*CWA 311(b)(2)(A), 40 CFR 116, 117
Discharge RQ = 5000 pounds (2270 kilograms)

BERYLLIUM FLOURIDE. [7787-49-7]
*CWA 311(b)(2)(A), 40 CFR 116, 117
Discharge RQ = 5000 pounds (2270 kilograms)

BERYLLIUM NITRATE. [7787-55-5], [13597-99-4]
*CWA 311(b)(2)(A), 40 CFR 116, 117
Discharge RQ = 5000 pounds (2270 kilograms)

BERYLLIUM OXIDE. [1304-56-9]
IARC—carcinogenic in animals

BERYLLIUM PHOSPHATE. [13598-15-7]
IARC—carcinogenic in animals

BERYLLIUM SULFATE. [13510-49-1]
IARC—carcinogenic in animals

BERYL ORE. [1302-52-9]
IARC—carcinogenic in animals

BHC (technical grades). *See* Lindane.
IARC—carcinogenic in animals
40 CFR 162.11, FIFRA-RPAR, Final Action Voluntary Cancelation
Current Status: 43 FR 31432 (7/21/78)
NTIS # PB80 216781
Criteria Possibly Met or Exceeded: Oncogenicity

4′4′′′-BIACETANILIDE. Syn: Diacetylbenzidine. [613-35-4]
OSHA Candidate List of Potential Occupational Carcinogens, 45 FR 53672, August 12, 1980
IARC—carcinogenic in animals

BICHROMATES. *See* Dichromates.

BICYCLO[2.2.1]HEPT-5-ENE-2,3-DICARBOXYLIC ACID, 1,4,5,6,7,7-HEXACHLORO-. [115-28-6]
*TSCA 8(a), 40 CFR 712, 47 FR 26992, June 22, 1982, Chemical Information Rule

BIFENOX. [42576-02-3]
FIFRA—Registration Standard—Issued June 1981 (herbicide)

2,2′-BIOXIRANE. *See* Alkyl epoxides. [1464-53-5]

ITSCA 8(a), 40 CFR 712, 47 FR 26992, June 22, 1982, Chemical Information Rule
*TSCA 8(d), Unpublished Health and Safety Studies Reporting, 40 CFR 716, 47 FR 387, September 2, 1982
*RCRA 40 CFR 261
EPA hazardous waste no. U085 (ignitable waste, toxic waste)

BIPHENYL. Syn: Diphenyl. [92-52-4]
*OSHA 29 CFR 1910.1000 Table Z-1
PEL(TWA): 0.2 ppm, 1mg/m^3
ACGIH-TLV
TWA: 0.2 ppm, 1.5 mg/m^3
STEL: 0.6 ppm, 4 mg/m^3
TSCA 4(e), ITC
Tenth Report of the Interagency Testing Committee to the administrator, Environmental Protection Agency, May 1982; 47 FR 22585, May 25, 1982
*TSCA 8(a), 40 CFR 712, 47 FR 26992, June 22, 1982, Chemical Information Rule
*TSCA 8(d), Unpublished Health and Safety Studies Reporting, 40 CFR 716, 47 FR 38800, September 2, 1982; 48 FR 13178, March 30, 1983
TSCA-CHIP available

1,1′-BIPHENYL. *See* Biphenyl. [92-52-4]

2-BIPHENYLAMINE HYDROCHLORIDE. Syn: 2-Aminobiphenyl. [2185-92-4]
NCI—carcinogenic in animals

(1,1′-BIPHENYL)-4,4′-BIS(DIAZONIUM), 3,3′-DIMETHOXY-.
See Bisazobiphenyl dyes. [20282-70-6]
TSCA 8(d), Unpublished Health and Safety Studies Reporting, 40 CFR 716, 47 FR 387, September 2, 1982

(1,1′-BIPHENYL)-4,4′-DIAMINE. [92-87-5]
*RCRA 40 CFR 261
EPA hazardous waste no. U021

(1,1'-BIPHENYL)-4,4'-DIAMINE, 3,3'-DICHLORO-. [91-94-1]
*RCRA 40 CFR 261
EPA hazardous waste no. U073

(1,1'-BIPHENYL)-4,4'-DIAMINE DIHYDROCHLORIDE. See Benzidine dihydrochloride. [531-85-1]

(1,1'-BIPHENYL)-4,4'-DIAMINE, 3,3'-DIMETHOXY-. [119-90-4]
*RCRA 40 CFR 261
EPA hazardous waste no. U091

(1,1'-BIPHENYL)-4,4'-DIAMINE, 3,3'-DIMETHYL)-. [119-93-7]
*RCRA 40 CFR 261
EPA hazardous waste no. U095

3,3',4,4'-BIPHENYL-TETRAMINE TETRAHYDROCHLORIDE. Syn: 3,3'-Diaminobenzidine tetrahydrochloride. [7411-49-6]
OSHA Candidate List of Potential Occupational Carcinogens, 45 FR 53672, August 12, 1980

BIS(1-AZIRIDINYL)MORPHOLINO-PHOSPHINE SULFIDE.
IARC—carcinogenic in animals

BISAZOBIPHENYL DYES derived from benzidine and its congeners, orthotoluidine(dimethylbenzidine) and dianisidine (dimethoxy-benzidine). See Individual dyes
*TSCA 8(d), Unpublished Health and Safety Studies Reporting, 40 CFR 716, 47 FR 387, September 2, 1982

BIS(2-CHLOROETHOXY)METHANE.
CWA 307(a)(1), Priority Pollutant
*RCRA 40 CFR 261
EPA hazardous waste no. U024

BIS(CHLOROETHYL)ETHER. [6986-48-7, 111-44-4]
OSHA Candidate List of Potential Occupational Carcinogens (EPA Assessment Group List), 45 FR 53672, August 12, 1980

CWA 307(a)(1), Priority Pollutant
IARC—carcinogenic in animals

N,N-BIS(2-CHLOROETHYL)-2-NAPHTHYLAMINE. [494-03-1]
IARC—carcinogenic in humans

BIS(2-CHLOROISOPROPYL) ETHER. [108-60-1]
*RCRA 40 CFR 261
EPA hazardous waste no. U027

1,2-BIS(CHLOROMETHOXY) ETHANE. [13483-18-6]
IARC—carcinogenic in animals

1,4'-BIS(CHLOROMETHOXYMETHYL) BENZENE. [56894-91-8]
IARC—carcinogenic in animals

BIS(CHLOROMETHYL)ETHER. [542-88-1]
IARC—carcinogenic in humans
*RCRA 40 CFR 261
EPA hazardous waste no. P016

BIS(2-CHLORO-1-METHYLETHYL)ETHER (BCME). [108-60-1]
NCI—carcinogenic in animals
TSCA-CHIP available

BIS(DIMETHYLTHIOCARBAMOYL)DISULFIDE. [137-26-8]
*RCRA 40 CFR 261
EPA hazardous waste no. U244

BIS(2-ETHYLHEXYL)PTHALATE. [117-81-7]
*RCRA 40 CFR 261
EPA hazardous waste no. U028

BIS(2-ETHYLHEXYL)TEREPHTHALATE. [6422-86-2]
TSCA 4(e), ITC
Eleventh Report of the Interagency Testing Committee to the administrator, Environmental Protection Agency, November 1982; 47 FR 54626, December 3, 1982

BISMUTH TELLURIDE, SE-DOPPED. [1304-82-1]

ACGIH-TLV
TWA: 10 mg/m^3
STEL: 20 mg/m^3

ACGIH-TLV
TWA: 5 mg/m^3
STEL: 10 mg/m^3

BISPHENOL A. [80-05-7]

NCI—may be carcinogenic in animals

TSCA-CHIP available

BISPHENOL A-DIGLYCIDYL ETHER.
See Propane, 2,2-bis(2,3-epoxypropoxyphenyl)-. [1675-54-3]

BKLFI-2.

FIFRA—Registration Standard—Issued December 1981 (disinfectant)

BLEACHING POWDER. See Calcium hypochlorite. [7778-54-3]

BLUE VRS. [129-17-9]

IARC—carcinogenic in animals

BOLLEN.

FIFRA—Data Call In, 45 FR 75488, November 14, 1980
Status: letter issued 8/29/81

BOLSTAR® (SULPROFOS).
[35400-43-2]

FIFRA—Registration Standard—Issued August 1981 (insecticide)

BORATES, TETRA-, SODIUM SALTS, ANHYDROUS. [1303-96-4]

ACGIH-TLV
TWA: 1 mg/m^3

BORATES, TETRA-, SODIUM SALTS, DECAHYDRATE. [1303-96-4]

ACGIH-TLV
TWA: 5 mg/m^3

BORATES, TETRA-, SODIUM SALTS, PENTAHYDRATE. [1303-96-4]

ACGIH-TLV
TWA: 1 mg/m^3

BORIC ACID and SALTS. [10043-35-3]

FIFRA—Registration Standard Development—Data Evaluation and Development of Regulatory Position (insecticide)

BORON HYDRIDES. See Decaborane and Diborane. [17702-41-9, 19287-45-7]

BORON OXIDE. [1303-86-2]

*OSHA 29 CFR 1910.1000 Table Z-1
PEL(TWA): 15 mg/m^3

ACGIH-TLV
TWA: 10 mg/m^3
STEL: 20 mg/m^3

BORON TRIBROMIDE. [10294-33-4]

ACGIH-TLV
TWA: 10 mg/m^3
STEL: 30 mg/m^3

BORON TRIFLUORIDE. [7636-07-2]

*OSHA 29 CFR 1910.1000 Table Z-1
PEL(Ceiling): 1 ppm, 3 mg/m^3

ACGIH-TLV
TWA: ceiling limit, 1 ppm, 3 mg/m^3

NIOSH Criteria Document (Pub. No. 77-122), NTIS Stock No. PB 274747, December 14, 1976

The standard is designed to protect the health and safety of workers for up to a 10-hour workday, 40-hour workweek, over a working lifetime. Sufficient technology exists to prevent adverse effects in workers, but techniques to measure airborne levels of boron trifluoride for compliance with an environmental limit are not adequate. Therefore, an environmental limit is not recommended herein, in part because of the unavailablity of adequate monitoring methods. Work practices and engineering controls are recommended for control of exposure since reliable environmental data will not be available. The standard will be subject to review and revision as more information is acquired.

Boron trifluoride is highly reactive in the presence of water vapor. A dense, white mist is formed when boron trifluoride reacts with the moisture present in the air. The hydration and hydrolysis of boron trifluoride appear to be rapid and extensive. The term "boron trifluoride" as used throughout this document refers to the unreacted gas, the products formed by the reaction of boron trifluoride gas upon release into the environment, or both. "Occupational exposure to boron trifluoride" is defined as working in areas where boron trifluoride is manufactured, used, or handled, or is evolved as a result of chemical processes.

NFPA—hazardous chemical
health—3, flammability—0, reactivity—1

BRILLIANT BLUE FCF (DIAMMONIUM and DISODIUM SALTS). [3844-45-9], [2650-18-2]

IARC—carcinogenic in animals

TSCA-CHIP available

BROMACIL and SALTS. [314-40-9]

ACGIH-TLV
TWA: 1 ppm, 10 mg/m^3
STEL: 2 ppm, 20 mg/m^3

FIFRA—Registration Standard Development—Data Evaluation and Development of Regulatory Position (herbicide)

BROMINE. [7726-95-6]

*OSHA 29 CFR 1910.1000 Table Z-1
PEL(TWA): 0.1 ppm, 0.7 mg/m^3

ACGIH-TLV
TWA: 0.1 ppm, 0.7 mg/m^3
STEL: 0.3 ppm, 2 mg/m^3

NFPA—hazardous chemical
health—4, flammability—0, reactivity—0,
oxy—oxidizing chemical

BROMINE and BROMINE COMPOUNDS. [74-83-9, 74-96-4, 75-25-2, 78-75-1, 96-12-8, 106-93-4, 107-04-0, 11-24-0, 557-91-5, 594-34-3, 7926-95-6]

TSCA-CHIP available

BROMINE CYANIDE. See Cyanogen bromide. [506-68-3]

BROMINE PENTALFLUORIDE. [7789-30-2]

ACGIH-TLV
TWA: 0.1 ppm, 0.7 mg/m^3
STEL: 0.3 ppm, 2 mg/m^3

NFPA—hazardous chemical
health—4, flammability—0, reactivity—3,
W—water may be hazardous in fire fighting,
oxy—oxidizing chemical, reacts explosively with water or steam to produce toxic and corrosive fumes

BROMINE TRIFLUORIDE. [7787-71-5]

NFPA—hazardous chemical
health—4, flammability—0, reactivity—3,
W—water may be hazardous in fire fighting,
oxy—oxidizing chemical, reacts explosively with water or steam to produce toxic and corrosive fumes

BROMOACETONE. See Propanone, 1-bromo-. [598-31-2]

BROMOANILINE. See Aniline and chloro-, bromo- and/or nitro- anilines. [106-40-1]

4-BROMOBENZENAMINE. See Aniline and chloro-, bromo- and/or nitro- anilines. [106-40-1]

BROMOCHLOROMETHANE. See Chlororobromomethane. [74-97-5]

2-BROMO-6-CHLORO-4-NITROANILINE. [99-29-6]

2-BROMO-4,6-DINITROANILINE. See Aniline and chloro-, bromo- and/or nitro- anilines. [1817-73-9]

2-BROMO-4,6-DINITRO-BENZENAMINE. See Aniline and chloro-, bromo-, and/or anilines. [1817-73-8]

BROMOETHANE. See Methyl bromide. [74-96-4]

BROMOFORM. [75-25-2]

*OSHA 29 CFR 1910.1000 Table Z-1

Skin
PEL(TWA): 0.5 ppm, 5 mg/m³
ACGIH-TLV
TWA: 0.5 ppm, 5 mg/m³
CWA 307(a)(1), 40 CFR 125, Priority Pollutant
*RCRA 40 CFR 261
EPA hazardous waste no. U225

1-BROMOPENTANE.
NFPA—hazardous chemical (flammable chemical)
health—1, flammability—3, reactivity—0

4-BROMOPHENYL PHENYL ETHER.
*RCRA 40 CFR 261
EPA hazardous waste no. U030

3-BROMOPROPENE-1. See Allyl bromide. [106-95-6]

3-BROMOPROPYNE. [106-96-7]
NFPA—hazardous chemical
health—4, flammability—3, reactivity—4

BRUCINE. [357-57-3]
*RCRA 40 CFR 261
EPA hazardous waste no. P018

BTF. See Benzotrifluoride. [98-08-8]

BUFENCARB. [2282-34-0]
FIFRA—Data Call In, 45 FR 75488, November 14, 1980
Status: agency decision reached
NFPA—hazardous chemical
health—2, flammability—4, reactivity—2

BUTADIENE (1,3-BUTADIENE). [106-99-0]
*OSHA 29 CFR 1910.1000 Table Z-1
PEL(TWA): 1000 ppm, 2200mg/m³
ACGIH-TLV
TWA: 1000 ppm, 2200 mg/m³
STEL: 1250 ppm, 2750 mg/m³
NFPA—hazardous chemical
health—2, flammability—4, reactivity—2

1,3-BUTADIENE, 2-CHLORO-. [126-99-8]
*TSCA 8(a), 40 CFR 712, 47 FR 26992, June 22, 1982, Chemical Information Rule
TSCA-CHIP available

1,3-BUTADIENE, 1,1,2,3,4,4-HEXACHLORO-. [87-68-3]
*TSCA 8(a), 40 CFR 712, 47 FR 26992, June 22, 1982, Chemical Information Rule
*RCRA 40 CFR 261
EPA hazardous waste no. U128

BUTANAL. See Butyraldehyde. [123-72-8]

BUTANAMIDE, *N,N*'-(3,3'-DIMETHYL[1,1'-BIPHENYL]-4,4'-DIYL)BIS[3-OXO-]. See Bisazobiphenyl dyes. [91-96-3]
*TSCA 8(d), Unpublished Health and Safety Studies Reporting, 40 CFR 716, 47 FR 387, September 2, 1982

1-BUTANAMINE, *N*-BUTYL-*N*-NITROSO-. [924-16-3]
*RCRA 40 CFR 261
EPA hazardous waste no. U172

BUTANE. [106-97-8]
ACGIH-TLV
TWA: 800 ppm, 1900 mg/m³
NFPA—hazardous chemical (flammable chemical)
health—1, flammability—4, reactivity—0

1,4-BUTANEDIOL DIMETHANESULFONATE (MYLERAN). [55-98-1]
IARC—carcinogenic in animals

BUTANETHIOL. See Butyl mercaptan. [109-79-5]
See Thiols, NIOSH Critera Document.

BUTANOIC ACID. See Butyric acid. [107-92-6]

BUTANOIC ACID, 4-[BIS(2-CHLORO-ETHYL)AMINO]BENZENE-.
[305-03-3]

*RCRA 40 CFR 261
EPA hazardous waste no. U035

1-BUTANOL. [71-36-3]

*RCRA 40 CFR 261
EPA hazardous waste no. U031 (ignitable waste)

2-BUTANOL. [78-92-2]

NFPA—hazardous chemical (flammable chemical)
health—1, flammability—3, reactivity—0

2-BUTANONE. See Methyl ethyl ketone (MEK). [78-93-3]

2-BUTANONE PEROXIDE. See Methyl ethyl ketone peroxide. [1338-23-4]

2-BUTENAL. See Crotonaldehyde. [123-73-9]

2-BUTENE, 1,4-DICHLORO-.
[764-41-0]

*RCRA 49 CFR 172.101
EPA hazardous waste no. U074 (ignitable waste, toxic waste)

BUTENES.

NFPA—hazardous chemical (flammable chemical)
health—1, flammability—4, reactivity—0

2-BUTOXYETHANOL (BUTYL CELLOSOLVE). [111-76-2]

*OSHA 29 CFR 1910.1000 Table Z-1

Skin

PEL(TWA): 50 ppm, 240 mg/m^3
ACGIH-TLV

Skin

TWA: 25 ppm, 120 mg/m^3
STEL: 75 ppm, 360 mg/m^3

BUTTER OF ARSENIC. See Arsenic chloride. [7784-34-1]

BUTYL ACETATE (n-BUTYL ACETATE). [123-86-4]

*OSHA 29 CFR 1910.1000 Table Z-1
PEL(TWA): 150 ppm, 710 mg/m^3
ACGIH-TLV
TWA: 150 ppm, 710 mg/m^3
STEL: 200 ppm, 950 mg/m^3
NFPA—hazardous chemical (flammable chemical
health—1, flammability—3, reactivity—0

*CWA 311(b)(2)(A), 40 CFR 116, 117
Discharge RQ = 5000 pounds (2270 kilograms)

sec-BUTYL ACETATE. [105-46-4]

*OSHA 29 CFR 1910.1000 Table Z-1
PEL(TWA): 200 ppm, 950 mg/m^3
ACGIH-TLV
TWA: 200 ppm, 950 mg/m^3
STEL: 250 ppm, 1190 mg/m^3

tert-BUTYL ACETATE. [540-88-5]

*OSHA 29 CFR 1910.1000 Table Z-1
PEL(TWA): 200 ppm, 950 mg/m^3
ACGIH-TLV
TWA: 200 ppm, 950 mg/m^3
STEL: 250 ppm, 1190 mg/m^3

BUTYL ACRYLATE. [141-32-2]

ACGIH-TLV
TWA: 10 ppm, 55 mg/m^3
NFPA—hazardous chemical
health—2, flammability—2, reactivity—2

BUTYL ALCOHOL. [71-36-3]

*OSHA 29 CFR 1910.1000 Table Z-1
PEL(TWA): 100 ppm, 300 mg/m^3
ACGIH-TLV

Skin

TWA(Ceiling): 50 ppm, 150 mg/m^3
*RCRA 40 CFR 261
EPA hazardous waste no. U031 (ignitable waste)

sec-BUTYL ALCOHOL. [78-92-2]

*OSHA 29 CFR 1910.1000 Table Z-1
PEL(TWA): 150 ppm, 450 mg/m^3

ACGIH-TLV
TWA: 100 ppm, 305 mg/m^3
STEL: 150 ppm, 455 mg/m^3

tert-BUTYL ALCOHOL. [75-65-0]
*OSHA 29 CFR 1910.1000 Table Z-1
PEL(TWA): 100 ppm, 300 mg/m^3
ACGIH-TLV
TWA: 100 ppm, 300 mg/m^3
STEL: 150 ppm, 450 mg/m^3
NFPA—hazardous chemical (flammable chemical)
health—1, flammability—3, reactivity—0

BUTYLAMINE. [109-73-9]
*OSHA 29 CFR 1910.1000 Table Z-1
Skin
PEL(Ceiling): 5 ppm, 15 mg/m^3
ACGIH-TLV
Skin
TWA: ceiling limit, 5 ppm, 15 mg/m^3
NFPA—hazardous chemical
health—2, flammability—3, reactivity—0
*CWA 311(b)(2)(A), 40 CFR 116, 117
Discharge RQ = 1000 pounds (454 kilograms)

BUTYLATE. [2008-41-5]
FIFRA—Data Call In, 45 CFR 75488, November 14, 1980
Status: letter issued 12/10/81
FIFRA—Registration Standard Development—Data Collection (herbicide)

t-BUTYL BENZALDEHYDE. [939-97-9]
TSCA-CHIP available

n-BUTYL BENZOATE. [136-60-7]
NFPA—hazardous chemical (flammable chemical)
health—1, flammability—1, reactivity—0

t-BUTYL BENZOIC ACID. [98-73-7]
TSCA-CHIP available

BUTYL BENZYL PHTHALATE.
[85-68-7]
NCI—may be carcinogenic in animals
CWA 307(a)(1), 40 CFR 125, Priority Pollutant

tert-BUTYL CHROMATE, as CrO$_3$
[1189-85-1]
*OSHA 29 CFR 1910.1000 Table Z-1
Skin
PEL(Ceiling): 0.1 mg/m^3 (as CrO$_3$)
ACGIH-TLV
Skin
TWA: ceiling limit, 0.1 mg/m^3

1,2-BUTYLENE OXIDE. [106-88-7]
NFPA—hazardous chemical
health—2, flammability—3, reactivity—2

BUTYL ETHER. See Dibutyl ether.
[142-96-1]

n-BUTYL GLYCIDYL ETHER (BGE).
[2426-08-6]
*OSHA 29 CFR 1910.1000 Table Z-1
PEL(TWA): 50 ppm, 270 mg/m^3
ACGIH-TLV
TWA: 25 ppm, 135 mg/m^3
See Glycidyl ethers, NIOSH Criteria Document.
TSCA 4(e)
Third Report of the TSCA Interagency Testing Committee to the administrator, Environmental Protection Agency, October 1978; 43 FR 50630, October 30, 1979

BUTYL GLYCOLYL BUTYL PHTHALATE.
[85-70-1]
TSCA 4(e), ITC
Seventh Report of the TSCA Interagency Testing Committee to the administrator, Environmental Protection Agency, October 1980; 45 FR 78432, November 25, 1980, EPA responded to the committee's recommendations for testing 46 FR 54487, November 2, 1981
*TSCA 8(d), Unpublished Health and Safety Studies Reporting, 40 CFR 716, 47 FR

38800, September 2, 1982; 48 FR 13178, March 30, 1983

t-BUTYL HYDROPEROXIDE. [75-91-2]
NFPA—hazardous chemical
health—1, flammability—4, reactivity—4, oxy—oxidizing chemical
TSCA-CHIP available

n-BUTYL LACTATE. [138-22-7]
ACGIH-TLV
TWA: 5 ppm, 25 mg/m^3

BUTYLLITHIUM in hydrocarbon solvents.
NFPA—hazardous chemical
health—3, flammability—4, reactivity—2,
W—water may be hazardous in fire fighting

BUTYL MERCAPTAN. [109-79-5]
*OSHA 29 CFR 1910.1000 Table Z-1
PEL(TWA): 10 ppm, 35 mg/m^3
ACGIH-TLV
TWA: 0.5 ppm, 1.5 mg/m^3

t-BUTYL PERBENZOATE. [614-45-9]
NFPA—hazardous chemical
health—1, flammability—3, reactivity—4, oxy—oxidizing chemical

t-BUTYL PEROXYACETATE (75% solution in benzene or mineral spirits). [107-71-1]
NFPA—hazardous chemical
health—2, flammability—3, reactivity—4, oxy—oxidizing chemical

t-BUTYL PEROXYPIVALATE (75% solution in mineral spirits). [927-07-1]
NFPA—hazardous chemical
health—0, flammability—3, reactivity—4, oxy—oxidizing chemical

o-sec-BUTYLPHENOL. [89-72-5]
ACGIH-TLV
Skin
TWA: 5 ppm, 30 mg/m^3

n-BUTYL PHTHALATE. [84-74-2]
*CWA 311(b)(2)(A), 40 CFR 116, 117
Discharge RQ = 100 pounds (45.4 kilograms)

t-BUTYL TOLUENE. [98-51-1]
*OSHA 29 CFR 1910.1000 Table Z-1
PEL(TWA): 10 ppm, 60 mg/m^3
ACGIH-TLV
TWA: 10 ppm, 60 mg/m^3
STEL: 20 ppm, 120 mg/m^3
TSCA-CHIP available

BUTYRALDEHYDE (NORMAL and ISO). [123-72-8-, 78-84-2]
NFPA—hazardous chemical
health—2, flammability—3, reactivity—0

BUTYRIC ACID. [107-92-6]
*CWA 311(b)(3)(A), 40 CFR 116, 117
Discharge RQ = 5000 pounds (2270 kilograms)
NFPA—hazardous chemical
health—2, flammability—2, reactivity—0

BUTYRIC ACID, 2-AMINO-4-(ETHYLTHIO)-, DL-. Syn: DL-ethionine. [67-21-0]
OSHA Candidate List of Potential Occupational Carcinogens, 45 FR 53672, August 12, 1980

BUTYRIC ACID, 2-AMINO-4-(ETHYLTHIO)-, L-. Syn: Ethionine. [13073-35-3]
OSHA Candidate List of Potential Occupational Carcinogens, 45 FR 53672, August 12, 1980

β-BUTYROLACTONE. [3068-88-0]
IARC—carcinogenic in animals

ISOBUTYRONITRILE. [78-82-0]
See Nitriles, NIOSH Criteria Document

n-BUTYRONITRILE. [109-74-0]
See Nitriles, NIOSH Criteria Document

C

CACODYLIC ACID. [75-60-5]
40 CFR 162.11, FIFRA-RPAR
Notice of Intent to Register Issued
Current Status: Insufficient evidence to issue a RPAR. Decision document in agency review. Returned to Registration Division in December 1981.
Criteria Possibly Met or Exceeded: Oncogenicity, Mutagenicity, Teratogenicity, Fetotoxicity, and Reproductive Effects
*RCRA 40 CFR 261
EPA hazardous waste no. U136

CADMIUM [7440-43-9]

NIOSH Criteria Document (Pub. No. 76-192), NTIS Stock No. PB 274237, August 23, 1976
"Cadmium" refers to elemental cadmium and all cadmium compounds. An "action level" is defined as half the time-weighted average concentration environmental limit of cadmium. "Occupational exposure to cadmium" is defined as exposure to cadmium at a concentration greater than the action level. Exposures at lower environmental concentrations will require adherence to Section 6(b) and 7(d). See the Criteria Document for details.

Occupational exposure to cadmium shall be controlled so that workers are not exposed to cadmium at a concentration greater than 40 μg Cd/m^3 determined as a time-weighted average (TWA) exposure concentration for up to a 10-hour workday, 40-hour workweek, or at a ceiling concentration greater than 200 μg Cd/m^3 for any 15-minute sampling period. The standard is designed to protect the health and safety of workers for up to a 10-hour workday, 40-hour workweek, over a working lifetime.

40 CFR 162.11, FIFRA-RPAR issued
Current Status: PD 1 published, 42 FR 56574 (10/26/77); comment period closed 2/10/78
Risk/benefit assessment is ongoing.
Criteria Possibly Met or Exceeded: Oncogenicity, Mutagenicity, Teratogenicity, and Fetotoxicity.
CWA 304(a)(1), 45 FR 79318, November 28, 1980, Water Quality Criteria

Freshwater Aquatic Life

For total recoverable cadmium the criterion (in micrograms per liter) to protect freshwater aquatic life as derived using the Guidelines is the numerical value given by e (1.05 [ln (hardness)] $-$ 8.53) as a 24-hour average and the concentration (in micrograms per liter) should not exceed the numerical value given by e (1.05[ln (hardness) $-$ 3.73) at any time. For example, for hardnesses of 50, 100, and 200 mg/liter as CaCO$_3$ the criteria are 0.012, 0.025, and 0.051 μg/liter, respectively, and the concentration of total recoverable cadmium should not exceed 1.5, 3.0, and 6.3 μg/liter respectively, at any time.

Saltwater Aquatic Life

For total recoverable cadmium the criterion to protect saltwater aquatic life as derived using the guidelines is 4.5 μg/liter as a 24-hour average and the concentration should not exceed 59 μg/liter at any time.

Human Health

The ambient water quality criterion for cadmium is recommended to be identical to the existing drinking water standard which is 10 μg/liter. Analysis of the toxic effects data resulted in a calculated level that is protective of human health against the ingestion of contaminated water and contaminated aquatic organisms. The calculated value is comparable to the present standard. For this reason a selective criterion based on exposure solely from consumption of 6.5 g of aquatic organisms was not derived.

CWA 307(a)(1), Priority Pollutant

*National Interim Primary Drinking Water Regulations, 40 CFR 141; 40 FR 59565, December 24, 1975; amended by 41 FR 28402, July 9, 1976; 44 FR 68641, November 29, 1979; corrected by 45 FR 15542, March 11, 1980; 45 FR 57342, August 27, 1980; 47 FR 18998, March 3, 1982; corrected by 47 FR 10998, March 12, 1982

Maximum contaminant level—0.010 mg/liter

CAA 112

Chemical proposed to be assessed as a Hazardous Air Pollutant by a House of Representatives bill to amend the CAA

CADMIUM and compounds. [7440-43-9]

OSHA Candidate List of Potential Occupational Carcinogens (EPA Carcinogen Assessment Group List), 45 FR 53672, August 12, 1980

CADMIUM ACETATE. [543-90-8]

*CWA 311(b)(2)(A), 40 CFR 116, 117
Discharge RQ = 100 pounds (4.54 kilograms)

CADMIUM BROMIDE. [7789-42-6]

*CWA 311(b)(2)(A), 40 CFR 116, 117
Discharge RQ = 100 pounds (45.4 kilograms)

CADMIUM CHLORIDE. [10108-64-2]

IARC—carcinogenic in animals

*CWA 311(b)(2)(A), 40 CFR 116, 117
Discharge RQ = 100 pounds (4.54 kilograms)

CADMIUM DUST. [7440-43-9]

*OSHA 1910.1000 Table Z-2
0.2 mg/m^3, 8-hour TWA
0.6 mg/m^3, acceptable ceiling concentration
ACGIH-TLV
TWA: 0.05 mg/m^3
STEL: 0.2 mg/m^3

CADMIUM FUME (CADMIUM OXIDE). [1306-19-0]

*OSHA 1910.1000 Table Z-2
0.1 mg/m^3, 8-hour TWA
0.3 mg/m^3, acceptable ceiling concentration
ACGIH-TLV
TWA: ceiling limit, 0.05 ppm
ACGIH-TLV (production as cadmium)
TWA: 0.05 mg/m^3
Industrial substance suspect of carcinogenic potential for humans—which is suspect of inducing cancer, based on either (1) limited epidemiologic evidence, exclusive of clinical reports of single cases, or (2) demonstration of carcinogenesis in one or more animal species by appropriate methods. Worker exposure by all routes should be carefully controlled to levels consistent with the animal and human experience data. Intended changes for 1982.

IARC—carcinogenic in animals (cadmium powder)

IARC—carcinogenic in humans (cadmium-using industries—possibly cadmium oxide)

CADMIUM SULFATE. [10124-36-4]

IARC—carcinogenic in animals

CADMIUM SULFIDE. [1306-23-6]

IARC—carcinogenic in animals

CALCIUM. [7440-70-2]

NFPA—hazardous chemical
health—1, flammability—1, reactivity—2,
W—water may be hazardous in fire fighting

CALCIUM ARSENATE. See Arsenic. [7778-44-1]

*CWA 311(b)(2)(A), 40 CFR 116, 117
Discharge RQ = 1000 pounds (454 kilograms)

CALCIUM ARSENITE. See Arsenic. [52740-16-6]

*CWA 311(b)(2)(A), 40 CFR 116, 117
Discharge RQ = 1000 pounds (454 kilograms)

CALCIUM BICHROMATE.
See Dichromates.

CALCIUM CARBIDE. [75-20-7]
NFPA—hazardous chemical
health—1, flammability—4, reactivity—2, W—water may be hazardous in fire fighting
*CWA 311(b)(2)(A), 40 CFR 116, 117
Discharge RQ = 5000 pounds (2270 kilograms)

CALCIUM CARBONATE (MARBLE). [1317-65-3]
ACGIH-TLV
TWA: Nuisance particulate, 30 mppcf or 10 mg/m^3 of total dust < 1% quartz, or 5 mg/m^3 respirable dust
STEL: 20 mg/m^3

CALCIUM CHLORATE. [10137-74-3]
NFPA—hazardous chemical
when nonfire: health—0, flammability—0, reactivity—2, oxy—oxidizing chemical
when fire: health—2, flammability—0, reactivity—2, oxy—oxidizing chemical

CALCIUM CHROMATE. [13765-19-0]
OSHA Candidate List of Potential Occupational Carcinogens (EPA Carcinogen Assessment Group List), 45 FR 53672, August 12, 1980
IARC—carcinogenic in animals
*CWA 311(b)(2)(A), 40 CFR 116, 117
Discharge RQ = 1000 pounds (454 kilograms)
*RCRA 40 CFR 261
EPA hazardous waste no. U032 (hazardous substance)

CALCIUM CYANAMIDE. [156-62-7]
ACGIH-TLV
TWA: 0.5 mg/m^3
STEL: 1 mg/m^3

CALCIUM CYANIDE. [592-01-8]
See Cyanide, hydrogen and cyanide salts, NIOSH Criteria Document.

NFPA—hazardous chemical
health—3, flammability—0, reactivity—0
*CWA 311(b)(2)(A), 40 CFR 116, 117
Discharge RQ = 1000 pounds (454 kilograms)
*RCRA 40 CFR 261
EPA hazardous waste no. P021 (hazardous substance)

CALCIUM DICHROMATE.
See Dichromates.

CALCIUM DODECYLBENZENESULFONATE [26264-06-2]
*CWA 311(b)(2)(A), 40 CFR 116, 117
Discharge RQ = 1000 pounds (454 kilograms)

CALCIUM HYDROXIDE. (LIME) [1305-62-0]
ACGIH-TLV
TWA: 5 mg/m^3
*CWA 311(b)(2)(A), 40 CFR 116, 117 (deleted)
Discharge RQ = 5000 pounds (2270 kilograms)

CALCIUM HYPOCHLORITE (BLEACHING POWDER). [7778-54-3]
NFPA—hazardous chemical
when nonfire: health—1, flammability—0, reactivity—2, oxy—oxidizing chemical
when fire: health—2, flammability—0, reactivity—2, oxy—oxidizing chemical
*CWA 311(b)(2)(A), 40 CFR 116, 117
Discharge RQ = 100 pounds (4.54 kilograms)

CALCIUM NAPTHENATE. [61789-36-4]
TSCA 4(e), ITC
Twelfth Report of the Interagency Testing Committee to the administrator, Environmental Protection Agency, May 1983; 48 FR 24443, June 1, 1983
*TSCA 8(a), 40 CFR 712, 48 FR 28443, June 22, 1983, Chemical Information Rule

CALCIUM OXIDE (QUICKLIME). [1305-78-8]
*OSHA 29 CFR 1910.1000 Table Z-1
PEL(TWA): 5 mg/m^3

ACGIH-TLV
TWA: 2 mg/m³
NFPA—hazardous chemical
health—1, flammability—0, reactivity—1
*CWA 311(b)(2)(A), 40 CFR 116, 117 (deleted)
Discharge RQ = 5000 pounds (2270 kilograms)

CALCIUM SILICATE.
ACGIH-TLV
TWA: Nuisance particulate, 30 mppcf or 10 mg/m³ or total dust < 1% quartz, or 5 mg/m³ respirable dust

CAMPHENE, OCTACHLORO-.
See Chlorinated camphene. [8001-35-2]

CAMPHOR, SYNTHETIC. [76-22-2]
*OSHA 29 CFR 1910.1000 Table Z-1
PEL(TWA): 2 mg/m³
ACGIH-TLV
TWA: 2 ppm, 12 mg/m³
STEL: 3 ppm, 18 mg/m³

CANTHARIDIN. [56-25-7]
IARC—carcinogenic in animals

CAPROLACTAM. [105-60-2]
ACGIH-TLV (dust)
TWA: 1 mg/m³
STEL: 3 mg/m³
ACGIH-TLV (vapor)
TWA: 5 ppm, 20 mg/m³
STEL: 10 ppm, 40 mg/m³

CAPTAFOL. [2425-06-1]
ACGIH-TLV

Skin

TWA: 0.1 mg/m³
FIFRA—Data Call In, 45 FR 75488, November 14, 1980
Status: letter issued 9/24/81

CAPTAN. [133-06-2]
ACGIH-TLV
TWA: 5 mg/m³
STEL: 15 mg/m³

NCI—carcinogenic in animals
40 CFR 162.11, FIFRA-RPAR issued
Current Status: PD 1 published, 45 FR 54938 (8/18/80); comment period closed 12/1/80: NTIS# PB81 109449. Risk/benefit assessment is ongoing.
Criteria Possibly Met or Exceeded: Oncogenicity, Mutagenicity, and Other Chronic Effects
*CWA 311(b)(2)(A), 40 CFR 116, 117
Discharge RQ = 10 pounds (4.54 kilograms)

CARBAMIC ACID, BIS(2-HYDROXYETHYL)DITHIO-, MONOPOTASSIUM SALT. See Potassium bis(2-hydroxyethyl)dithiocarbamate. [23746-34-1]

CARBAMIC ACID, ETHYL ESTER. [51-79-6]
*RCRA 40 CFR 261
EPA hazardous waste no. U238

CARBAMIC ACID, METHYLNITROSO-, ETHYL ESTER. [615-53-2]
*RCRA 40 CFR 261
EPA hazardous waste no. U178

CARBAMIDE, N-ETHYL-N-NITROSO-. [759-73-9]
*RCRA 40 CFR 261
EPA hazardous waste no. U176

CARBAMIDE, N-METHYL-N-NITROSO-. [684-93-5]
*RCRA 40 CFR 261
EPA hazardous waste no. U177

CARBAMIDE, THIO-. [62-56-6]
*RCRA 40 CFR 261
EPA hazardous waste no. U219

CARBAMIMIDOSELENOIC ACID.
*RCRA 40 CFR 261
EPA hazardous waste no. P103

CARBAMOYL CHLORIDE, DIMETHYL-. [79-44-7]
*RCRA 40 CFR 261
EPA hazardous waste no. U097

CARBAMYLHYDRAZINE. *See* Semicarbazide. [57-56-7]

CARBAMYLHYDRAZINE HYDROCHLORIDE. *See* Semicarbazide monohydrochloride [563-41-7]

CARBARYL (SEVIN®). [63-25-2]
*OSHA 29 CFR 1910.1000 Table Z-1
PEL(TWA): 5 mg/m^3
ACGIH-TLV
TWA: 5 mg/m^3
STEL: 10 mg/m^3
NIOSH Criteria Document (Pub. No. 77-107), NTIS Stock No. PB 273801, September 30, 1976
"Carbaryl" is the generic name for the 1-naphthyl ester of *N*-methylcarbamic acid or 1-naphthyl *N*-methylcarbamate. "Action level" is defined as one-half the recommended time-weighted average (TWA) environmental exposure limit for carbaryl. "Occupational exposure to carbaryl" is defined as exposure to airborne carbaryl at concentrations greater than the action level. Exposure to carbaryl at concentrations less than or equal to the action level shall not require adherence to the recommended standard, except for Sections 3, 4(a,b), and 7(b). See the Criteria Document for details.

Occupational exposure to carbaryl shall be controlled so that no employee is exposed to carbaryl at concentrations greater than 5 mg/m^3 in air determined as a TWA concentration for up to a 10-hour work shift, 40-hour workweek.

The standard is designed to protect the health and safety of employees for up to a 10-hour work shift, 40-hour workweek, over a working lifetime.

40 CFR 162.11, FIFRA-RPAR, Notice of Intent to Register Issued
Current Status: 45 FR 81869 (12/12/80). The agency has returned this compound to the registration process with the stipulation that the following measures be considered: (1) a FIFRA Section 3(c)(2)(B) action that registrants remedy data limitations identified in pre-RPAR review and, (2) appropriate label changes be made to minimize exposure to carbaryl. Both measures are being considered in the registration review of this chemical.
Criteria Possibly Met or Exceeded: Oncogenicity, Mutagenicity, and Teratogenicity
FIFRA—Registration Standard Development—Data Evaluation and Development of Regulatory Position (insecticide)
*CWA 311(b)(2)(A), 40 CFR 116, 117
Discharge RQ = 100 pounds (4.54 kilograms)

CARBAZOLE, 3-AMINO-9-ETHTYL-.
Syn: 3-Amino-*N*-ethylcarbazole. [132-32-1]
OSHA Candidate List of Potential Occupational Carcinogens, 45 FR 53672, August 12, 1980
NCI—carcinogenic in animals
TSCA-CHIP available

CARBOFURAN. [1563-66-2]
ACGIH-TLV
TWA: 0.1 mg/m^3
FIFRA—Data CAll In, 45 FR 75488, November 14, 1980
Status: agency decision reached, no additional chronic toxicological data required at this time
*CWA 311(b)(2)(A), 40 CFR 116, 117
Discharge RQ = 10 pounds (4.54 kilograms)

CARBOFURAN INTERMEDIATES. *See* individual carbofurans.
TSCA 4(e), ITC
Eleventh Report of the Interagency Testing Committee to the administrator, Environmental Protection Agency, November 1982; 47 FR 54626, December 3, 1982, recommended but not designated for response within 12 months

CARBOLIC ACID. *See* Phenol. [108-95-2]

CARBON BISULFIDE. *See* Carbon disulfide. [75-15-0]

CARBON BLACK. [7440-44-0]

*OSHA 29 CFR 1910.1000 Table Z-1
PEL(TWA): 3.5 mg/m^3

ACGIH-TLV
TWA: 3.5 mg/m^3
STEL: 7 mg/m^3

NIOSH Criteria Document (Pub. No. 78-204), NTIS Stock No. PB 81-225625, September 1978

The term "carbon black" refers to material consisting of more than 85% elemental carbon in the form of near-spherical colloidal particles and coalesced particle aggregates of colloidal size obtained by partial combustion or thermal decomposition of hydrocarbons. If carbon black contains polycyclic aromatic hydrocarbons (PAHs), that is, if it contains cyclohexane-extractable substances at a concentration greater than 0.1%, "occupational exposure to carbon black" is defined as any work involving any contact, airborne or otherwise, with this substance. If carbon black contains cyclohexane-extractable substances at a concentration of 0.1% or less, "occupational exposure to carbon black" is defined as any work involving exposure to carbon black at a concentration greater than half the recommended environmental limit of 3.5 mg/m^3; exposure to this carbon black at lower concentrations will not require adherence to all sections except Sections 2(a), 3(a), 6(b-e), 7, and 8(a). See the Criteria Document for details.

Particulate polycyclic organic material (PPOM), polynuclear aromatic hydrocarbons (PNAs), and PAHs are terms frequently encountered in the literature and sometimes used interchangeably to describe various products of the petroleum and petrochemical industries. Some of these aromatic hydrocarbons, such as 3,4-benzpyrene, pyrene, and 1,2-benzpyrene are formed during carbon black manufacture and their adsorption on the carbon black could pose a risk of cancer after exposure to the carbon black.

Occupational exposure to carbon black shall be controlled so that employees are not exposed to carbon black at a concentration greater than 3.5 mg/m^3 or air, or to PAHs at a concentration greater than 0.1 mg, measured as the cyclohexane-extractable fraction, per cubic meter of air (0.1 mg/m^3), determined as time-weighted average (TWA) concentrations for up to a 10-hour work shift in a 40-hour workweek.

The standard is designed to protect the health and provide for the safety of employees for up to a 10-hour work shift, 40-hour workweek, over a working lifetime.

TSCA-CHIP available

CARBON DIOXIDE. [124-38-9]

*OSHA 29 CFR 1910.1000 Table Z-1
PEL(TWA): 5000 ppm, 9000 mg/m^3

ACGIH-TLV
TWA: 5000 ppm, 9000 mg/m^3
STEL: 15000 ppm, 27000 mg/m^3

NIOSH Criteria Document (Pub. No. 76-194), NTIS Stock No. PB 266597, August 10, 1976

"Occupational exposure to carbon dioxide" is defined as exposure at a concentration greater than the time-weighted average (TWA) environmental limit. "Overexposure to carbon dioxide" is defined as any exposure at a concentration sufficient to produce signs of respiratory difficulty or central nervous system effects. Exposure to carbon dioxide at or below the TWA environmental limit will not require adherence to all sections except for Sections 1, 3, 5, 6, and the first paragraph of Section 7. See the Criteria Document for details.

Employee exposure to carbon dioxide shall be controlled so that the environmental limit does not exceed 10,000 ppm parts of air (1%) by volume (approximately 18,000 mg/m^3 of air) determined as a TWA concentration for up to a 10-hour work shift in a 40-hour workweek, with a ceiling concentration of 30,000 ppm parts of air (3%) by volume (approximately 54,000 mg/m^3 of air) as determined by a sampling period not to exceed 10 minutes.

The standard is designed to protect the health and safety of workers for up to a 10-hour work shift in a 40-hour workweek over a normal working lifetime.

CARBON DISULFIDE. Syn: carbon bisulfide. [75-15-0]
*OSHA 1910.1000 Table Z-2
20 ppm, 8-hour TWA
30 ppm, acceptable ceiling concentration
100 ppm/30 min, acceptable maximum peak
ACGIH-TLV
Skin
TWA: 10 ppm, 30 mg/m^3
NIOSH Criteria Document (Pub. No. 77-156), NTIS Stock No. PB 274199, May 4, 1977

The term "carbon disulfide" refers to either vaporized or liquid carbon disulfide. Synonyms for carbon disulfide include carbon bisulfide, carbon sulfide, and dithiocarbonic anhydride. "Occupational exposure to airborne carbon disulfide" is defined as exposure to airborne carbon disulfide at or above half the recommended time-weighted average (TWA) concentration limit or contact of skin or eyes with liquid carbon disulfide. Where there is no occupational exposure to carbon disulfide, adherence is required to Sections 3, 4(a), 4(b), 5, 6, 7, and 8 only. See Criteria Document for details.

Employee exposure to carbon disulfide shall be controlled so that no worker is exposed to carbon disulfide at a concentration greater than 3 mg of carbon disulfide per cubic meter of air (1 ppm parts of air by volume) determined as a TWA concentration for up to a 10-hour work shift in a 40-hour workweek, or to more than 30 mg carbon disulfide per cubic meter of air (10 ppm) as a ceiling concentration for any 15-minute period.

The standard is designed to protect the health and provide for the safety of employees for up to a 10-hour work shift, 40-hour workweek, over a working lifetime.

FIFRA—Data Call In, 45 FR 75488, November 14, 1980
Status: letter issued 7/2/81
NFPA—hazardous chemical
health—2, flammability—3, reactivity—0
*CWA 311(a)(2)(A), 40 CFR 116, 117
Discharge RQ = 5000 pounds (2270 kilograms)

*RCRA 40 CFR 261
EPA hazardous waste no. P022 (hazardous substance)

CARBONIC ACID, DITHALLIUM SALT. [6533-73-9]
*RCRA 49 CFR 172.101
EPA hazardous waste no. U215 (ignitable waste)

CARBONIN (VITAVAX®). [5234-68-4]
FIFRA—Registration Standard—Issued August 1981 (fungicide)

CARBON MONOXIDE. [630-08-0]
*OSHA 29 CFR 1910.1000 Table Z-1
PEL(TWA): 50 ppm, 55 mg/m^3
ACGIH-TLV
TWA: 50 ppm, 55 mg/m^3
STEL: 400 ppm, 440 mg/m^3
NIOSH Criteria Document (Pub. No. 73-11000), NTIS Stock No. PB 212629, August 3, 1972

Occupational exposure to carbon monoxide shall be controlled so that no worker shall be exposed at a concentration greater than 35 ppm determined as a time-weighted average (TWA) exposure for an 8-hour workday, as measured with a portable, direct reading, hopcalite-type carbon monoxide meter calibrated against known concentrations of CO, or with gas detector tube units certified under Title 42 of the Code of Federal Regulations, Part 84.

No level of carbon monoxide to which workers are exposed shall exceed a ceiling concentration of 200 ppm.

Because of the well-defined relationship between smoking and the concomitant exposure to CO in inhaled smoke the recommended standard may not provide the same degree of protection to those workers who smoke as it will to nonsmokers. Likewise, under conditions of reduced ambient oxygen concentration, such as would be encountered by workers at very high altitudes (e.g., 5000–8000 feet above sea level), the permissible exposure stated in the recommended standard should be appropriately lowered to compensate for loss in the oxygen-carrying capacity of the blood. In addition, workers with physical impairments that interfere with normal oxygen delivery to the tissue (e.g., emphysema, anemia, coronary heart disease) will not be provided the same degree of protection as the healthy worker population.

The recommended standard is designed to protect the safety and health of workers who are performing a normal 8-hour per day, 40-hour per week work assignment.

*CAA §109(b); Part C §160–178; 49 CFR 50, National Ambient Air Quality Standard (NAAQS) (not to be exceeded more than one per year)

CARBONOCHLORIDIC ACID, METHYL ESTER. [79-22-1]

*RCRA 40 CFR 261

EPA hazardous waste no. U156 (ignitable waste, toxic waste)

	$\mu g/m^3$	
	Primary	Secondary
8 hr	10,000	Same as primary
1 hr	40,000 (35 ppm)	Same as primary

CARBON OXYFLUORIDE. [353-50-4]

*RCRA 40 CFR 261

EPA hazardous waste no. U033 (reactive waste, toxic waste)

CARBON TETRABROMIDE.
[558-13-4]

ACGIH-TLV

TWA: 0.1 ppm, 1.4 mg/m^3

STEL: 0.3 ppm, 4 mg/m^3

CARBON TETRACHLORIDE. [56-23-5]

*OSHA 1910.1000 Table Z-1

10 ppm, 8-hour TWA

25 ppm, acceptable ceiling concentration

200 ppm/5 minutes in any 4 hours, acceptable maximum peak

ACGIH-TLV

Skin

TWA: 5 ppm, 30 mg/m^3

STEL: 20 ppm, 125 mg/m^3

Industrial substance suspect of carcinogenic potential for humans—which is suspect of inducing cancer, based on either (1) limited epidemiologic evidence, exclusive of clinical reports of single cases, or (2) demonstration of carcinogenesis in one or more animal species by appropriate methods. Worker exposure by all routes should be carefully controlled to levels consistent with the animal and human experience data.

NIOSH Criteria Document, revised. No NIOSH publication number is assigned. Available only from NIOSH. No date is given in the revision. 1976 is the date of the original document.

"Occupational exposure to carbon tetrachloride" is defined as exposure to carbon tetrachloride in any establishment where carbon tetrachloride is used, manufactured, or stored. Exposure to carbon tetrachloride under any of the above conditions will require adherence to all sections of the standard.

Carbon tetrachloride shall be controlled in the workplace so that the concentration of carbon tetrachloride is not greater than 2 ppm (12.6 mg/m^3) of breathing zone air in a 45-liter air sample taken over a period not to exceed 1 hour in duration.

The standard is designed to protect the health and provide for the safety of workers for up to a 10-hour workday, 40-hour workweek over a working lifetime.

IARC—carcinogenic in animals

40 CFR 162.11, FIFRA-RPAR issued

Current Status: PD 1 published, 45 FR 68534 (10/15/80); comment period closed 11/24/80: NTIS# PB80 121782. Risk/benefit assessment is ongoing.

Criteria Possibly Met or Exceeded: Oncogenicity and Toxic Effects on Liver and Kidney

NFPA—hazardous chemical

health—3, flammability—0, reactivity—0

TSCA-CHIP available

CWA 304(a)(1), 45 FR 79318, November 28, 1980, Water Quality Criteria

Freshwater Aquatic Life

The available data for carbon tetrachloride indicate that acute toxicity to freshwater aquatic life occurs at concentrations as low as 35,200 µg/liter and would occur at lower concentrations among species that are more sensitive than those tested. No data are available concerning the chronic toxicity of carbon tetrachloride to sensitive freshwater aquatic life.

Saltwater Aquatic Life

The available data for carbon tetrachloride indicate that acute toxicity to saltwater aquatic life occurs at concentrations as low as 50,000 µg/liter and would occur at lower concentrations among species that are more sensitive than those tested. No data are available concerning the chronic toxicity of carbon tetrachloride to sensitive saltwater aquatic life.

Human Health

For the maximum protection of humans health from the potential carcinogenic effects due to exposure of carbon tetrachloride through ingestion of contaminated water and contaminated aquatic organisms the ambient water concentration should be zero based on the nonthreshold assumption for this chemical. However, zero level may not be attainable at the present time. Therefore, the levels that may result in incremental increase of cancer risk over the lifetime are estimated at 10^{-5}, 10^{-6}, and 10^{-7}. The corresponding criteria are 4.0 µg/liter, 0.40 µg/liter, and 0.04 µg/liter, respectively. If the above estimates are made for consumption of aquatic organisms only, excluding consumption of water, the levels are 69.4 µg/liter, 6.94 µg/liter, and 0.69 µg/liter, respectively. Other concentrations representing different risk levels may be calculated by use of the guidelines. The risk estimate range is presented for information purposes and does not represent an agency judgment on an "acceptable" risk level.

CWA 307(a)(1), Priority Pollutant

*CWA 311(b)(2)(A), 40 CFR 116, 117
Discharge RQ = 5000 pounds (2270 kilograms)

CAA 112
Chemical proposed to be assessed as a Hazardous Air Pollutant by a House of Representatives bill to amend the CAA

*RCRA 40 CFR 261
EPA hazardous waste no. U211 (hazardous substance)

CARBONYL CHLORIDE.
See Phosgene. [75-44-5]

See Phosgene, NIOSH Criteria Document.

*RCRA 40 CFR 261
EPA hazardous waste no. P095 (hazardous substance)

CARBONYL FLUORIDE. [353-50-4]

ACGIH-TLV
TWA: 2 ppm, 5 mg/m³
STEL: 5 ppm, 15 mg/m³

*RCRA 40 CFR 261
EPA hazardous waste no. U033 (reactive waste, toxic waste)

CARBOPHENOTHION. [786-19-6]

FIFRA—Data Call In, 45 FR 75488, November 14, 1980
Status: agency decision reached

CARRAGEENANS, NATIVE.

IARC—carcinogenic in animals

CATECHOL. Syn: Pyrocatechol. [120-80-9]

ACGIH-TLV
TWA: 5 ppm, 20 mg/m³

CAUSTIC ARSENIC CHLORIDE. See Arsenic chloride. [7784-34-1]

CAUSTIC OIL OF ARSENIC.
See Arsenic chloride. [7784-34-1]

CAUSTIC POTASH. See Potassium hydroxide. [1310-58-3]

CAUSTIC SODA. See Sodium hydroxide. [1310-73-2]

CELLULOID. See Cellulose nitrate.

CELLULOSE (PAPER FIBER).
[9004-34-6]

ACGIH-TLV
TWA: Nuisance particulate, 30 mmpcf or 10

mg/m³ of total dust < 1% quartz, or 5 mg/m³ respirable dust
STEL: 20 mg/m³

CELLULOSE NITRATE (not explosive grade). [9004-70-0]

NFPA—hazardous chemical
when nonfire: health—0, flammability—3, reactivity—3
when fire: health—2, flammability—3, reactivity—3

CEMENT, PORTLAND.

ACGIH-TLV
30 mppcf

CESIUM BICHROMATE.

See Chromium (VI), NIOSH Criteria Document.

CESIUM CHROMATE.

See Chromium (VI), NIOSH Criteria Document.

CESIUM HYDROXIDE. [21351-79-1]

ACGIH-TLV
TWA: 2 mg/m³

CHLORACETYL CHLORIDE. See Chloroacetyl chloride. [79-04-9]

CHLORAL.

*RCRA 40 CFR 261
EPA hazardous waste no. U034

CHLORABEN. [133-90-4]

NCI—carcinogenic in animals
FIFRA—Data Call In, 45 FR 75488, November 14, 1980
Status: letter being drafted
FIFRA—Registration Standard—Issued August 1981 (herbicide)

CHLORAMBUCIL. [305-03-3]

OSHA Candidate List of Potential Occupational Carcinogens (EPA Carcinogen Assessment Group List), 45 FR 53672, August 12, 1980
IARC—carcinogenic in animals
*RCRA 40 CFR 261
EPA hazardous waste no. U035

CHLORAMPHENICOL. [56-75-7]

IARC—carcinogenic in humans

CHLORANIL. [188-75-2]

40 CFR 162.11, FIFRA-RPAR, Final Action
Current Status: Voluntary Cancellation, 42 FR 3702 (1/19/77)
Criteria Possibly Met or Exceeded: Oncogenicity

CHLORDANE (1,2,4,5,6,7,8,8-OCTACHLORO-4,7-METHANO-3a,4,7,7a-TETRAHYDROINDANE).
[57-74-9]

*OSHA 29 CFR 1910.1000 Table Z-1

Skin

PEL(TWA): 0.5 mg/m³
ACGIH-TLV

Skin

TWA: 0.5 mg/m³
STEL: 2 mg/m³
(alpha and gamma isomers):
OSHA Candidate List of Potential Occupational Carcinogens (EPA Carcinogen Assessment Group List), 45 FR 53672, August 12, 1980
NCI—carcinogenic in animals
40 CFR 162.11, FIFRA-RPAR, Notice of Intent to Cancel/Suspend Issued
Current Status: Notice of Intent to Cancel 39 FR 41298 (11/26/74). (Cancellation of most uses.)
Criteria Possibly Met or Exceeded: Oncogenecity and Reductions in Nontarget and Endangered Species
CWA 304(a)(1), 45 FR 79318, November 28, 1980, Water Quality Criteria

Freshwater Aquatic Life

For chlordane the criterion to protect freshwater aquatic life as derived using the guidelines is 0.0043 μg/liter at any time.

Saltwater Aquatic Life

For chlordane the criterion to protect saltwater aquatic life as derived using the guidelines is 0.0040 µg/liter as a 24-hour average and the concentration should not exceed 2.4 µg/liter at any time.

Human Health

For the maximum protection of human health from the potential carcinogenic effects due to exposure to chlordane through ingestion of contaminated water and contaminated aquatic organisms, the ambient water concentration should be zero based on the nonthreshold assumption for this chemical. However, zero level may not be attainable at the present time. Therefore, the levels that may result in incremental increase of cancer risk over the lifetime are estimated at 10^{-5}, 10^{-6}, and 10^{-7}. The corresponding criteria are 4.6 ng/liter, 0.46 ng/liter and 0.046 ng/liter, respectively. If the above estimates are made for consumption of aquatic organisms only, excluding consumption of water, the levels are 4.8 ng/liter 0.48 ng/liter, and 0.048 ng/liter, respectively. Other concentrations representing different risk levels may be calculated by use of the guidelines. The risk estimate range is presented for information purposes and does not represent an agency judgment on an "acceptable" risk level.

*CWA 311(b)(2)(A), 40 CFR 116, 117
Discharge RQ = 1 pound (0.454 kilogram)
CWA 307(a)(1), Priority Pollutant

*RCRA 49 CFR 172.101
EPA hazardous waste no. U036 (hazardous substance)

CHLORDECONE. *See* Kepone.
[143-50-0]

CHLORDIMEFORM. [6164-98-3]
FIFRA—Data Call In, 45 FR 75488, November 14, 1980
Status: letter being drafted

CHLOREN *See* Dichloroethyl ether.
[111-44-4]

CHLORENDIC ACID. [115-28-6]
TSCA 4(e), ITC
Ninth Report of the TSCA Interagency Testing Committee to the administrator, Environmental Protection Agency, October 1981; 47 FR 5456, February 5, 1982, removed from the Priority List because the EPA administrator has responded to the committee's prior recommendation for testing of the chemical, 47 FR 44878, October 12, 1982
*TSCA 8(d), Unpublished Health and Safety Studies Reporting, 40 CFR 716, 47 FR 38800, September 2, 1982; 48 FR 13178, March 30, 1983

CHLORIDE.
*National Secondary Drinking Water Regulations, 40 CFR 143; 44 FR 42198, July 19, 1979, effective January 19, 1981
Maximum contaminant level—250 mg/liter
Color—15 color units

CHLORINATED BENZENES, MONO- and DI-. [95-50-1, 106-46-7, 108-90-7, 541-73-1]
*TSCA 8(d), Unpublished Health and Safety Studies Reporting, 40 CFR 716, 47 FR 387, September 2, 1982

CHLORINATED BENZENES, TRI-, TETRA-, and PENTA-. [87-61-6, 95-94-3, 108-70-3, 120-82-1, 608-93-5, 634-66-2, 634-90-2]
*TSCA 8(d), Unpublished Health and Safety Studies Reporting, 40 CFR 716, 47 FR 387, September 2, 1982
CWA 304(a)(1), 45 FR 79318, November 28, 1980, Water Quality Criteria

Freshwater Aquatic Life

The available data for chlorinated benzenes indicate that acute toxicity to freshwater aquatic life occurs at concentrations as low as 250 µg/liter and would occur at lower concentrations among species that are more sensitive than those tested. No data are available concerning the chronic toxicity of the more toxic of the chlorinated benzenes to sensitive freshwater aquatic life but toxicity occurs at concentrations as low as 50 µg/liter for a fish species exposed for 7.5 days.

Saltwater Aquatic Life

The available data for chlorinated benzenes indicate that acute toxicity to freshwater aquatic life occurs at concentrations as low as 250 µg/liter and would occur at lower concentrations among species that are more sensitive than those tested. No data are available concerning the chronic toxicity of the more toxic of the chlorinated benzenes at sensitive freshwater aquatic life but toxicity occurs at concentrations as low as 50 µg/liter for a fish species exposed for 7.5 days.

Saltwater Aquatic Life

The available data for chlorinated benzenes indicate that acute and chronic toxicity to saltwater aquatic life occur at concentrations as low as 160 and 129 µg/liter, respectively, and would occur at lower concentrations among species that are more sensitive than those tested.

Human Health

For the maximum protection of human health from the potential carcinogenic effects due to exposure to hexachlorobenzene through ingestion of contaminated water and contaminated aquatic organisms, the ambient water concentration should be zero based on the nonthreshold assumption for this chemical. However, zero level may not be attainable at the present time. Therefore, the levels that may result in incremental increase of cancer risk over the lifetime are estimated at 10^{-5}, 10^{-6}, and 10^{-7}. The corresponding recommended criteria are 7.2 ng/liter, 0.72 ng/liter, and 0.072 ng/liter, respectively. If the above estimates are made for consumption of aquatic organisms only, excluding consumption of water, the levels are 7.4 ng/liter, 0.74 ng/liter, and 0.074 ng/liter, respectively.

For the protection of human health from the toxic properties of 1,2,4,5-tetrachlorobenzene ingested through water and contaminated aquatic organisms, the ambient water criterion is determined to be 38 µg/liter.

For the protection of human health from the toxic properties of 1,2,4,5-tetrachlorobenzene ingested through contaminated aquatic organisms alone, the ambient water criterion is determined to be 48 µg/liter.

For the protection of human health from the toxic properties of pentachlorobenzene ingested through water and contaminated aquatic organisms, the ambient water criterion is determined to be 74 µg/liter.

For the protection of human health from the toxic properties of pentachlorobenzene ingested through contaminated aquatic organisms alone, the ambient water criterion is determined to be 85 µg/liter.

Using the present guidelines, a satisfactory criterion cannot be derived at this time due to the insufficiency in the available data for trichlorobenzene.

For comparison purposes, two approaches were used to derive criterion levels for monochlorobenzene. Based on available toxicity data, for the protection of public health, the derived level is 488 µg/liter. Using available organoleptic data, for controlling undesirable taste and odor quality of ambient water, the estimated level is 20 µg/liter. It should be recognized that organoleptic data as a basis for establishing a water quality criterion have limitations and have no demonstrated relationship to potential adverse human health effects.

CHLORINATED CAMPHENE.

[8001-35-2]

*OSHA 29 CFR 1910.1000 Table Z-1

Skin

PEL(TWA): 0.5 mg/m^3

ACGIH-TLV

Skin

TWA: 0.5 mg/m^3
STEL: 1 mg/m^3
*RCRA 40 CFR 261
EPA hazardous waste no. P123

CHLORINATED DIPHENYL OXIDE.

[55720-99-5]

*OSHA 29 CFR 1910.1000 Table Z-1
PEL(TWA): 0.5 mg/m^3
ACGIH-TLV
TWA: 0.5 mg/m^3
STEL: 2 mg/m^3

CHLORINATED ETHANES.

OSHA Candidate List of Potential Occupational Carcinogens (EPA Carcinogen Assessment Group List), 45 FR 53672, August 12, 1980

CWA 304(a)(1), 45 FR 79318, November 28, 1980, Water Quality Criteria

Freshwater Aquatic Life

The available freshwater data for chlorinated ethanes indicate that toxicity increases greatly with increasing chlorination, and that acute toxicity occurs at concentrations as low as 118,000 µg/liter for 1,2-dichloroethane, 18,000 µg/liter for two trichloroethanes, 9320 µg/liter for two tetrachloroethanes, 7240 µg/liter for pentachloroethane, and 980 µg/liter for hexachloroethane. Chronic toxicity occurs at concentrations as low as 20,000 µg/liter for 1,2-dichloroethane, 9400 µg/liter for 1,1,2-trichloroethane, 2400 µg/liter for 1,1,2,2-tetrachloroethane, 1100 µg/liter for pentachloroethane, and 540 µg/liter for hexachloroethane. Acute and chronic toxicity would occur at lower concentrations among species that are more sensitive than those tested.

Saltwater Aquatic Life

The available saltwater data for chlorinated ethanes indicate that toxicity increases greatly with increasing chlorination and that acute toxicity to fish and invertebrate species occurs at concentrations as low as 113,000 µg/liter for 1,2-dichloroethane, 31,200 µg/liter for 1,1,1-trichloroethane, 9020 µg/liter for 1,1,2,2-tetrachloroethane, 390 µg/liter for pentachloroethane, and 940 µg/liter for hexachloroethane. Chronic toxicity occurs at concentrations as low as 281 µg/liter for pentachloroethane. Acute and chronic toxicity would occur at lower concentrations among species that are more sensitive than those tested.

Human Health

For the maximum protection of human health from the potential carcinogenic effects due to exposure of 1,2-dichloroethane through ingestion of contaminated water and contaminated aquatic organisms, the ambient water concentration should be zero based on the nonthreshold assumption for this chemical. However, zero level may not be attainable at the present time. Therefore, the levels that may result in incremental increase of cancer risk over the lifetime are estimated at 10^{-5}, 10^{-6}, and 10^{-7}. The corresponding, criteria are 9.4 µg/liter, 0.94 µg/liter, and 0.094 µg/liter, respectively. If the above estimates are made for consumption of aquatic organisms only, excluding consumption of water, the levels are 2430 µg/liter, 243 µg/liter, and 24.3 µg/liter, respectively. Other concentrations representing different risk levels may be calculated by use of the guidelines. The risk estimate range is presented for information purposes and does not represent an agency judgment on an "acceptable" risk level.

For the protection of human health from the toxic properties of 1,1,1-trichloroethane ingested through water and contaminated aquatic organisms, the ambient water criterion is determined to be 18.4 mg/liter.

For the protection of human health from the toxic properties of 1,1,1-trichloroethane ingested through contaminated aquatic organisms alone, the ambient water criterion is determined to be 1.03g/liter.

For the maximum protection of human health from the potential carcinogenic effects due to exposure of 1,1,2-trichloroethane through ingestion of contaminated water and contaminated aquatic organisms, the ambient water concentration should be zero based on the nonthreshold assumption for this chemical. However, zero level may not be attainable at the present time. Therefore, the levels that may result in incremental increase of cancer risk over the lifetime are estimated at 10^{-5}, 10^{-6}, and 10^{-7}. The corresponding criteria are 6.0 µg/liter, 0.6 µg/liter, and 0.06 µg/liter, respectively. If the above estimates are made for consumption of aquatic organisms only, excluding consumption of water, the levels are 418 µg/liter, 41.8 µg/liter, and 4.18 µg/liter, respectively. Other concentrations representing different risk levels may be calculated by use of the guidelines. The risk estimate range is presented for information purposes and does not represent an agency judgment on an "acceptable" risk level.

For the maximum protection of human health from the potential carcinogenic effects due to exposure of 1,1,2,2-tetrachloroethane through ingestion of contaminated water and contaminated aquatic organisms, the ambient water concentration should be zero based on the nonthreshold assumption for this chemical. However, zero level may not be attainable at the present time. Therefore, the levels that may result in incremental increase of cancer risk over the lifetime are estimated at 10^{-5}, 10^{-6}, and 10^{-7}. The corresponding criteria are 1.7 µg/liter, 0.17 µg/liter, and 0.017 µg/liter, respectively. If the above estimates are made for consumption of aquatic organisms only, excluding consumption of water, the levels are 107 µg/liter, 10.7 µg/liter, and 1.07 µg/liter, respectively. Other concentrations representing different risk levels may be calculated by use of the guidelines. The risk estimate range is presented for information purposes and does not represent an agency judgment on an "acceptable" risk level.

Using the present guidelines, a satisfactory criterion cannot be derived at this time due to the insufficiency in the available data for 1,1-dichloroethane.

Using the present guidelines, a satisfactory criterion cannot be derived at this time due to the insufficiency in the available data for 1,1,1,2-tetrachloroethane.

Using the present guidelines, a satisfactory criterion cannot be derived at this time due to the insufficiency in the available data for pentachloroethane.

For the maximum protection of human health from the potential carcinogenic effects due to exposure of hexachloroethane through ingestion of contaminated water and contaminated aquatic organisms, the ambient water concentration should be zero based on the nonthreshold assumption for this chemical. However, zero level may not be attainable at the present time. Therefore the levels that may result in incremental increase of cancer risk over the lifetime are estimated at 10^{-5}, 10^{-6}, and 10^{-7}. The corresponding criteria are 19 µg/liter, 1.9 µg/liter, and 0.19 µg/liter, respectively. If the above estimates are made for consumption of aquatic organisms only, excluding consumption of water, the levels are 87.4 µg/liter, 8.74 µg/liter, and 0.87 µg/liter, respectively. Other concentrations representing different risk levels may be calculated by use of the guidelines. The risk estimate range is presented for information purposes and does not represent an agency judgment on an "acceptable" risk level.

Using the present guidelines, a satisfactory criterion cannot be derived at this time due to the insufficiency in the available data for monochloroethane.

CHLORINATED LIME. *See* Calcium hypochlorite. [7778-54-3]

CHLORINATED NAPHTHALENES. *See* chlorinated derivatives of napthalene (empirical formula $C_{10}H_xCl_y$ where $x + y = 8$). [90-13-1, 1321-64-8, 1321-65-9]

TSCA 4(e), ITC
This category consists of chlorinated derivates of naphthalene (empirical formula $C_{10}H_xCl_y$ where $x + y = 8$). Second Report of the TSCA Interagency Testing Committee to the administrator, Environmental Protection Agency, April 1978; 43 FR 16684, April 19, 1978, 44 FR 28095, responded to by the EPA administrator, May 14, 1979, EPA responded to the committee's recommendations for testing 46 FR 54491, November 2, 1981

*TSCA 8(d), Unpublished Health and Safety Studies Reporting, 40 CFR 716, 47 FR 387, September 2, 1982

CWA 304(a)(1), 45 FR 79318, November 28, 1980, Water Quality Criteria

Freshwater Aquatic Life

The available data for chlorinated naphthalenes indicate that acute toxicity to freshwater aquatic life occurs at concentrations as low as 1600 µg/liter and would occur at lower concentrations among species that are more sensitive than those tested. No data are available concerning the chronic toxicity of chlorinated naphthalenes at sensitive freshwater aquatic life.

Saltwater Aquatic Life

The available data for chlorinated naphthalenes indicate that acute toxicity to saltwater aquatic life occurs at concentrations as low as 7.5 µg/liter and would occur at lower concentrations among species that are more sensitive than those tested. No data are available concerning the chronic toxicity of chlorinated naphthalenes to sensitive saltwater aquatic life.

Human Health

Using the present guidelines, a satisfactory criterion cannot be derived at this time due to the insufficiency in the available data for chlorinated naphthalenes.

CHLORINATED PARAFFINS.

Chlorinated paraffin oils and chlorinated paraffin waxes, with chlorine content of 35–70% by weight. [61788-76-9,63449-39-8, 68920-70-7]

TSCA 4(e), ITC
This category is comprised of a series of mixtures of chlorination products of materials known commercially as paraffin oils or paraffin waxes; those having a chlorine content of 35–70% by weight are included.
Initial Report to the administrator, Environmental Protection Agency, TSCA Interagency Testing Committee, October 1, 1977; 42 FR 55026, October 12, 1977, responded to by the EPA administrator, 43 FR 50134, October 26, 1978, EPA responded to the committee's recommendations for testing, 47 FR 1017, January 8, 1982

*TSCA 8(d), Unpublished Health and Safety Studies Reporting, 40 CFR 716, 47 FR 387, September 2, 1982

CHLORINATED PHENOLS.

CWA 304(a)(1), 45 FR 79318, November 28, 1980, Water Quality Criteria

Freshwater Aquatic Life

The available freshwater data for chlorinated phenols indicate that toxicity generally increases with increasing chlorination, and that acute toxicity occurs at concentrations as low as 30 µg/liter for 4-chloro-3-methylphenol to greater than 500,000 µg/liter for other compounds. Chronic toxicity occurs at concentrations as low as 970 µg/liter for 2,4,6-trichlorophenol. Acute and chronic toxicity would occur at lower concentrations among species that are more sensitive than those tested.

Saltwater Aquatic Life

The available saltwater data for chlorinated phenols indicate that toxicity generally increases with increasing chlorination and that acute toxicity occurs at concentrations as low as 440 µg/liter for 2,3,4,6-tetrachlorophenol and 29,700 µg/liter for 4-chlorophenol. Acute toxicity would occur at lower concentrations among species that are more sensitive than those tested. No data are available concerning the chronic toxicity of chlorinated phenols to sensitive saltwater aquatic life.

Human Health

Sufficient data are not available for 3-monochlorophenol to derive a level that would protect against the potential toxicity of this compound. Using available organoleptic data, for controlling undesirable taste and odor quality of ambient water, the estimated level is 0.1 µg/liter. It should be recognized that organoleptic data is a basis for establishing a water quality criterion have limitations and have no demonstrated relationship to potential adverse human health effects.

Sufficient data are not available for 4-monochlorophenol to derive a level that would protect against the potential toxicity of this compound. Using available organoleptic data, for controlling undesirable taste and odor quality of ambient water, the estimated level is 0.1 µg/liter. It should be recognized that organoleptic data as a basis for establishing a water quality criterion have limitations and have no demonstrated relationship to potential adverse human health effects.

Sufficient data are not available for 2,3-dichlorophenol to derive a level that would protect against the potential toxicity of this compound. Using available organoleptic data, for controlling undesirable taste and odor quality of ambient water, the estimated level is 0.04 µg/liter. It should be recognized that organoleptic data as a basis for establishing a water quality criterion have limitations and have no demonstrated relationship to potential adverse human health effects.

Sufficient data are not available for 2,5-dichlorophenol to derive a level which would protect against the potential toxicity of this compound. Using available organoleptic data, for controlling undesirable taste and odor quality of ambient water, the estimated level is 0.5 µg/liter. It should be recognized that organoleptic data as a basis for establishing a water quality criterion have limitation and have no demonstrated relationship to potential adverse human health effects.

Sufficient data are not available for 2,6-dichlorophenol to derive a level that would protect against the potential toxicity of this compound. Using available organoleptic data, for controlling undesirable taste and odor quality of ambient water, the estimated level is 0.2 µg/liter. It should be recognized that organoleptic data as a basis for establishing a water quality criterion have limitations and have no demonstrated relationship to potential adverse human health effects.

Sufficient data are not available for 3,4-dichlorophenol to derive a level that would protect against the potential toxicity of this compound. Using available organoleptic data, for controlling undesirable taste and odor quality of ambient water, the estimated level is 0.3 µg/liter. It should be recognized that organoleptic data as a basis for establishing a water quality criterion have limitations and have no demonstrated relationship to potential adverse human health effects.

Sufficient data are not available for 2,3,4,6-tetrachlorophenol to derive a level that would protect against the potential toxicity of this compound. Using available organoleptic data, for controlling undesirable taste and odor quality of ambient water, the estimated level is 1 µg/liter. It should be recognized that organoleptic data as a basis for establishing a water quality criterion have limitations and have no demonstrated relationship to potential adverse human health effects.

For comparison purposes, two approaches were used to derive criterion levels for 2,4,5-trichlorophenol. Based on available toxicity data, for the protection of public health, the derived level is 2.6 µg/liter. Using available organoleptic data, for controlling undesirable taste and odor quality of ambient water, the estimated level is 1.0 µg/liter. It should be recognized that organoleptic data as a basis for establishing a water quality criterion have limitations and have no demonstrated relationship to potential adverse human health effects.

For the maximum protection of human health from the potential carcinogenic effects due to exposure of 2,4,6-trichlorophenol through ingestion of contaminated water and contaminated aquatic organisms, the ambient water concentration should be zero based on the nonthreshold assumption for this chemical. However, zero level may not be attainable at the present time. Therefore, the levels that may result in incremental increase of cancer risk over the lifetime are estimated at 10^{-5}, 10^{-6}, and 10^{-7}. The corresponding criteria are 12 µg/liter, 1.2 µg/liter, and 0.12 µg/liter, respectively. If the above estimates are made for consumption of aquatic organisms only, excluding consumption of water, the levels are 36 µg/liter, 3.6 µg/liter, and 0.36 µg/liter, respectively. Other concentrations representing different risk levels may be calculated by use of the guidelines. The risk estimate range is presented for information purposes and does not represent an agency judgment on an "acceptable" risk level.

Using available organoleptic data, for controlling undesirable taste and odor quality of ambient water, the estimated level is 2 µg/liter. If should be recognized that organoleptic data as a basis for establishing a water quality criterion have limitations and have no demonstrated relationship to potential adverse human health effects.

Sufficient data are not available for 2-methyl-4-chlorophenol to derive a level that would protect against any potential toxicity of this compound. Using available organoleptic data, for controlling undesirable taste and odor quality of ambient water, the estimated level is 1800 µg/liter. It should be recognized that organoleptic data as a basis for establishing a water quality criterion have limitations and have no demonstrated relationship to potential adverse human health effects.

Sufficient data are not available for 3-methyl-4-chlorophenol to derive a level that would protect against the potential toxicity of this compound. Using available organoleptic data, for controlling undesirable taste and odor quality of ambient water, the estimated level is 3000 µg/liter. It should be recognized that organoleptic data as a basis for establishing a water quality criterion have limitations and have no demonstrated relationship to potential adverse human health effects.

Sufficient data are not available for 3-methyl-6-chlorophenol to derive a level that would protect against the potential toxicity of this compound. Using available organoleptic data, for controlling undesirable taste and odor quality of ambient water, the estimated level is 20 µg/liter. It should be recognized that organoleptic data as a basis for establishing a water quality criterion have limitations and have no demonstrated relationship to potential adverse human health effects.

CHLORINE. [7782-50-5]

*OSHA 29 CFR 1910.1000 Table Z-1
PEL(Ceiling): 1 ppm, 3 mg/m^3

ACGIH-TLV
TWA: 1 ppm, 3 mg/m^3
STEL: 3 ppm, 9 mg/m^3

NIOSH Criteria Document (Pub. No. 76-170), NTIS Stock No. PB 266367, May 25, 1976

"Chlorine" is defined as liquid or gaseous molecular chlorine. "Occupational exposure to chlorine" is defined as exposure to airborne concentrations of chlorine at or above one-half of the recommended workplace environmental limit. Adherence only to Sections 3, 4(a) (1, 3, and 4), 4(b) (6, 7, 9, 10), 5, 6, and 7 is required when workplace environmental concentrations of chlorine are less than one-half of the recommended workplace environmental limit.

Exposure to chlorine shall be controlled so that no worker is exposed to chlorine at an airborne concentration greater than 0.5 parts of chlorine per million parts of air (0.5 ppm) for any 15-minute sampling period. This shall be designated as a ceiling concentration.

The standard is designed to protect the health and safety of the worker for up to a 10-hour workday, 40-hour workweek, over a working lifetime.

*CWA 311(b)(2)(A), 40 CFR 116, 117
Discharge RQ = 10 pounds (4.54 kilograms)
NFPA—hazardous chemical
health—3, flammability—0, reactivity—0, oxy—oxidizing chemical

CHLORINE CYANIDE. [506-77-4]

*RCRA 40 CFR 261
EPA hazardous waste no. P033 (hazardous substance)

CHLORINE DIOXIDE. [10049-04-4]

*OSHA 29 CFR 1910.1000 Table Z-1
PEL(TWA): 0.1 ppm, 0.3 mg/m^3

ACGIH-TLV
TWA: 0.1 ppm, 0.3 mg/m^3
STEL: 0.3 ppm, 0.9 mg/m^3

CHLORINE TRIFLUORIDE. [7790-91-2]

*OSHA 29 CFR 1910.1000 Table Z-1
PEL(Ceiling): 0.1 ppm, 0.4 mg/m^3

ACGIH-TLV
TWA: ceiling limit, 0.1 ppm, 0.4 mg/m^3

NFPA—hazardous chemical
health—4, flammability—0, reactivity—3, W—water may be hazardous in fire fighting, oxy—oxidizing chemical. Reacts explosively with water or steam to produce toxic and corrosive fumes.

CHLORMADINONE ACETATE.
[302-22-7]

IARC—carcinogenic in animals

CHLORNAPHAZINE. [494-03-1]
*RCRA 40 CFR 261
EPA hazardous waste no. U026

CHLOROACETALDEHYDE. [107-20-0]
*OSHA 29 CFR 1910.1000 Table Z-1
PEL(Ceiling): 1 ppm, 3 mg/m^3
ACGIH-TLV
TWA: ceiling limit, 1 ppm, 3 mg/m^3
*RCRA 40 CFR 261
EPA hazardous waste no. P023

alpha-CHLOROACETOPHENONE (PHENACYCHLORIDE). [532-27-4]
*OSHA 29 CFR 1910.1000 Table Z-1
PEL(TWA): 0.05 ppm, 0.3 mg/m^3
ACGIH-TLV
TWA: 0.05 ppm, 0.3 mg/m^3

CHLOROACETYL CHLORIDE.
[79-04-9]
ACGIH-TLV
TWA: 0.05 ppm, 0.2 mg/m^3
NFPA—hazardous chemical
health—3, flammability—0, reactivity—0

CHLOROALKYL ETHERS.
OSHA Candidate List of Potential Occupational Carcinogens (EPA Carcinogen Assessment Group List), 45 FR 53672, August 12, 1980

CWA 304(a)(1), 45 FR 79318, November 28, 1980, Water Quality Criteria

Freshwater Aquatic Life
The available data for chloroalkyl ethers indicate that acute toxicity to freshwater aquatic life occurs at concentrations as low as 238,000 µg/liter and would occur at lower concentrations among species that are more sensitive than those tested. No definitive data are available concerning the chronic toxicity of chloroalkyl ethers to sensitive freshwater aquatic life.

Saltwater Aquatic Life
No saltwater organisms have been tested with any chloroalkyl ether and no statement can be made concerning acute and chronic toxicity.

Human Health
For the maximum protection of human health from the potential carcinogenic effects due to exposure of bis-(chloromethyl)-ether through ingestion of contaminated water and contaminated aquatic organisms, the ambient water concentration should be zero based on the nonthreshold assumption for this chemical. However, zero level may not be attainable at the present time. Therefore, the levels that may result in incremental increase of cancer risk over the lifetime are estimated at 10^{-5}, 10^{-6}, and 10^{-7}. The corresponding criteria are 0.038 ng/liter, 0.0038 ng/liter, and 0.00038 ng/liter, respectively. If the above estimates are made for consumption of aquatic organisms only, excluding consumption of water, the levels are 18.4 ng/liter, 1.84 ng/liter, and 0.184 ng/liter, respectively. Other concentrations representing different risk levels may be calculated by use of the guidelines. The risk estimate range is presented for information purposes and does not represent an agency judgment on an "acceptable" risk level.

For the maximum protection of human health from the potential carcinogenic effects due to exposure of bis(2-chloroethyl) ether through ingestion of contaminated water and contaminated aquatic organisms, the ambient water concentration should be zero based on the nonthreshold assumption for this chemical. However, zero level may not be attainable at the present time. Therefore, the levels that may result in incremental increase of cancer risk over the lifetime are estimated at 10^{-5}, 10^{-6}, and 10^{-7}. The corresponding criteria are 0.3 µg/liter, 0.03 µg/liter, and 0.003 µg/liter, respectively. If the above estimates are made for consumption of aquatic organisms only, excluding consumption of water, the levels are 13.6 µg/liter, 1.36 µg/liter, 0.136 µg/liter respectively. Other concentrations representing different risk levels may be calculated by use of the guidelines. The risk estimate range is presented for information purposes and does not represented an agency judgment on an "acceptable" risk level.

For the protection of human health from the toxic properties of bis(2-chloroisopropyl) ether ingested through water and contaminated aquatic organisms, the ambient water criterion is determined to be 34.7 μg/liter.

For the protection of human health from the toxic properties of bis(2-chloroisopropyl) ether ingested through contaminated aquatic organisms alone, the ambient water criterion is determined to be 4.36 mg/liter.

GAMMA-CHLOROALLYL CHLORIDE. See Dichloropropene. [542-75-6]

m-CHLOROANILINE. See Aniline and chloro-, bromo-, and/or nitro- anilines. [108-42-9]

o-CHLOROANILINE. See Aniline and chloro-, bromo-, and/or nitro- anilines. [95-51-2]

p-CHLOROANILINE. See Aniline and chloro-, bromo-, and/or nitro- anilines. [106-47-8]
NCI—may be carcinogenic in animals
*RCRA 40 CFR 261
EPA hazardous waste no. P024

CHLOROBENZENE (MONOCHLOROBENZENE).
[108-90-7]
*OSHA 29 CFR 1910.1000 Table Z-1
PEL(TWA): 75 ppm, 350 mg/m^3
ACGIH-TLV
TWA: 75 ppm, 350 mg/m^3
NFPA—hazardous chemical
health—2, flammability—3, reactivity—0
CWA 307(a)(1), Priority Pollutant
*CWA 311(b)(2)(A), 40 CFR 116, 117
Discharge RQ = 100 pounds (45.4 kilograms)
CAA 112

Chemical proposed to be assessed as a Hazardous Air Pollutant by a House of Representatives bill to amend the CAA
*RCRA 49 CFR 172.101
EPA hazardous waste no. U037 (hazardous substance)

2-CHLOROBENZENAMINE.
See Aniline and chloro-, bromo-, and/or nitro- anilines. [95-51-2]

3-CHLOROBENZENAMINE.
See Aniline and chloro-, bromo-, and/or nitro- anilines. [108-42-9]

4-CHLOROBENZENAMINE.
See Aniline and chloro-, bromo-, and/or nitro anilines. [106 47 8]

3-CHLOROBENZENAMINE HYDROCHLORIDE. See Aniline and chloro-, bromo-, and/or nitro- anilines. [141-85-5]

4-CHLORO-1,2-BENZENEDIAMINE.
See o-Phenylenediamine, 4-chloro-. [95-83-0]

CHLOROBENZILATE. [510-15-6]
OSHA Candidate List of Potential Occupational Carcinogens (EPA Carcinogen Assessment Group List), 45 FR 53672, August 12, 1980
IARC—carcinogenic in animals
NCI—carcinogenic in animals
FIFRA—Registration Standard Development—Data Collection (insecticide)
FIFRA—Data Call In, 45 FR 75488, November 14, 1980
Status: letter being drafted
40 CFR 162.11, FIFRA-RPAR completed
Current Status: PD 1 published, 41, FR 21517 (5/26/76); comment period closed 7/26/76.
PD 2/3 and Notice of Proposed Determination published, 43 FR 29824 (7/11/78); comment period closed 8/10/78: NTIS# PB80 213887. Final Determination based on PD 2/3 published, 43, FR 29824 (7/11/78).

PD 4 and Notice of Intent to Cancel published, 44 FR 9547 (2/13/79).
Criteria Possibly Met or Exceeded: Oncogenicity and Testicular Effects

4-CHLOROBENZOTRIFLUORIDE.
[98-56-6]

TSCA 4(e), ITC
Ninth Report of the TSCA Interagency Testing Committee to the administrator, Environmental Protection Agency, October 1981; 47 FR 5456, February 5, 1982

*TSCA 8(d), Unpublished Health and Safety Studies Reporting, 40 CFR 716, 47 FR 38800, September 2, 1982; 48 FR 13178, March 30, 1983

1-(p-CHLOROBENZOYL)-5-METHOXY-2-METHYLINDOLE-3-ACETIC ACID. [53-86-1]

*RCRA 40 CFR 261
EPA hazardous waste no. U245

o-CHLOROBENZYLIDENE MALONONITRILE (OCBM). [2698-41-1]

*OSHA 29 CFR 1910.1000 Table Z-1
PEL(TWA): 0.05 ppm, 0.4 mg/m^3
ACGIH-TLV

Skin

TWA: ceiling limit, 0.05 ppm, 0.4 mg/m^3

CHLOROBROMOMETHAN. Syn: Bromochloromethane. [74-97-5]

*OSHA 29 CFR 1910.1000 Table Z-1
PEL(TWA): 200 ppm, 1050 mg/m^3
ACGIH-TLV
TWA: 200 ppm, 1050 mg/m^3
STEL: 250 ppm, 1300 mg/m^3
CWA 307(a)(1) Priority Pollutant

2-CHLORO-1,3-BUTADIENE.
See beta-Chloroprene. [126-99-8]

CHLOROCHROMIC ANHYDRIDE.
See Chromyl chloride. [14977-61-8]

p-CHLORO-m-CRESOL. [59-50-7]
CWA 307(a)(1) Priority Pollutant

*RCRA 40 CFR 261
EPA hazardous waste no. U039

4-CHLORO-m-CRESOL.
See Chloro-m-cresol. [59-50-7]

CHLORODIBROMOMETHANE.
[124-48-1]
CWA 307(a)(1) Priority Pollutant

CHLORODIETHYLALUMINUM.
NFPA—hazardous chemical
health—3, flammability—3, reactivity—3,
W—water may be hazardous in fire fighting

CHLORODIETHYLSILANE.
NFPA—hazardous chemical
health—3, flammability—3, reactivity—0

CHLORODIFLUOROMETHANE.
[75-45-6]

ACGIH-TLV
TWA: 1000 ppm, 3500 mg/m^3
STEL: 1250 ppm, 4375 mg/m^3

2-CHLORO-4,6-DINITROANILINE. See Aniline and chloro-, bromo-, and/or nitro- anilines. [3531-19-9]

2-CHLORO-4,6-DINITROBENZENAMINE. See Aniline and chloro-, bromo-, and/or nitro- anilines. [3531-19-9]

4-CHLORO-2,6-DINITROBENZENAMINE. See Aniline and chloro-, bromo-, and/or nitro- anilines. [5388-62-5]

1-CHLORO-2,4-DINITROBENZENE. [97-00-7]
NFPA—hazardous chemical
health—3, flammability—1, reactivity—4

CHLORODIPHENYL (42% chlorine).
[53449-21-9]

*OSHA 29 CFR 1910.1000 Table Z-1

Skin

PEL(TWA): 1 mg/m^3
ACGIH-TLV

Skin

TWA: 1 mg/m^3
STEL: 2 mg/m^3

CHLORODIPHENYL (54% chlorine). [11097-69-1]
*OSHA 29 CFR 1910.1000 Table Z-1

Skin

PEL(TWA): 0.5 mg/m^3
ACGIH-TLV

Skin

TWA: 0.5 mg/m^3
STEL: 1 mg/m^3

1-CHLORO-2,3-EPOXYPROPANE. See Epichlorohydrin. [106-89-9]

CHLOROETHANE. See Ethyl chloride. [75-00-3]

2-CHLOROETHANOL. See Ethylene chlorohydrin. [107-07-3]

CHLOROETHENE. See Vinyl chloride. [75-01-4]

CHLOROETHYLENE. See Vinyl chloride. [75-01-4]

2-CHLOROETHYL VINYL ETHER. [110-75-8]
CWA 307(a)(1) Priority Pollutant
*RCRA 40 CFR 261
EPA hazardous waste no. U042

CHLOROFLUOROALKANES (fully halogenated). See Chlorofluorocarbons.

CHLOROFLUOROCARBONS.

*TSCA 40 CFR 712 and 762, 43 FR 11318, March 17, 1978;
43 FR 29001, July 5, 1978; 43 FR 55241, November 27, 1978; 43 FR 59500, December 21, 1978; 45 FR 43721, June 30, 1980; 45 FR 78688, November 26, 1980; 46 FR 5981, January 21, 1981

CHLOROFORM. [67-66-3]
*OSHA 29 CFR 1910.1000 Table Z-1
PEL(Ceiling): 50 ppm, 240 mg/m^3
ACGIH-TLV
TWA: 10 ppm, 50 mg/m^3
STEL: 50 ppm, 225 mg/m^3
Industrial substance suspect of carcinogenic potential for humans—which is suspect of inducing cancer, based on either (1) limited epidemiologic evidence, exclusive of clinical reports of single cases, or (2) demonstration of carcinogenesis in one or more animal species by appropriate methods. Worker exposure by all routes should be carefully controlled to levels consistent with the animal and human experience data.

NIOSH Criteria Document, revised. No NIOSH publication number is assigned. Available only from NIOSH. No date is given in the revision. 1975 is the date of the original document.
"Occupational exposure to chloroform" is defined as exposure to chloroform in any establishment where chloroform is used, manufactured, or stored. Exposure to chloroform under any of the above conditions will require adherence to all sections of the standard.

Chloroform shall be controlled in the workplace so that the concentration of chloroform is not greater than 2 ppm (9.78 mg/m^3) of breathing zone air in a 45-liter air sample taken over a period not to exceed 1 hour in duration.

The standard is designed to protect the health and provide for the safety of workers for up to a 10-hour workday, 40-hour workweek over a working lifetime.

NIOSH Current Intelligence Bulletin No. 9—cancer-related bulletin

OSHA Candidate List of Potential Occupational Carcinogens (EPA Carcinogen Assessment Group List), 45 FR 53672, August 12, 1980

NCI—carcinogenic in animals

40 CFR 162.11, FIFRA-RPAR issued
Current Status: PD 1 published, 41 FR 14588 (4/6/76); comment period closed 7/

23/76. Risk/benefit assessment is ongoing. Criteria Possibly Met or Exceeded: Oncogenicity

NFPA—hazardous chemical
health—2, flammability—0, reactivity—0

CWA 304(a)(1), 45 FR 79318, November 28, 1980, Water Quality Criteria

Freshwater Aquatic Life

The available data for chloroform indicate that acute toxicity to freshwater aquatic life occurs at concentrations as low as 29,900 µg/liter and would occur at lower concentrations among species that are more sensitive than the three tested species. Twenty-seven-day LC_{50} values indicate that chronic toxicity occurs at concentrations as low as 1240 µg/liter and could occur at lower concentrations among species or other life stages that are more sensitive than the earliest life cycle stage of the rainbow trout.

Saltwater Aquatic Life

The data base for saltwater species is limited to one test and no statement can be made concerning acute or chronic toxicity.

Human Health

For the maximum protection of human health from the potential carcinogenic effects due to exposure of chloroform through ingestion of contaminated water and contaminated aquatic organisms, the ambient water concentration should be zero based on the nonthreshold assumption for this chemical. Howener, zero level may not be attainable at the present time. Therefore, the levels that may result in incremental increase of cancer risk over the lifetime are estimated at 10^{-5}, 10^{-6}, and 10^{-7}. The corresponding criteria are 1.90 µg/liter, 0.19 µg/liter, and 0.019 µg/liter, respectively. If the above estimates are made for consumption of aquatic organisms only, excluding consumption of water, the levels are 157 µg/liter, 15.7 µg/liter, and 1.57 µg.liter, respectively. Other concentrations representing different risk levels may be calculated by use of the guidelines. The risk estimate range is presented for information purposes and does not represent an agency judgment on an "acceptable" risk level.

CWA 307(a)(1), Priority Pollutant

*CWA 311(b)(2)(A), 40 CFR 116, 117
Discharge RQ = 5000 pounds (2270 kilograms)

CAA 112
Chemical proposed to be assessed as a Hazardous Air Pollutant by a House of Representatives bill to amend the CAA

*RCRA 40 CFR 261
EPA hazardous waste no. U044 (hazardous substance)

CHLOROHEXANE.

NFPA—hazardous chemical (flammable chemical)
health—0, flammability—3, reactivity—0

alpha-CHLOROHYDRIN (3-CHLORO-1,2-PROPANEDIOL). [96-24-2]

TSCA-CHIP available

CHLOROISOCYANURIC ACID.
[2782-57-2]

NFPA—hazardous chemical
health—3, flammability—0, reactivity—2

bis(2-CHLOROISOPROPYL)ETHER.
[108-60-1]

CWA 307(a)(1) Priority Pollutant

OSHA Candidate List of Potential Occupational Carcinogens (EPA Carcinogen Assessment Group List), 45 FR 53672, August 12, 1980

CHLOROMETHANE. See Methyl chloride. [74-87-3]

bis-CHLOROMETHYL ETHER.
[542-88-1]

*OSHA Standard 29 CFR 1910.1008
Cancer-Suspect Agent Shall not apply to solid or liquid mixtures containing less than 0.1% by weight or volume of bis-chloromethyl ether.

ACGIH-TLV
TWA: 0.001 ppm, 0.005 mg/m^3
Human carcinogen—no exposure or contact by any route—respiratory, skin or oral, as detected by the most sensitive methods—shall be permitted. The worker should be

properly equipped to insure virtually no contact with the carcinogen.

CWA 307(a)(1) Priority Pollutant

CHLOROMETHYL METHYL ETHER.
[107-30-2]

OSHA Candidate List of Potential Occupational Carcinogens (EPA Carcinogen Assessment Group List), 45 FR 53672, August 12, 1980

[possibly associated with bis(chloromethyl) ether]:
IARC—carcinogenic in humans
*RCRA 40 CFR 261
EPA hazardous waste no. U046

N,N-BIS(2-CHLOROMETHYL)-2-NAPTHYLAMINE.

OSHA Candidate List of Potential Occupational Carcinogens (EPA Carcinogen Assessment Group List), 45 FR 53672, August 12, 1980

CHLOROMETHYLOXIRANE.
See Epichlorohydrin. [106-89-8]

3-(CHLOROMETHYL)PYRIDINE (HYDROCHLORIDE). [6959-48-4]

NCI—carcinogenic in animals

2-CHLORONAPHTHALENE. [91-58-7]

CWA 307(a)(1) Priority Pollutant
*RCRA 40 CFR 261
EPA hazardous waste no. U047

beta-CHLORONAPHTHALENE. See 2-Chloronapthalene. [91-58-7]

CHLORONEB. [2675-77-6]

FIFRA—Registration Standard—Issued September 1980, NTIS# PB81 123804 (fungicide)

2-CHLORO-4-NITROANILINE. See Aniline and chloro-, bromo-, and/or nitro-anilines. [121-87-9]

4-CHLORO-2-NITROANILINE. See Aniline and chloro-, bromo-, and/or nitro-anilines. [89-63-4]

2-CHLORO-4-NITROBENZENAMINE.
See Aniline and chloro-, bromo-, and/or nitro- anilines. [121-87-9]

2-CHLORO-5-NITROBENZENAMINE.
See Aniline and chloro-, bromo-, and/or nitro- anilines. [6283-25-67]

4-CHLORO-2-NITROBENZENAMINE.
See Aniline and chloro-, bromo-, and/or nitro- anilines. [89-63-4]

4-CHLORO-3-NITROBENZENAMINE.
See Aniline and chloro-, bromo-, and/or nitro- anilines. [635-22-3]

2-CHLORONITROBENZENE. [88-73-3]

TSCA-CHIP available

4-CHLORONITROBENZENE.
[100-00-5]

TSCA-CHIP available

CHLORONITROBENZENES (*meta* or *ortho*). [88-73-3, 121-73-3]

NFPA—hazardous chemical
health—3, flammability—1, reactivity—1

1-CHLORO-1-NITROPROPANE.
[600-25-9]

*OSHA 29 CFR 1910.1000
PEL(TWA): 20 ppm, 100 mg/m^3
ACGIH-TLV
TWA: 2 ppm, 10 mg/m^3

CHLOROPENTAFLUOROETHANE.
[76-15-3]

ACGIH-TLV
TWA: 1000 ppm, 6320 mg/m^3

1-CHLOROPENTANE. [543-59-9]

NFPA—hazardous chemical (flammable chemical)
health—1, flammability—3, reactivity—0

4-CHLOROPHENE-1,3-DIAMINE. See *m*-Phenylenediamine, 4-chloro-.
[5131-60-2]

2-CHLOROPHENOL. [95-57-8]

CWA 304(a)(1), 45 FR 79318, November 28, 1980, Water Quality Criteria

74 o-CHLOROPHENOL

Freshwater Aquatic Life

The available data for 2-chlorophenol indicate that acute toxicity to freshwater aquatic life occurs at concentrations as low as 4380 µg/liter and would occur at lower concentrations among species that are more sensitive than those tested. No definitive data are avialable concerning the chronic toxicity of 2-chlorophenol to sensitive freshwater aquatic life but flavor impairment occurs in one species of fish at concentrations as low as 2000 µg/liter.

Saltwater Aquatic Life

No saltwater organisms have been tested with 2-chlorophenol and no statement can be made concerning acute and chronic toxicity.

Human Health

Sufficient data are not available for 2-chlorophenol to derive a level that would protect against the potential toxicity of this compound. Using available organoleptic data, for controlling undesirable taste and odor quality of ambient water, the estimated level is 0.1 µg/liter. It should be recognized that organoleptic data as a basis for establishing a water quality criterion have limitations and have no demonstrated relationship to potential adverse human health effects.

CWA 307(a)(1) Priority Pollutant

*RCRA 40 CFR 261
EPA hazardous waste no. U048

o-CHLOROPHENOL.
See 2-chlorophenol. [95-57-8]

4-CHLORO-m-PHENYLENEDIAMINE.
[5131-60-2]

NCI—carcinogenic in animals

4-CHLORO-o-PHENYLENEDIAMINE.
[95-83-0]

NCI—carcinogenic in animals

4-CHLOROPHENYL PHENYL ETHER.

CWA 307(a)(1) Priority Pollutant

1-(o-CHLOROPHENYL)THIOUREA.
[5344-82-1]

*RCRA 40 CFR 261
EPA hazardous waste no. P026

CHLOROPICRIN. [76-06-2]

*OSHA 29 CFR 1910.1000 Table Z-1
PEL(TWA): 0.1 ppm, 0.7 mg/m^3

ACGIH-TLV
TWA: 0.1 ppm, 0.7 mg/m^3
STEL: 0.3 ppm, 2 mg/m^3

FIFRA—Registration Standard Development—Data Evaluation and Development of Regulatory Position (insecticide)

NFPA—hazardous chemical
health—4, flammability—0, reactivity—3

beta-CHLOROPRENE. [126-99-8]
See Chloroprene.

CHLOROPRENE (2-CHLORO-1,3-BUTADIENE). [126-99-8]

*OSHA 29 CFR 1910.1000 Table Z-1

Skin

PEL(TWA): 25 ppm, 90 mg/m^3

ACGIH-TLV

Skin

TWA: 10 ppm, 45 mg/m^3

NIOSH Criteria Document(Pub. No. 77-210), NTIS Stock No. PB 274777, August 9, 1977

Synonyms for chloroprene include 2-chloro-1,3-butadiene, 2-chloroprene, and beta-chloroprene.

"Occupational exposure to chloroprene" is defined as work in any establishment where chloroprene is manufactured, stored, handled, used, or otherwise present. Occupational exposure to chloroprene shall require adherence to all sections of the standard.

The employer shall control exposure to chloroprene so that no employee is ever exposed at a concentration greater than 3.6 mg/m^3 of air (1 ppm), determined as a ceiling concentration for any 15-minute sampling period during a 40-hour workweek. The schedule for such sampling shall be determined by a professional industrial hygienist in accordance with good industrial hygiene practice.

The standard is designed to protect the health and provide for the safety of employees for up to a 10-hour work shift, 40-hour workweek, over a working lifetime.

NIOSH Current Intelligence Bulletin No. 1—cancer-related bulletin

CAA 112
Chemical proposed to be assessed as a Hazardous Air Pollutant by a House of Representatives bill to amend the CAA

3-CHLOROPROPENE. See Allyl chloride. [107-05-1]

3-CHLOROPROPIONITRILE. [542-76-7]

*RCRA 40 CFR 261
EPA hazardous waste no. P027

CHLOROPROPYLENE OXIDE. See Epichlorohydrin. [106-89-8]

o-CHLOROSTRYENE. [1331-28-8]

ACGIH-TLV
TWA: 50 ppm, 285 mg/m^3
STEL: 75 ppm, 430 mg/m^3

CHLOROSULFONIC ACID. See Chlorosulfuric acid. [7790-94-5]

CHLOROSULFURIC ACID. [7790-94-5]

NFPA—hazardous chemical
health—3, flammability—0, reactivity—2, W—water may be hazardous in fire fighting, oxy—oxidizing chemical. Reacts explosively with water or steam to produce toxic and corrosive fumes.

*CWA 311(b)(2)(A), 40 CFR 116, 117
Discharge RQ = 1000 pounds (454 kilograms)

CHLOROTHALONIL. [1897-45-6]

NCI—carcinogenic in animals
FIFRA—Data Call In, 45 FR 75488, November 14, 1980
Status: letter issued 9/14/81
FIFRA—Registration Standard Development—Data Collection (fungicide)

alpha-CHLOROTOLUENE.
See Benzyl chloride. [100-44-7]

o-CHLOROTOLUENE. See 2-Chlorotoluene. [95-49-8]

2-CHLOROTOLUENE. [95-49-8]

ACGIH-TLV

Skin

TWA: 50 ppm, 250 mg/m^3
STEL: 75 ppm, 375 mg/m^3

TSCA 4(e), ITC
Eighth Report of the TSCA Interagency Testing Committee to the administrator, Environmental Protection Agency, April 1981; 46 FR 28138, May 22, 1981, EPA responded to committee's recommendations for testing, 45 FR 18172, April 28, 1982

*TSCA 8(d), Unpublished Health and Safety Studies Reporting, 40 CFR 716, 47 FR 38800, September 2, 1982; 48 FR 13178, March 30, 1983

4-CHLORO-o-TOLUIDINE (HYDROCHLORIDE). [3165-93-3]

NCI—carcinogenic in animals

*RCRA 40 CFR 261
EPA hazardous waste no. U049

5-CHLORO-o-TOLUIDINE. [95-79-4]

NCI—carcinogenic in animals

2-CHLORO-6-(TRICHLOROMETHYL) PYRIDINE. See Nitrapyrin. [1929-82-4]

CHLORPYRIFOS. [2921-88-2]

ACGIH-TLV

Skin

TWA: 0.2 mg/m^3
STEL: 0.6 mg/m^3

FIFRA—Data Call In, 45 FR 75488, November 14, 1980
Status: agency decision reached, no additional chronic toxicological data required at this time

*CWA 311(b)(2)(A), 40 CFR 116, 117
Discharge RQ = 1 pound (0.454 kilogram)

CHROME PIGMENT.
NIOSH Current Intelligence Bulletin No. 4—cancer-related bulletin

CHROMIC ACETATE. [1066-30-4]
*CWA 311(b)(2)(A), 40 CFR 116, 117
Discharge RQ = 1000 pounds (454 kilograms)

CHROMIC ACID. [11115-74-5]
*OSHA 1910.1000 Table Z-2
1 mg/10 m^3 Acceptable ceiling concentration
NIOSH Criteria Document (Pub. No. 73-11021), NTIS Stock No. PB 222221, July 17, 1973

"Chromic acid" is defined to mean chromium trioxide [chromium(VI) oxide, or chromic acid anhydride] and aqueous solutions thereof. "Occupational exposure to chromic acid" is defined as exposure above half the recommended workroom environmental standard.

Occupational exposure to chronic acid shall be controlled so that no worker is exposed either to: (1) a concentration of chromic acid greater than 0.05 mg as chromium trioxide per cubic meter of air determined as a time-weighted average exposure for an 8-hour workday, 40-hour workweek; or (2) a ceiling concentration in excess of 0.1 mg as chromium trioxide per cubic meter as determined by a sampling time of 15 minutes. The standard is designed to protect the health and safety of workers for an 8-hour workday, 40-hour workweek over a working lifetime.

*CWA 311(b)(2)(A), 40 CFR 116, 117
Discharge RQ = 1000 pounds (454 kilograms)
NFPA—hazardous chemical
health—3, flammability—0, reactivity—1, oxy—oxidizing chemical.

CHROMIC ACID, CALCIUM SALT. [13765-19-0]
*RCRA 40 CFR 261
EPA hazardous waste no. U032 (hazardous substance)

CHROMIC CHROMATE. [24613-89-6]
IARC—carcinogenic in animals

CHROMIC SULFATE. [10101-53-8]
*CWA 311(b)(2)(A), 40 CFR 116, 117
Discharge RQ = 1000 pounds (454 kilograms)

CHROMIC TRIOXIDE.
See Chromic acid. [11115-74-5]

CHROMITE ORE processing (CHROMATE), as Cr.
ACGIH-TLV
TWA: 0.05 mg/m^3
Human carcinogen—substance associated with industrial processes, recognized to have carcinogenic or cocarcinogenic potential

CHROMIUM. [7440-47-3]
ACGIH-TLV—Chromium(III) compounds, as Cr
TWA: 0.5 mg/m^3
ACGIH-TLV—Chromium(II) compounds as Cr
TWA: 0.5 mg/m^3
ACGIH-TLV—Chromium(VI) compounds, as Cr, water soluble
TWA: 0.05 mg/m^3
ACGIH-TLV—Chromium(VI) compounds, as Cr, certain water insolubles
TWA: 0.05 mg/m^3
Human carcinogen—associated with industrial processes, recognized to have carcinogenic or cocarcinogenic potential.

Chromium(VI), NIOSH Criteria Document (Pub. No. 76-129), NTIS Stock No. PB 248595, December 1, 1975
For the purpose of this standard, "chromium(VI)" is defined as the chromium in all materials in the +6 (hexavalent) state.

There are two recommended standards for chromium(VI). (See Chromic acid, NIOSH Criteria Document.) One addresses occupational exposure to a group of noncarcinogenic, but otherwise hazardous materials, while the other pertains to occupations and workplaces where there is exposure to other chromium(VI) materials associated with an increased incidence of lung cancer.

On the basis of the chemical analysis of airborne chromium(VI) materials, there is no practical means of distinguishing between these two groups of chromium(VI) materials. Until the airborne chromium(VI) in a particular workplace is demonstrated by the employer to be of the type considered to be noncarcinogenic, all airborne chromium(VI) shall be considered to comprise carcinogenic materials.

Based on current evidence, "noncarcinogenic chromium(VI)" is the chromium(VI) in monochromates and bichromates (dichromates) of hydrogen, lithium, sodium, potassium, rubidium, cesium, and ammonium, and chromium(VI) oxide (chromic acid anhydride). "Carcinogenic chromium(VI)" comprises any and all chromium(VI) materials not included in the noncarcinogenic group. "Occupational exposure to carcinogenic chromium(VI)" is defined as exposure to airborne chromium(VI) at concentrations greater than one-half of the workplace environmental limit for carcinogenic chromium(VI). "Occupational exposure to noncarcinogenic chromium(VI)" is defined as exposure to airborne chromium(VI) at concentrations greater than one-half of the workplace environmental limit for noncarcinogenic chromium(VI). Exposure to chromium(VI) at concentrations less than one-half of the workplace environmental limit will not require adherence to all sections of the standard except for Sections 3(a,b,c,d), 4a, 5, 6(b,c,e,f), and 7. See the Criteria Document for details

Carcinogenic chromium(VI) shall be controlled in the workplace so that the airborne workplace concentration of chromium(VI), sampled and analyzed according to the procedures in Appendices I and II in the Criteria Document, is not greater than 1 μg Cr(VI)/m^3 of breathing zone air.

Noncarcinogenic chromium(VI) shall be controlled in the workplace so that the airborne workplace concentration is not greater than 25 μg Cr(VI)/m^3 of breathing zone air determined as a time-weighted average (TWA) exposure for up to a 10-hour workday, 40-hour workweek, and is not greater than 50 μg Cr(VI)/m^3 of breathing zone air as determined by any 15-minute sample.

Procedures for sampling and analysis of chromium(VI) in air shall be as provided in Appendices I and II, or by any method shown to be equivalent in precision, accuracy, and sensitivity to the methods specified.

The standard is designed to protect the health and safety of workers for up to a 10-hour workday, 40-hour workweek over a working lifetime.

OSHA Candidate List of Potential Occupational Carcinogens, (EPA Carcinogen Assessment Group List), 45 FR 53672, August 12, 1980

CWA 304(a)(1), 45 FR 79318, November 28, 1980, Water Quality Criteria

Freshwater Aquatic Life

For total recoverable hexavalent chromium the criterion to protect freshwater aquatic life as derived using the guidelines is 0.29 μg/liter as a 24-hour average and the concentration should not exceed 21 μg/liter at any time.

For freshwater aquatic life the concentration (in micrograms per liter) of total recoverable trivalent chromium should not exceed the numerical value given by $e(1.08[\ln(\text{hardness})] + 3.48)$ at any time. For example, at hardnesses of 50, 100, and 200 mg/liter as CaCo$_3$ the concentration of total recoverable trivalent chromium should not exceed 2200, 4700, and 9000 μg/liter, respectively, at any time. The available data indicate that chronic toxicity to freshwater aquatic life occurs at concentrations as low as 44 μg/liter and would occur at lower concentrations among species that are more sensitive than those tested.

Saltwater Aquatic Life

For total recoverable hexavalent chromium the criterion to protect saltwater aquatic life as derived using the guidelines is 18 μg/liter as a 24-hour average and the concentration should not exceed 1260 μg/liter at any time.

For total recoverable trivalent chromium, the available data indicate that acute toxicity to saltwater aquatic life occurs at concentrations as low as 10,300 µg/liter and would occur at lower concentrations among species that are more sensitive than those tested. No data are available concerning the chronic toxicity of trivalent chromium to sensitive saltwater aquatic life.

Human Health

For the protection of human health from the toxic properties of chromium(III) ingested through water and contaminated aquatic organisms, the ambient water criterion is determined to be 170 mg/liter.

For the protection of human health from the toxic properites of chromium(III) ingested through contaminated aquatic organisms alone, the ambient water criterion is determined to be 3433 mg/liter.

The ambient water quality criterion for total chromium(VI) is recommended to be identical to the existing drinking water standard which is 50 µg/liter. Analysis of the toxic effects data resulted in a calculated level that is protective of human health against the ingestion of contaminated water and contaminated aquatic organisms. The calculated value is comparable to the present standard. For this reason a selective criterion based on exposure solely from consumption of 6.5 g of aquatic organisms was not derived.

CWA 307(a)(1), Priority Pollutant

*National Interim Primary Drinking Water Regulations, 40 CFR 141; 40 FR 59565, December 24, 1975; amended by 41 FR 28402, July 9, 1976; 44 FR 68641, November 29, 1979; corrected by 45 FR 15542, March 11, 1980; 45 FR 57342, August 27, 1980; 47 FR 18998, March 3, 1982; corrected by 47 FR 10998, March 12, 1982

Maximum contaminant level—0.05 mg/liter

CHROMIUM (chromate-producing industries). [7440-47-3]

IARC—carcinogenic in humans

CHROMIUM, METAL and INSOLUBLE SALTS. [7440-47-3]

*OSHA 29 CFR 1910.1000 Table Z-1
PEL(TWA): 1 mg/m^3
ACGIH-TLV
TWA: 0.5 mg/m^3

CHROMIUM, SOLUBLE CHROMIC, CHROMOUS SALT. [7440-47-3]

*OSHA 29 CFR 1910.1000 Table Z-1
PEL(TWA): 0.5 mg/m^3 (as Cr)
ACGIH-TLV
TWA: 0.5 mg/m^3

CHROMIUM OXYCHLORIDE.
See Chromyl chloride. [14977-61-8]

CHROMOUS CHLORIDE. [10049-05-5]

*CWA 311(b)(2)(A), 40 CFR 116, 117
Discharge RQ = 1000 pounds (454 kilograms)

CHROMYL CHLORIDE. [14977-61-8]

ACGIH-TLV (1982 addition)
TWA: 0.025 ppm, 0.15 mg/m^3

NFPA—hazardous chemical
health—3, flammability—0, reactivity—1

CHRYSENE. See Coal tar pitch volatiles. [218-01-9]

ACGIH-TLV
TWA: Industrial substance suspect of carcinogenic potential for humans—which is suspect of inducing cancer, based on either (1) limited epidemiologic evidence, exclusive of clinical reports of single cases, or (2) demonstration of carcinogenesis in one or more animal species by appropriate methods. Worker exposure by all routes should be carefully controlled to levels consistent with the animal and human experience data.

OSHA Candidate List of Potential Occupational Carcinogens (EPA Carcinogen Assessment Group List), 45 FR 53672, August 12, 1980

IARC—carcinogenic in animals

CWA 307(a)(1), Priority Pollutant
*RCRA 40 CFR 261
EPA hazardous waste no. U050

CHRYSOIDINE. [532-82-1]
IARC—carcinogenic in animals

CHRYSOTILE (SERPENTINE).
See Asbestos. [12001-29-5]
*TSCA 8(d), Unpublished Health and Safety Studies Reporting, 40 CFR 716, 47 FR 387, September 2, 1982

C.I. 76070. See p-Phenylenediamine, 2-nitro-. [5307-14-2]

C.I. 76555. See Phenol, 4-amino-2-nitro-. [119-34-6]

C.I. ACID YELLOW 73. [518-47-8]
NCI—carcinogenic in animals

C.I. AZOICDIAZO COMPONENT 11.
Syn: o-Toluidine, 4-chloro-, hydrochloride. [3165-93-3]
OSHA Candidate List of Potential Occupational Carcinogens, 45 FR 53672, August 12, 1980

C.I. DIRECT BLACK 38, DIRECT BLUE 6, and DIRECT BROWN 95: BENZIDINE DERIVED DYES. [1937-37-1, 2602-46-2, 16071-86-6]
NIOSH Current Intelligence Bulletin No. 24—cancer-related bulletin

C.I. DIRECT BLACK 38, DISODIUM SALT. Syn: 2,7-Naphthalenedisulfonic acid, 4-amino-3-((4'-((2,4-diaminophenyl)azo)(1,1'-biphenyl)-4-yl)azo)5-hydroxy-5-(phenylazo), disodium salt. See Bisazobiphenol dyes. [1937-37-1]
OSHA Candidate List of Potential Occupational Carcinogens, 45 FR 53672, August 12, 1980
*TSCA 8(a), 40 CFR 712, 47 FR 26992, June 22, 1982, Chemical Information Rule
*TSCA 8(d) Unpublished Health and Safety Studies Reporting, 40 CFR 716, 47 FR 387, September 2, 1982
NCI—carcinogenic in animals

C.I. DIRECT BLUE 6, TETRASODIUM SALT. Syn: 2,7-Naphthalenedisulfonic acid, 3,3'-((4,4'-biphenylene)bis(azo)bis(5-amino-4-hydroxy-, tetrasodium salt). See Bisazobiphenol dyes. [2602-46-2]
OSHA Candidate List of Potential Occupational Carcinogens, 45 FR 53672, August 12, 1980
*TSCA 8(a), 40 CFR 712, 47 FR 26992, June 22, 1982, Chemical Information Rule
*TSCA 8(d) Unpublished Health and Safety Studies Reporting, 40 CFR 716, 47 FR 387, September 2, 1982
NCI—carcinogenic in animals

C.I. DIRECT BLUE 218. See Bisazobiphenyl dyes. [10401-50-0]
*TSCA 8(d) Unpublished Health and Safety Studies Reporting, 40 CFR 716, 47 FR 387, September 2, 1982

C.I. DIRECT BROWN 74. See Bisazobiphenyl dyes. [8014-91-3]
*TSCA 8(d) Unpublished Health and Safety Studies Reporting, 40 CFR 716, 47 FR 387, September 2, 1982

C.I. DIRECT BROWN 95. Syn: Copper, (5-((4'-((2,5-dihydroxy-4-((2-hydroxy-5-sulfophenyl)azo)phenyl)azo)(1,1'-biphenyl)-4yl)azo)2-hydroxybenzoato(2-))-, disodium salt). [16071-86-6]
OSHA Candidate List of Potential Occupational Carcinogens, 45 FR 53672, August 12, 1980
NCI—carcinogenic in animals

C.I. DIRECT BROWN 154, DISODIUM SALT. See Bisazobiphenyl dyes. [6360-54-9]
*TSCA 8(d), Unpublished Health and Safety Studies Reporting, 40 CFR 716, 47 FR 387, September 2, 1982

C.I. DIRECT ORANGE 6, DISODIUM SALT. See Bisazobiphenyl dyes. [6637-88-3]

*TSCA 8(d), Unpublished Health and Safety Studies Reporting, 40 CFR 716, 47 FR 387, September 2, 1982

C.I. DISPERSE BLACK 6, DIHYDROCHLORIDE. Syn: Benzidine, 3,3'-dimethoxy, dihydrochloride. [20325-40-0]

OSHA Candidate List of Potential Occupational Carcinogens, 45 FR 53672, August 12, 1980

C.I. DISPERSE ORANGE 11. Syn: Anthraquinone, 1-amino-2-methyl-. [32-28-0]

OSHA Candidate List of Potential Occupational Carcinogens, 45 FR 53672, August 12, 1980

C.I. DISPERSE YELLOW 3. [2832-40-8]

NCI—carcinogenic in animals

TSCA-CHIP available

C.I. SOLVENT ORANGE 2. Syn: 1-(o-Tolylazo)-beta-naphthol. [2646-17-5]

OSHA Candidate List of Potential Occupational Carcinogens, 45 FR 53672, August 12, 1980

C.I. SOLVENT YELLOW 1. Syn: Aniline, p-(phenylazo). [60-09-3]

OSHA Candidate List of Potential Occupational Carcinogens, 45 FR 53672, August 12, 1980

C.I. SOLVENT YELLOW 3. Syn: o-Toluidine, 4-(o-tolylazo)-. [97-56-3]

OSHA Candidate List of Potential Occupational Carcinogens, 45 FR 53672, August 12, 1980

C.I. SOLVENT YELLOW 14. [842-07-9]

NCI—carcinogenic in animals

C.I. SOLVENT YELLOW 34. Syn: Aniline, 4,4'-(imidocarbonyl)bis(N,N-dimethyl-). [492-80-8]

OSHA Candidate List of Potential Occupational Carcinogens, 45 FR 53672, August 12, 1980

C.I. VAT YELLOW 4. [128-66-5]

NCI—carcinogenic in animals

CINNAMYL ANTHRANILATE. [87-29-6]

NCI—carcinogenic in animals

CITRUS RED NO. 2. [6358-53-8]

OSHA Candidate List of Potential Occupational Carcinogens (EPA Carcinogen Assessment Group List), 45 FR 53672, August 12, 1980

IARC—carcinogenic in animals

CLOPIDOL. [2871-90-6]

ACGIH-TLV
TWA: 10 mg/m^3
STEL: 20 mg/m^3

COAL DUST.

ACGIH-TLV
2 mg/m^3 (respirable dust fraction < 5% quartz), if > 5% quartz, use respirable mass formula

COAL GASIFICATION PLANTS.

NIOSH Criteria Document (Pub. No. 78-191), NTIS Stock No. PB 80 164874, September 1978

These criteria and the recommended standard apply to the exposure of employees to toxicants and hazardous operating conditions in commercial coal gasification plants. As used herein, the term "commercial coal gasification plant" refers to any plant using coal to produce a gas that will be sold as a source of energy or otherwise utilized for commercial purposes. These criteria and the recommended standard pertain principally to the types of plants whose technology, construction, and utilization are anticipated around the year 1985. The term "toxicants" applies to all raw materials, products, and by-products of coal gasification processes that may produce a toxic effect; toxicants include, but are not limited to, asphyxiants, irritants, nuisance particulates, poisons, and carcinogens. The following terms are used interchangeably with the term "toxicant(s)": toxic compound(s), toxic material(s), toxic gas(es), hazardous material(s), and hazardous agent(s). The term "hazardous operating conditions" refers to conditions that may impair the health of, or cause physical injury to, employees.

For all sections of the recommended standard except workplace monitoring, the terms "occupational exposure" and "employee exposure" are defined as any contact with any toxicant(s) in the work environment. For purposes of workplace monitoring, these terms are defined as in the existing federal standards (29 CFR 1910) except where NIOSH has used different language, in which case the NIOSH definition applies.

No attempt has been made herein to develop permissible levels of exposure to toxic substances specific to coal gasification plants. It is recommended that applicable existing federal occupational exposure limits (29 CFR 1910, Subpart Z) be enforced, except where NIOSH has recommended a reduction in the existing federal limit or where there is no existing federal limit, in which cases the NIOSH recommendations should apply. Valid and reproducible techniques for measuring exposure are available to industry and government agencies. Furthermore, existing technology is adequate to permit compliance with the recommended standard.

The recommended standard is designed to protect the health and provide for the safety of employees for up to a 10-hour work shift, 40-hour workweek, during a working lifetime. Compliance with all sections of the recommended standard should prevent or greatly reduce the adverse effects of toxicants or hazardous conditions on the health of employees and provide for their safety.

A tabular summary is presented in Table 1 of the regulatory status of various hazardous agents that are potentially present in coal gasification plants.

Most of Table 1 is based on NIOSH criteria documents. For chemical substances not covered by NIOSH recommended standards, the current federal occupational exposure standard is listed (49): 40. Title 29 of the Code of Federal Regulations (CFR), Part 1910.1000, Subpart Z, 1976.

a. Up to an 8-hour time-weighted average unless otherwise noted
b. Up to a 10-hour time-weighted average unless otherwise noted

The recommended standard is designed to protect the health and provide for the safety of employees for up to a 10-hour work shift, 40-hour workweek, during a working lifetime.

COAL TAR PITCH VOLATILES (benzene soluble fraction) (Anthracene, BaP, PHENANTHRENE, ACRIDINE, CHRYSENE, PYRENE) [8007-45-2]

*OSHA 29 CFR 1910.1000 Table Z-1
PEL(TWA): 0.2 mg/m^3
Interpretation of term: 29 CFR 1910.1002 (revised 48 FR 2768, January 21, 1983)
Coal tar pitch volatiles include the fused polycyclic hydrocarbons which volatize from the distillation residues of coal, petroleum (excluding asphalt), wood, and other organic matter. Asphalt (CAS 8052-42-4, and CAS 64742-93-4) is not covered under the "coal tar pitch volatiles" standard.

ACGIH-TLV
TWA: 0.2 mg/m^3
Human carcinogen—associated with industrial processes, recognized to have carcinogenic or cocarcinogenic potential

COAL TAR. See Coal tar products, NIOSH Criteria Document. [8007-45-2]

NIOSH Criteria Document (Pub. No. 78-107), NTIS Stock No. PB 276917, September 29, 1977
The term "coal tar products," as used in this recommended standard, includes coal tar and two of the fractionation products of coal tar, creosote and coal tar pitch, derived from the carbonization of bituminous coal. Coal tar, coal tar pitch, and creosote derived from bituminous coal often contain identifiable components which by themselves are carcino-

Table 1. Summary of Health Effects of Agents Potentially Present in Coal Gasification Plants.

Agent	Current Federal Occupational Exposure Standard(s)	NIOSH Recommendation for Permissible Exposure Limit(s)
Ammonia	50 ppm (34.8 $\mu g/m^3$)	50 ppm ceiling (34.8 mg/m^3) (5 min)
Arsenic, inorganic	0.5 mg/m^3	2 $\mu g/m^3$ ceiling (15 min)
Benzene	1 ppm (3.2 mg/m^3); 5 ppm maximum ceiling (15 min)	1 ppm ceiling (3.2 mg/m^3) (60 min)
Beryllium	2 $\mu g/m^3$; 5 $\mu g/m^3$ acceptable ceiling: 25 $\mu g/m^3$ maximum ceiling (30 min)	0.5 $\mu g/m^3$ (130 min)
Cadmium	0.1 mg/m^3; fumes = 0.3 mg/m^3 ceiling (erroneously published as 3 mg/m^3); 0.2 mg/m^3; dust = 0.6 mg/m^3 ceiling	40 $\mu g/m^3$; 200 $\mu g/m^3$ ceiling (15 min)
Carbon dioxide	5000 ppm (9000 mg/m^3)	10,000 ppm (18,000 mg/m^3; 30,000 ppm ceiling (54,000 mg/m^3) (10 min)
Carbon monoxide	50 ppm (55 mg/m^3)	35 ppm (40 mg/m^3); 200 ppm ceiling (229 mg/m^3)
Carbonyl sulfide	None	None
Chromium (VI)	100 $\mu g/10\ m^3$ ceiling	1 $\mu g/m^3$ for carcinogenic Cr (VI); 25 $\mu g/m^3$ for other chromium; 50 $\mu g/m^3$ ceiling (15 min)
Coal dust	2.4 mg/m^3, if respirable dust fraction less than 5% SiO_2; if respirable fraction is more than 5% SiO_2, respirable mass formula is (10 mg/m^3)/(% SiO_2 + 2)	None
Coal tar products	0.2 mg/m^3 (for benzene-soluble fraction of tar pitch volatiles)	0.1 mg/m^3 (cyclohexane-extractable fraction of coal tar, coal tar pitch, creosote, or mixtures)
Coke-oven emissions	150 $\mu g/m^3$	Work practices to minimize exposure to emissions
Cresol	5 ppm (22 mg/m^3) (skin)	10 mg/m^3
Cyanide, hydrogen and cyanide salts	10 ppm (alkali cyanides); 5 mg CN/m^3 (cyanide)	5 mg CN/m^3 ceiling (4.7 ppm) (10 min)
Fluorides, inorganic	2.5 mg/m^3	2.5 mg/m^3
Hot environments	None	Variable (sliding scale)
Hydrogen chloride	5 ppm (7 mg/m^3)	None
Hydrogen sulfide	20 ppm acceptable ceiling; 50 ppm maximum ceiling (10 min)	15 mg/m^3 ceiling (approximately 10 ppm) (10 min)
Isopropyl ether	500 ppm (2100 mg/m^3)	None
Lead, inorganic	0.2 mg/m^3	Less than 100 $\mu g/m^3$
Manganese	5 mg/m^3 ceiling	None
Mercury, inorganic	0.1 mg/m^3 ceiling	0.05 mg/m^3
Methanol	200 ppm (260 mg/m^3)	200 ppm (262 mg/m^3); 800 ppm (1048 mg/m^3) ceiling (15 min)
Nickel, inorganic and compounds	1 mg/m^3 (metal and soluble compounds as Ni)	15 $\mu g\ Ni/m^3$

Table 1. (Continued)

Agent	Current Federal Occupational Exposure Standard(s)	NIOSH Recommendation for Permissible Exposure Limit(s)
Nickel carbonyl	0.001 ppm (0.007 mg/m^3)	0.001 ppm (0.007 mg/m^3)
Nitrogen oxides	NO_2: 5 ppm (9 mg/m^3) NO: 25 ppm (30 mg/m^3)	NO_2: 1 ppm (1.8 mg/m^3) ceiling (15 min); NO: 25 ppm (30 mg/m^3)
Noise	90 dBA	85 dBA (8-hour TWA); 115 dBA ceiling
Phenol	5 ppm (skin)	20 mg/m^3 (5.2 ppm); 60 mg/m^3 (15.6 ppm) ceiling (15 min)
Selenium	0.2 mg/m^3 compounds as Se	None
Silica, crystalline	$(250)/(\% SiO_2 + 5)$ in mppcf, or (10 mg/m^3 $(\% SiO_2 + 2)$ respirable quartz	50 μg/m^3 respirable free silica
Sulfur dioxide	5 ppm (13 mg/m^3)	0.5 ppm (1.3 mg/m^3)
Toluene	200 ppm; 300 ppm acceptable ceiling; 500 ppm maximum peak above acceptable ceiling (10 min)	100 ppm (376 mg/m^3); 200 ppm (750 mg/m^3) ceiling (10 min)
Vanadium	Vanadium pentoxide: dust = 0.5 mg/m^3 ceiling; fume = 0.1 mg/m^3 ceiling	Vanadium compounds: 0.05 mg/m^3 ceiling (15 min)
Xylene	100 ppm (435 mg/m^3)	100 ppm (434 mg/m^3); 200 ppm (868 mg/m^3) ceiling (10 min)

genic, such as benzo(a)-pyrene, benzanthracene, chrysene, and phenanthrene. Other chemicals from coal tar products such as anthracene, carbazole, fluoranthene, and pyrene may also cause cancer, but these causal realtionships have not been adequately documented. "Occupational exposure to coal tar products" is defined as any contact with coal tar, coal tar pitch, or creosote in the work environment.

From the epidemiologic and experimental toxicologic evidence on coal tar, coal tar pitch, and creosote, NIOSH has concluded that they are carcinogenic and can increase the risk of lung and skin cancer in workers. Therefore, the permissible exposure limit recommended is the lowest concentration that can be reliably detected by the recommended method of environmental monitoring. While compliance with this limit should substantially reduce the incidence of cancer produced by coal tar products, no absolutely safe concentration can be established for a carcinogen at this time. The environmental limit is proposed to reduce the risk, and the employer should regard it as the upper boundary of exposure and make every effort to keep exposure as low as is technically feasible.

Occupational exposure to coal tar products shall be controlled so that employees are not exposed to coal tar, coal tar pitch, creosote, or mixtures of these substances at a concentration greater than 0.1 m^3 of the cyclohexane-extractable fraction of the sample, determined as a time-weighted average (TWA) concentration for up to a 10-hour work shift in a 40-hour workweek.

The standard is designed to protect the health and provide for the safety of employees for up to a 10-hour work shift, 40-hour workweek, over a working lifetime.

40 CFR 162.11, FIFRA-RPAR issued
Current Status: PD 1 published, 43 FR 48154 (10/18/78); comment period closed 9/30/78; NTIS# PB80 312879. PD 2/3 (wood uses only) completed and Notice of Determination published in 46 FR 13020 (2/19/81); comment period closed 5/20/81. Final decision is being developed.

Criteria Possibly Met or Exceeded: Oncogenicity and Mutagenicity

COBALT, as Co METAL, DUST and FUME. [7440-48-4]

*OSHA 29 CFR 1910.1000 Table Z-1
PEL(TWA): 0.1 mg/m^3
ACGIH-TLV
TWA: 0.1 mg/m^3
Intended changes for 1982:
TWA: 0.05 mg/m^3
STEL: 0.1 mg/m^3

COBALT NAPHTHENATE. [61789-51-3]

TSCA 4(e), ITC
Twelfth Report of the Interagency Testing Committee to the administrator, Environmental Protection Agency, May 1983; 48 FR 24443, June 1, 1983
*TSCA 8(a), 40 CFR 712, 48 FR 28443, June 22, 1983, Chemical Information Rule
TSCA-CHIP available

COBALTOUS BROMIDE. [7789-43-7]

*CWA 311(b)(2)(A), 40 CFR 116, 117
Discharge RQ = 1000 pounds (454 kilograms)

COBALTOUS FORMATE. [544-18-3]

*CWA 311(b)(2)(A), 40 CFR 116, 117
Discharge RQ = 1000 pounds (454 kilograms)

COBALTOUS NITRATE. [10141-05-6]

NFPA—hazardous chemical
when nonfire: health—0, flammability—0, reactivity—0, oxy—oxidiziing chemical
when fire: health—1, flammability—0, reactivity—0, oxy—oxidizing chemical

COBALTOUS SULFAMATE. [14017-41-5]

*CWA 311(b)(2)(A), 40 CFR 116, 117
Discharge RQ = 1000 pounds (454 kilograms)

COD LIVER OIL. [8001-69-2]

NFPA—hazardous chemical (flammable chemical)
health—0, flammability—1, reactivity—0

COKE OVEN EMISSIONS.

*OSHA Standard 29 CFR 1910.1029, Cancer Hazard Permissible exposure limit. The employer shall assure that no employee in the regulated area is exposed to coke oven emissions at concentrations greater than 150 µg/m^3 of air, averaged over any 8-hour period.

NIOSH Criteria Document (Pub. No. 73-11016), NTIS Stock No. PB 216167, February 28, 1973
No environmental limit is proposed because sufficient data are not available at present with which to determine a safe environmental level. Recommendations are made for protection of the worker through a combination of periodic medical examinations, the use of respiratory protective devices, and operating procedures or "work practice."

These recommendations are not intended to supplant the existing OSHA standard for coal tar pitch volatiles as published in the *Federal Register,* part 1910.93, Volume 37, dated October 18, 1972.

OSHA Candidate List of Potential Occupational Carcinogens, (EPA Carcinogen Assessment Group List), 45 FR 53672, August 12, 1980

CAA 112
Chemical proposed to be assessed as a Harzardous Air Pollutant by a House of Representatives bill to amend the CAA

COLLODION. See Cellulose nitrate. [9004-70-0]

COPPER. [7440-50-8]

CWA 304(a)(1), 45 FR 79318, November, 28, 1980, Water Quality Criteria

Freshwater Aquatic Life

For total recoverable copper the criterion to protect freshwater aquatic life as derived using the guidelines is 5.6 µg/liter as a 24-hour average and the concentration (in micrograms per liter) should not exceed the numerical value given by $e(0.94[\ln(\text{hardness})] - 1.23)$ at any time. For example, at hardnesses of 50, 100, and 200 mg/liter

CaCO$_3$ the concentration of total recoverable copper should not exceed 12, 22, and 43 µg/liter at any time.

Saltwater Aquatic Life

For total recoverable copper the criterion to protect saltwater aquatic life as derived using the guidelines is 4.0 µg/liter as a 24-hour average and the concentration should not exceed 23 µg/liter at any time.

Human Health

Sufficient data are not available for copper to derive a level that would protect against the potential toxicity of this compound. Using available organoleptic data, for controlling undesirable taste and odor qualtiy of ambient water, the estimated level is 1 mg/liter. It should be recognized that organoleptic data as a basis for establishing a water quality criterion have limitations and have no demonstrated relationship to potential adverse human health effects.

CWA 307(a)(1), Priority Pollutant
*National Secondary Drinking Water Regulations, 40 CFR 143; 44 FR 42198, July 19, 1979, effective January 19, 1981
Maximum contaminant level—1 mg/liter, corrosivity—noncorrosive

COPPER, DUSTS AND MISTS.
[7440-50-8]

*OSHA 29 CFR 1910.1000
PEL(TWA): 1 mg/m^3
ACGIH-TLV
TWA: 1 mg/m^3
STEL: 2 mg/m^3

COPPER, FUME. [7440-50-8]

*OSHA 29 CFR 1910.1000 Table Z-1
PEL(TWA): 0.1 mg/m^3
ACGIH-TLV
TWA: 0.2 mg/m^3

COPPER ACETOARSENITE.
[12002-03-8]

40 CFR 162.11, FIFRA-RPAP, Final Action, Voluntary Cancellation
Current Status: 42 FR 18422 (4/7/77)
Criteria Possibly Met or Exceeded: Oncogenicity, Mutagenicity, and Teratogenicity

COPPER ARSENATE, BASIC (all products). [12002-03-8]

40 CFR 162.11, FIFRA-RPAR, Final Action, Voluntary Cancellation
Current Status: 42 FR 19422 (4/7/77)
Criteria Possibly Met or Exceeded: Oncogenicity

COPPER CARBONATE. [12069-69-1]

FIFRA—Data Call In, 45 FR 75488, November 14, 1980
Status: agency decision reached, data call in requirements (chronic feeding, oncogenicity, reproduction, and teratology) waived.

COPPER (BORDEAUX MIXTURE).
[8011-63-0]

FIFRA—Data Call In, 45 FR 75488, November 14, 1980
Status: agency decision reached, data call in requirements (chronic feeding, oncogenicity, reproduction, and teratology) waived.

COPPER CYANIDES.
[544-92-3, 14763-77-0]

*RCRA 40 CFR 261
EPA hazardous waste no. P029

COPPER µ-[[6,6'-[(3,3'-DIHYDROXY[1,1'-BIPHENYL]-4,4'-DIYL)BIS(AZO)]]BIS[4-AMINO-5-HYDROXY-1,3-NAPHTHALENEDISULFONATO]] (4-)]]DI-, TETRASODIUM SALT.

$C_{32}H_{20}Cu_2N_6O_{16}S_4$·4Na. See Bisazobiphenyl dyes. [16143-79-6]

*TSCA 8(d), Unpublished Health and Safety Studies Reporting, 40 CFR 716, 47 FR 387, September 2, 1982

COPPER [µ-[7-[[3,3'-DIHYDROXY-4'-[[1-HYDROXY-6-(PHENYLAMINO)-3-SULFO-2-NAPHTHALENYL]AZO][1,1'-BIPHENYL]-4-YL]AZO]-8-HYDROXY-1,6-NAPHTHALENEDISULFONATO (4-)]]DI-, TRISODIUM SALT. See Bisazobiphenyl dyes. [6656-03-7]

*TSCA 8(d), Unpublished Health and Safety Studies Reporting, 40 CFR 716, 47 FR 387, September 2, 1982

COPPER [5-[[4'-[[2,5-DIHYDROXY-4-[(2-HYDROXY-5-SULFOPHENYL)AZO]PHENYL-AZO][1.1'BIPHENYL]-4-YL]AZO]-2-HYDROXYBENZOATO(2-)]DISODIUM SALT. See Bisazobiphenyl dyes.
[16071-86-6]

*TSCA 8(d), Unpublished Health and Safety Studies Reporting, 40 CFR 716, 47 FR 387, September 2, 1982

COPPER NAPHTHENATE. [1338-02-9]
FIFRA—Data Call In, 45 FR 75488, November 14, 1980
Status: agency decision reached, data call in requirements (chronic feeding, oncogenicity, reproduction, and teratology) waived.

COPPER NITRATE. See Cupric nitrate.
[10031-43-3]

COPPER OXIDE. [1317-39-1]
FIFRA—Data Call In, 45 FR 75488, November 14, 1980
Status: agency decision reached, data call in requirements (chronic feeding, oncogenicity, reproduction, and teratology) waived.

COPPER OXYCHLORIDE. [1332-40-7]
FIFRA—Data Call In, 45 FR 75488, November 14, 1980
Status: agency decision reached, data call in requirements (chronic feeding, oncogenicity, reproduction, and teratology) waived.

COPPER QUINOLINOLATE.
[10380-28-6]
FIFRA—Data Call In, 45 FR 75488, November 14, 1980
Status: agency decision reached, data call in requirements (chronic feeding, oncogenicity, reproduction, and teratology) waived.

COPPER SULFATE BASIC. [7758-99-8]
FIFRA—Data Call In, 45 FR 75488, November 14, 1980
Status: agency decision reached, data call in requirements (chronic feeding, oncogenicity, reproduction, and teratology) waived.

COPPER SULFATE MONOHYDRATE.
FIFRA—Data Call In, 45 FR 75488, November 14, 1980
Status: agency decision reached, data call in requirements (chronic feeding, oncogenicity, reproduction, and teratology) waived.

CORN OIL. [8001-30-7]
NFPA—hazardous chemical (flammable chemical)
health—0, flammability—1, reactivity—0

COTTON DUST.
OSHA Standard 29 CFR 1910.1043
Permissible exposure limits. (1) The employer shall assure that not employee who is exposed to cotton dust in yarn manufacturing is exposed to airborne concentrations of lintfree respirable cotton dust greater than 200 $\mu g/m^3$ mean concentration, averaged over an 8-hour period, as measured by a vertical elutriator or a method of equivalent accuracy and precision.

(2) The employer shall assure that no employee who is exposed to cotton dust in the textile processes known as slashing and weaving is exposed to airborne concentrations of lintfree respirable cotton dust greater than 750 $\mu g/m^3$ mean concentration, averaged over an 8-hour period, as measured by a vertical elutriator or a method of equivalent accuracy and precision.

(3) The employer shall assure that no employee who is exposed to cotton dust (except for exposures in yarn manufacturing and slashing and weaving) is exposed to airborne concentrations of lintfree respirable cotton dust greater than 500 $\mu g/m^3$ mean concentrations, averaged over an 8-hour period, as measured by a vertical elutriator or a method of equivalent accuracy and precision.

Shall not apply to: (i) the harvesting of cotton; (ii) the ginning of cotton (see 1910.1046); (iii) maritime operations (see 1915–1918); (iv) the handling or processing of woven or knitted materials; and(v) the handling or processing of washed cotton.

NIOSH Criteria Document (Pub. No. 75-118), September 26, 1974

"Exposure to cotton dust" includes any work with cotton that results in airborne cotton dust; "cotton dust" is defined as dust generated into the atmosphere as a result of the processing of cotton fibers combined with any naturally occurring materials such as stems, leaves, bracts, and inorganic matter which may have accumulated on the cotton fibers during the growing or harvesting period. Any dust generated from processing of cotton through the weaving of fabric in textile mills and dust generated in other operations or manufacturing processes using new or waste cotton fibers or cotton fiber by-products from textile mills is considered cotton dust. The recommended standard does not apply to dust generated from the handling or processing of woven materials.

Although there are significant gaps in knowledge of the etiology of byssinosis, the standard is designed to compensate for these gaps and to provide the greatest feasible degree of health protection for exposed workers over their working lifetime. The significant thrust of the standard for cotton dust is medical management and surveillance, administrative controls, and work practices. Since no definitive environmental level can assure complete health protection, none is recommended in this document. However, to ensure that effective engineering controls are implemented and dust concentrations reduced, an environmental standard should be fixed. The concentration should be set at the lowest level feasible in order to reduce the prevalence and severity of byssinosis. Byssinosis is a disease characterized by chronic bronchitis sometimes complicated by emphysema or asthma.

COTTON DUST IN COTTON GINS.

*OSHA Standard 29 CFR 1910.1046
Applies to the control of employee exposure to cotton dust in cotton gins

COTTON DUST(RAW).

*OSHA 29 CFR 1910.1000 Table Z-1
PEL(TWA): 1 mg/m^3
Lintfree dust as measured by the vertical elutriator:
ACGIH-TLV
TWA: 0.2 mg/m^3
STEL: 0.6 mg/m^3

COTTONSEED OIL.

NFPA—hazardous chemical (flammable chemical)
health—0, flammability—1, reactivity—0

COUMAPHOS. [56-72-4]

FIFRA—Registration Standard—Issued August 1981 (insecticide)
*CWA 311(b)(2)(A), 40 CFR 116, 117
Discharge RQ = 10 pounds (4.54 kilograms)

COUMARIN. [91-64-5]

IARC—carcinogenic in animals

CRAG® HERBICIDE. [136-78-7]

*OSHA 29 CFR 1910.1000 Table Z-1
PEL(TWA): 15 mg/m^3

CREOSOTE. [8001-58-9]

See Coal tar products, NIOSH Criteria Document.

OSHA Candidate List of Potential Occupational Carcinogens (EPA Carcinogen Assessment Group List), 45 FR 53672. August 12, 1980

40 CFR 162.11, FIFRA-RPAR issued
Current Status: Same as above (Coal Tar/Creosote/Coal Tar Neutral Oil PD 1 and 2/3 combined). PD 2/3 (wood uses only) completed and Notice of Determination published in 46 13020 (2/19/81); comment period closed 5/20/81. Final decision is being developed.
Criteria Possibly Met or Exceeded: Oncogenicity and Mutagenicity

*RCRA 40 CFR 261
EPA hazardous waste no. U051

m-CRESIDINE. [102-50-1]

NCI—carcinogenic in animals

p-CRESIDINE. See o-Anisidine, 5-methyl-. [120-71-8]

CRESOL (all isomers). [1319-77-3]

*OSHA 29 CFR 1910.1000 Table Z-1

Skin

PEL(TWA): 5 ppm, 22 mg/m^3

ACGIH-TLV

Skin

TWA: 5 ppm, 55 mg/m^3

NIOSH Criteria Document (Pub. No. 78-133), NTIS Stock No. is not available. February 1978

In this document, the term "cresol" applies to the ortho, meta, or para isomer of the aromatic organic compound $CH_3C_6H_4OH$ or to any combination of the three isomers in a mixture. Examples of commercial mixtures that often contain cresol are the cresylic acids, which are generally defined as mixtures of cresol, xylenols, and phenol in which 50% of the material boils above 204°C. The criteria and recommendation for cresol will apply to cresylic acid mixtures that contain cresol.

"Occupational exposure to cresol," because of the systematic effects, absorption through the skin on contact, and possible dermal irritation, is defined as work in any area where cresol is produced, processed, stored, or otherwise used. The "action level" is defined as one-half the recommended time-weighted average (TWA) environmental limit. Adherence to all provisions of the standard is required if an employee is occupationally exposed to airborne cresol at concentrations in excess of the action level. If the employee is occupationally exposed at concentrations equal to or below the action level, then all sections of the recommended standard except Sections 4(c)(2) and 8(a) shall be complied with because adverse effects can be produced by skin and eye contact.

When skin contact is prevented, exposure to cresol shall be controlled so that no employee is exposed to cresol at a concentration greater than 10 mg/m^3 of air (2.3 ppm by volume), determined as a time-weighted average (TWA) concentration for up to a 10-hour work shift and 40-hour workweek.

The standard is designed to protect the health and provide for the safety of employees for up to a 10-hour work shift, 40-hour workweek, over a working lifetime.

*CWA 311(b)(2)(A), 40 CFR 116, 117

Discharge RQ = 1000 pounds (454 kilograms)

*RCRA 40 CFR 261

EPA hazardous waste no. U052 (hazardous substance)

See below.

***m*-CRESOL.** [108-39-4]

See Cresol, NIOSH Criteria Document.

TSCA 4(e), ITC

Initial Report to the administrator, Environmental Protection Agency, TSCA Interagency Testing Committee, October 1, 1977; 42 FR 55026, October 12, 1977, responded to be the EPA administrator, 43 FR 50134, October 26, 1978

***o*-CRESOL.** [95-48-7]

See Cresol, NIOSH Criteria Document

TSCA 4(e), ITC

Initial Report to the administrator, Environmental Protection Agency, TSCA Interagency Testing Committee, October 2, 1977; 42 FR 55026, October 12, 1977, responded to by the EPA administrator, 43 FR 50134, October 26, 1978

***p*-CRESOL.** [106-44-5]

See Cresol, NIOSH Criteria Document.

TSCA 4(e), ITC

Initial Report to the administrator, Environmental Protection Agency, TSCA Interagency Testing Committee, October 1, 1977; 42 FR 55026, October 12, 1977, responded to by the EPA administrator, 43 FR 50134, October 26, 1978.

CRESOL (*ortho-, meta-, para-*). [95-48-7, 106-44-5, 108-39-4]

NFPA—hazardous chemical

when *ortho:* health—3, flammability—2, reactivity—0

when *meta* and *para:* health—3, flammability—1, reactivity—0

*TSCA 8(d), Unpublished Health and Safety Studies Reporting, 40 CFR 716, 47 FR 387, September 2, 1982

CAA 112
Chemical proposed to be assessed as a Hazardous Air Pollutant by a House of Representives bill to amend the CAA

para-CRESYL GLYCIDYL ETHER. [2186-24-5]

TSCA 4(e), ITC
Third Report of the TSCA Interagency Testing Committee to the administrator, Environmental Protection Agency, October 1978; 43 FR 50630, October 30, 1978

CRESYLIC ACID. See Cresol. [1319-77-3]

CROCIDOLITE (RIEBECKITE). See Asbestos. [53799-46-5]

*TSCA 8(d), Unpublished Health and Safety Studies Reporting, 40 CFR 716, 47 FR 387, September 2, 1982

CROTONALDEHYDE. [123-73-9]
*OSHA 29 CFR 1910.1000 Table Z-1
PEL(TWA): 2 ppm, 6 mg/m^3

ACGIH-TLV
TWA: 2 ppm, 6 mg/m^3
STEL: 6 ppm, 18 mg/m^3
NFPA—hazardous chemical
health—3, flammability—3, reactivity—2
*CWA 311(b)(2)(A), 40 CFR 116, 117
Discharge RQ = 100 pounds (45.4 kilograms)
*RCRA 40 CFR 261
EPA hazardous waste no. U053 (hazardous substance)

CRUFOMATE. [299-86-5]
ACGIH-TLV
TWA: 5 mg/m^3
STEL: 20 mg/m^3

CRYOGENINE. See Semicarbazide, 1-phenyl-. [103-03-7]

CRYOLITE. [15096-52-3]
FIFRA—Registration Standard Development—Data Collection (insecticide)

FIFRA—Data Call In, 45 FR 75488, November 14, 1980
Status: letter issued 7/28/81

CUMENE. [98-82-8]
*OSHA 29 CFR 1910.1000 Table Z-1
PEL(TWA): 50 ppm, 245 mg/m^3
ACGIH-TLV

Skin

TWA: 50 ppm, 245 mg/m^3
STEL: 75 ppm, 365 mg/m^3
NFPA—hazardous chemical
health—2, flammability—3, reactivity—0
*RCRA 40 CFR 261
EPA hazardous waste no. U055 (ignitable waste)

CUMENE HYDROPEROXIDE. [80-15-9]

NFPA—hazardous chemical
health—1, flammability—2, reactivity—4, oxy—oxidizing chemical
TSCA-CHIP available

CUMOL. See Cumene. [98-82-8]

CUPFERRON. See Hydroxylamine, N-Nitroso-N-phenyl-, ammonium salt. [135-20-6]

CUPRIC ACETATE. [142-71-2]
*CWA 311(b)(2)(A), 40 CFR 116, 117
Discharge RQ = 100 pounds (45.4 kilograms)

CUPRIC ACETOARSENITE. [12002-03-8]
*CWA 311(b)(2)(A), 40 CFR 116, 117
Discharge RQ = 100 pounds (45.4 kilograms)

CUPRIC CHLORIDE. [7447-39-4]
*CWA 311(b)(2)(A), 40 CFR 116, 117
Discharge RQ = 10 pounds (4.54 kilograms)

CUPRIC NITRATE. [3251-23-8]
NFPA—hazardous chemical
when nonfire: health—0, flammability—0, reactivity—0, oxy—oxidizing chemical

when fire: health—1, flammability—0, reactivity—0, oxy—oxidizing chemical

*CWA 311(b)(2)(A), 40 CFR 116, 117
Discharge RQ = 100 pounds (45.4 kilograms)

CUPRIC OXALATE. [5893-66-3]

*CWA 311(b)(2)(A), 40 CFR 116, 117
Discharge RQ = 100 pounds (45.4 kilograms)

CUPRIC SULFATE. [7758-98-7]

*CWA 311(b)(2)(A), 40 CFR 116, 117
Discharge RQ = 10 pounds (4.54 kilograms)

CUPRIC SULFATE AMMONIATED. [10380-29-7]

*CWA 311(b)(2)(A), 40 CFR 116, 117
Discharge RQ = 100 pounds (45.4 kilograms)

CUPRIC TARTRATE. [815-82-7]

*CWA 311(b)(2)(A), 40 CFR 116, 117
Discharge RQ = 100 pounds (45.4 kilograms)

Cutting Fluid:
TSCA-CHIP available

CYANAMIDE. [420-04-2]

ACGIH-TLV
TWA: 2 mg/m^3

CYANAZINE. [21725-46-2]

FIFRA—Data Call In, 45, FR 75488, November 14, 1980
Status: letter being drafted
FIFRA—Registration Standard Development—Data Collection (herbicide)

CYANIDE. [151-50-8, 143-33-9]

*OSHA 29 CFR 1910.1000 Table Z-1

Skin

PEL(TWA): 5 mg/m^3 (as CN)

ACGIH-TLV

Skin

TWA: 5 mg/m^3

NIOSH Criteria Document (Pub. No. 77-108), NTIS Stock No. PB 266230, March 4, 1977

These criteria and recommended standards apply to occupational exposure of workers to HCN and cyanide salts. Synonyms for HCN are hydrocyanic acid, prussic acid, and formonitrile. For the purpose of this document, cyanide salts are defined as sodium cyanide (NaCN), potassium cyanide (KCN), or calcium cyanide [Ca(CN)$_2$]. The word "cyanide" or the symbol "CN" is used to designate salts as well as hydrogen cyanide.

The "action level" for hydrogen cyanide or a cyanide salt is defined as one-half the corresponding recommended ceiling environmental exposure limit. "Occupational exposure" to these compounds is defined as exposure to airborne concentrations greater than the corresponding action levels.

The criteria and recommended standards apply to any area in which HCN, any of the cyanide salts or materials containing any of them, alone or in combination with other substances is used, produced, packaged, processed, mixed, blended, handled, stored in large quantities, or applied. Exposure to cyanide at concentrations less than or equal to the respective action levels will not require adherence to the recommended standards except for Sections 2–7 and 8(c).

Employee exposure to HCN shall be controlled so as not to exceed 5 mg/m^3 of air expressed as CN (4.7 ppm), determined as a ceiling concentration based on a 10-minute sampling period.

Employee exposure to cyanide salts shall be controlled so as not to exceed 5 mg/m^3 of air expressed as CN, determined as a ceiling value based upon a 10-minute sampling period.

Whenever the air is analyzed for cyanide salts, a concurrent analysis shall be made for HCN. Neither of the respective ceiling values may be exceeded nor may the combined values exceed 5 mg/m^3 measured as CN during a 10-minute sampling period.

The standards are designed to protect the health of workers for up to a 10-hour workday, 40-hour workweek, over a working lifetime.

CWA 304(a)(1), 45 FR 79318, November 28, 1980, Water Quality Criteria

Freshwater Aquatic Life

For free cyanide (sum of cyanide present as HCN and CN$^-$, expressed as CN) the criterion to protect freshwater aquatic life as derived using the guidelines is 3.5 µg/liter as a 24-hour average and the concentration should not exceed 52 µg/liter at any time.

Saltwater Aquatic Life

The available data for free cyanide (sum of cyanide present as HCN and CN$^-$, expressed as CN) indicate that acute toxicity to saltwater aquatic life occurs at concentrations as low as 30 µg/liter and would occur at lower concentrations among species that are more sensitive than those tested. If the acute-chronic ratio for saltwater organisms is similar to that for freshwater organisms, chronic toxicity would occur at concentrations as low as 2.0 µg/liter for the tested species and at lower concentrations among species that are more sensitive than those tested.

Human Health

The ambient water quality criterion for cyanide is recommended to be identical to the existing drinking water standard which is 200 µg/liter. Analysis of the toxic effects data resulted in a calculated level that is protective of human health against the ingestion of contaminated water and contaminated aquatic organisms. The calculated value is comparable to the present standard For this reason a selective criterion based on exposure solely from consumption of 6.5 g of aquatic organisms was not derived.

*CWA 307(a)(1), Priority Pollutant

*RCRA 40 CFR 261
EPA hazardous waste no. P030

CYANOACETIC ACID. [372-09-8]

NFPA—hazardous chemical
when nonfire: health—0, flammability—1, reactivity—0
when fire: health—3, flammability—1, reactivity—0

CYANOGEN. [460-19-5]

ACGIH-TLV
TWA: 10 ppm, 20 mg/m^3
NFPA—hazardous chemical
health—4, flammability—4, reactivity—2
*RCRA 40 CFR 261
EPA hazardous waste no. P031

CYANOGEN BROMIDE. [506-68-3]

NFPA—hazardous chemical
health—3, flammability—0, reactivity—2
*RCRA 40 CFR 261
EPA hazardous waste no. U246

CYANOGEN CHLORIDE. [506-77-4]

ACGIH-TLV
TWA: Ceiling limit, 0.3 ppm, 0.6 mg/m^3
*CWA 311(b)(2)(A), 40 CFR 116, 117
Discharge RQ = 10 pounds (4.54 kilograms)
*RCRA 40 CFR 261
EPA hazardous waste no. P033 (hazardous substance)

CYANURIC ACID and CHLORINATED DERIVATIVES. [87-90-1, 108-80-5, 2244-21-5, 2624-17-1, 2782-57-2, 2893-78-9]

TSCA-CHIP available

CYACASIN. [14901-08-7]

OSHA Candidate List of Potential Occupational Carcinogens (EPA Carcinogen Assessment Group List), 45 FR 53672, August 12, 1980
IARC—carcinogenic in animals

CYCLOCHLOROTINE. [12663-46-6]

IARC—carcinogenic in animals

CYCLODODECANE, 1,2,5,6,9,10-HEXABROMO-. [3194-55-6]

*TSCA 8(a), 40 CFR 712, 47 FR 26992, June 22, 1982, Chemical Information Rule

CYCLODODECANE, HEXABROMO-. [25637-99-4]

*TSCA 8(a), 40 CFR 712, 47 FR 26992, June 22, 1982, Chemical Information Rule

1,4-CYCLOHEXADIENEDIONE. See
2,5-Cyclohexadiene-1,4-dione. [106-51-4]

2,5-CYCLOHEXADIENE-1,4-DIONE. [106-51-4]

*TSCA 8(a), 40 CFR 712, 47 FR 26992, June 22, 1982, Chemical Information Rule

*RCRA 40 CFR 261
EPA hazardous waste no. U197

CYCLOHEXANE. [110-82-7]

*OSHA 29 CFR 1910.1000 Table Z-1
PEL(TWA): 300 ppm, 1050 mg/m^3

ACGIH-TLV
TWA: 300 ppm, 1050 mg/m^3
STEL: 375 ppm, 1300 mg/m^3

NFPA—hazardous chemical (flammable chemical)
health—1, flammability—3, reactivity—0

*CWA 311(b)(2)(A), 40 CFR 116, 117
Discharge RQ = 100 pounds (45.4 kilograms)

*RCRA 40 CFR 261
EPA hazardous waste no. U056 (hazardous substance, ignitable waste)

CYCLOHEXANE, 1,2,3,4,5,6-HEXACHLORO-. Syn: Benzene hexachloride. [608-73-1]

OSHA Candidate List of Potential Occupational Carcinogens, 45 FR 53672, August 12, 1980

*RCRA 49 CFR 172.101
EPA hazardous waste no. U127

CYCLOHEXANETHIOL.
See Thiols, NIOSH Criteria Document.

CYCLOHEXANOL. [108-93-0]

*OSHA 29 CFR 1910.1000 Table Z-1
PEL(TWA): 50 ppm, 200 mg/m^3

ACGIH-TLV
TWA: 50 ppm, 200 mg/m^3

CYCLOHEXANONE. [108-94-1]

ACGIH-TLV
TWA: 25 ppm, 100 mg/m^3
STEL: 100 ppm, 400 mg/m^3

See Ketones, NIOSH Criteria Document.

NFPA—hazardous chemical
health—1, flammability—2, reactivity—0

TSCA 4(e), ITC
Fourth Report of the Interagency Testing Committee to the administrator, Environmental Protection Agency, April 1979; 44 FR 31866, June 1, 1979

*TSCA 8(a), 40 CFR 712, 47 FR 26992, June 22, 1982, Chemical Information Rule

*TSCA 8(d), Unpublished Health and Safety Studies Reporting, 40 CFR 716, 47 FR 387, September 2, 1982

*RCRA 40 CFR 261
EPA hazardous waste no. U057 (ignitable waste)

CYCLOHEXENE. [110-83-8]

*OSHA 29 CFR 1910.1000 Table Z-1
PEL(TWA): 300 ppm, 1015 mg/m^3

ACGIH-TLV
TWA: 300 ppm, 1015 mg/m^3

2-CYCLOHEXEN-1-ONE, 3,5,5-TRIMETHYL-. [78-59-1]

*TSCA 8(a), 40 CFR 712, 47 FR 26992, June 22, 1982, Chemical Information Rule

CYCLOHEXIMIDE. [66-81-9]

FIFRA—Registration Standard Development—Data Evaluation and Development of Regulatory Position (fungicide)

CYCLOHEXYL ALCOHOL. [108-93-0]

NFPA—hazardous chemical (flammable chemical)
health—1, flammability—2, reactivity—0

CYCLOHEXYLAMINE. [101-83-7, 108-91-8, 947-92-2, 3129-92-8, 3882-06-2, 5473-16-5, 20227-92-3, 20136-64-5, 34067-50-0, 34139-62-3]

ACGIH-TLV

Skin

TWA: 10 ppm, 40 mg/m^3

NFPA—hazardous chemical
health—2, flammability—3, reactivity—0

TSCA-CHIP available

CYCLONITE. [121-82-4]

ACGIH-TLV

Skin

TWA: 1.5 mg/m^3
STEL: 3 mg/m^3

CYCLOPENTADIENE. [542-92-7]

*OSHA 29 CFR 1910.1000 Table Z-1
PEL(TWA): 75 ppm, 200 mg/m^3

ACGIH-TLV
TWA: 75 ppm, 200 mg/m^3
STEL: 150 ppm, 400 mg/m^3

1,3-CYCLOPENTADIENE, 1,2,3,4,5,5-HEXACHLORO-. [77-47-4]

*TSCA 8(a), 40 CFR 712, 47 FR 26992, June 22, 1982, Chemical Information Rule
*RCRA 40 CFR 261
EPA hazardous waste no. U130 (hazardous substance)

CYCLOPENTANE. [287-92-3]

ACGIH-TLV
TWA: 600 ppm, 1720 mg/m^3
STEL: 900 ppm, 2580 mg/m^3

NFPA—hazardous chemical (flammable chemical)
health—1, flammability—3, reactivity—0

CYCLOPHOSPHAMIDE. [50-18-0]

OSHA Candidate List of Potential Occupational Carcinogens, 45 FR 53672, August 12, 1980
IARC—carcinogenic in humans
*RCRA 40 CFR 261
EPA hazardous waste no. U058

CYCLOPROPANE. [75-19-4]

See Waste anesthetic gases and vapors, NIOSH Criteria Document.

NFPA—hazardous chemical (flammable chemical)
health—1, flammability—4, reactivity—0

CYHEXATIN. [13121-70-5]

ACGIII-TLV
TWA: 5 mg/m^3
STEL: 10 mg/m^3

CYPRAZINE. [22936-86-3]

FIFRA—Data Call In, 45 FR 75488, November 14, 1980
Status: letter being drafted
FIFRA—Registration Standard Development—Data Collection (herbicide)

CYTEMBENA. [21739-91-3]

NCI—carcinogenic in animals

D

2,4-D (2,4-DICHLOROPHENOXYACETIC ACID). [94-75-7]

*OSHA 29 CFR 1910.1000 Table Z-1
PEL(TWA): 10 mg/m^3

ACGIH-TLV
TWA: 10 mg/m^3
STEL: 20 mg/m^3

*CWA 311(b)(2)(A), 40 CFR 116, 117
Discharge RQ = 100 pounds (45.4 kilograms)

*National Interim Primary Drinking Water Regulations, 40 CFR 141; 40 FR 59565, December 24, 1975; amended by 41 FR 28402, July 9, 1976; 44 FR 68641, November 29, 1979; corrected by 45 FR 15542, March 11, 1980; 45 FR 57342, August 27, 1980; 47 FR 18998, March, 3, 1982; corrected by 47 FR 10998, March 12, 1982
Maximum contaminant level—0.1 mg/liter

*RCRA 40 CFR 261
EPA hazardous waste no. U240 (hazardous substance)

2,4-D, SALTS and ESTERS. [94-75-7]

*RCRA 40 CFR 261
EPA hazardous waste no. U240 (hazardous substance)

2,4-D ESTER. [94-11-1, 94-79-1, 94-80-4, 1320-18-9, 1928-38-7, 1928-61-6, 1929-73-3, 2971-38-2, 25168-26-7, 53467-11-1]

*CWA 311(b)(2)(A), 40 CFR 116, 117
Discharge RQ = 100 pounds (45.4 kilograms)

DACARBAZINE. *See* Imidazole-4-carboxamide, 5-(3,3'-dimethyl-1-triazeno)-. [4342-03-4]

DAMINOZIDE. *See* Succinic acid, mono(2,2-dimethylhydrazide)-. [1596-84-5]

FIFRA—Registration Standard Development Data Evaluation and Development of Regulatory Position (growth regulator)

D AND C RED NO. 9. [5160-02-1]

NCI—carcinogenic in animals
TSCA-CHIP available

DAPSONE. *See* Aniline, 4,4'-sulfonylidi. [80-08-0]

DAUNOMYCIN. [20830-81-3]

OSHA Candidate List of Potential Occupational Carcinogens (EPA Carcinogen Assessment Group List), 45 FR 53672, August 12, 1980

IARC—carcinogenic in animals

*RCRA 40 CFR 261
EPA hazardous waste no. U059

2,4-DB. [94-82-6]

FIFRA—Data Call In, 45 FR 75488, November 14, 1980
Status: letter being drafted

DBCP. *See* 1,2-Dibromochloropropane. [96-12-8]

DDD. *See* DDE. [72-54-8]

DDE. Syn: DDD, 4,4'-DD, 4,4'-DDE, *p,p'*-DDD, *p,p'*-DDE. [72-54-8]

OSHA Candidate List of Potential Occupational Carcinogens, 45 FR 53672, August 12, 1980

IARC—carcinogenic in animals
NCI—carcinogenic in animals

40 CFR 162.11, FIFRA-RPAR, Notice of Intent to Cancel/Suspend Issued
Current Status: 36 FR 5254 (3/18/71) (cancelled)

CWA 307(a)(1), 40 CFR 125, Priority Pollutant

*RCRA 40 CFR 172.101
EPA hazardous waste no. U060 (hazardous substance)

DDT (DICHLORODIPHENYLTRICHLOROETHANE). [50-29-3]

*OSHA 29 CFR 1910.1000 Table Z-1

Skin

PEL(TWA): 1 mg/m^3

ACGIH-TLV
TWA: 1 mg/m^3
STEL: 3 mg/m^3

OSHA Candidate List of Potential Occupational Carcinogens (EPA Carcinogen Assessment Group List), 45 FR 53672, August 12, 1980

IARC—carcinogenic in animals

40 CFR 162.11, FIFRA-RPAR, Notice of Intent to Cancel/Suspend Issued
Current Status: 37 FR 13369 (7/7/72) (cancellation of most uses)

CWA 307(a)(1), Priority Pollutant

*CWA 311(b)(2)(A), 40 CFR 116, 117
Discharge RQ = 1 pound (0.454 kilogram)

*RCRA 40 CFR 261
EPA hazardous waste no. U061 (hazardous substance)

See below.

DDT and METABOLITES. [50-29-3]

CWA 304(a)(1), 45 FR 79318, November 28, 1980, Water Quality Criteria

Freshwater Aquatic Life

For DDT and its metabolites the criterion to protect freshwater aquatic life as derived using the guidelines is 0.0010 μg/liter as a 24-hour average and the concentration should not exceed 1.1 μg/liter at any time.

The available data for TDE indicate that acute toxicity to freshwater aquatic life occurs at concentrations as low as 0.6 μg/liter and would occur at lower concentrations among species that are more sensitive than those tested. No data are available concerning the chronic toxicity of TDE to sensitive freshwater aquatic life.

The available data for DDE indicate that acute toxicity to freshwater aquatic life occurs at concentrations as low as 1050 μg/liter and would occur at lower concentrations among species that are more sensitive than those tested. No data are available concerning the chronic toxicity of DDE to sensitive freshwater aquatic life.

Saltwater Aquatic Life

For DDT and its metabolites the criterion to protect saltwater aquatic life as derived using the guidelines is 0.0010 μg/liter as a 24-hour average and the concentration should not exceed 0.13 μg/liter at any time.

The available data for TDE indicate that acute toxicity to saltwater aquatic life occurs at concentrations as low as 3.6 μg/liter and would occur at lower concentrations among species that are more sensitive than those tested. No data are available concerning the chronic toxicity of TDE to sensitive saltwater aquatic life.

The available data for DDE indicate that acute toxicity to saltwater aquatic life occurs at concentrations as low as 14 μg/liter and would occur at lower concentrations among species that are more sensitive than those tested. No data are available concerning the chronic toxicity of DDE to sensitive saltwater aquatic life.

Human Health

For the maximum protection of human health from the potential carcinogenic effects due to exposure of DDT through ingestion of contaminated water and contaminated aquatic organisms, the ambient water concentration should be zero based on the nonthreshold assumption for this chemical. However, zero level may not be attainable at the present time. Therefore, the levels that may result in incremental increase of cancer risk over the lifetime are estimated at 10^{-5}, 10^{-6}, and 10^{-7}. The corresponding criteria are 0.24 ng/liter, 0.024 ng/liter, and 0.0024 ng/liter, respectively. If the above estimates are made for consumption of aquatic organisms only, excluding consumption of water, the levels are 0.24 ng/liter, 0.024 ng/liter, and 0.0024 ng/liter, respectively. Other concentrations representing different risk levels may be calculated by use of the guidelines. The risk estimate range is presented for information purposes and does not represent an agency judgment on an "acceptable" risk level.

DDVP. [62-73-7]
*OSHA 29 CFR 1910.1000 Table Z-1

Skin

PEL(TWA): 1 mg/m^3

DEAK. See Dichlorodiethylaluminum.

DECABORANE. [17702-41-9]
*OSHA 29 CFR 1910.1000 Table Z-1

Skin

PEL(TWA): 0.05 ppm, 0.3 mg/m^3
ACGIH-TLV

Skin

TWA: 0.05 ppm, 0.3 mg/m^3
STEL: 0.15 ppm, 0.9 mg/m^3
NFPA—hazardous chemical
health—3, flammability—2, reactivity—1

DECABORON TETRADECAHYDRIDE.
See Decaborane. [17702-41-9]

DECACHLORODIPHENYL.
See Polychlorinated biphenyls, NIOSH Criteria Document.

DECACHLOROOCTAHYDRO-1,3,4-METHENO-2H-CYCLOBUTA[c,d]PENTALEN-2-ONE. [143-50-0]
*RCRA 40 CFR 261
EPA hazardous waste no. U142 (hazardous substance)

n-DECANE. [124-18-5]
NFPA—hazardous chemical (flammable chemical)
health—0, flammability—2, reactivity—0

1-DECANETHIOL. See Thiols, NIOSH Criteria Document.

DECANOL. [112-30-1]
NFPA—hazardous chemical
health—0, flammability—2, reactivity—0

DEET. [134-62-3]
FIFRA—Registration Standard—Issued December 1980, NTIS# 540/R5 81-0004 (insecticide)

DEMENTON®. [8065-48-3]
*OSHA 29 CFR 1910.1000 Table Z-1

Skin

PEL(TWA): 0.1 mg/m^3
ACGIH-TLV

Skin

TWA: 0.01 ppm, 0.1 mg/m^3
STEL: 0.03 ppm, 0.3 mg/m^3
FIFRA—Data Call In, 45 FR 75488, November 14, 1980
Status: letter issued 6/13/81

β-2'-DEOXY-6-THIOGUANOSINE MONOHYDRATE. [1888-71-7]
NCI—carcinogenic in animals

DIACETONE ALCOHOL (4-HYDROXY-4-METHYL-2-PENTANONE). [123-42-2]
*OSHA 29 CFR 1910.1000 Table Z-1
PEL(TWA): 50 ppm, 240 mg/m^3
ACGIH-TLV
TWA: 50 ppm, 240 mg/m^3
STEL: 75 ppm, 360 mg/m^3
See Ketones, NIOSH Criteria Document.

DIACETYLBENZIDINE. See 4',4'''-Biacetanilide. [613-35-4]

N,N-DIACETYLBENZIDINE. See 4',4'''-Biacetanilide. [613-35-4]

DIACETYL PEROXIDE (25% solution). [110-22-5]

NFPA—hazardous chemical,
health—1, flammability—2, reactivity—4

DIALIFOR. [10311-84-9]

FIFRA—Registration Standard—Issued July 1981 (insecticide)

DIALLATE. [2303-16-4]

OSHA Candidate List of Potential Occupational Carcinogens (EPA Carcinogen Assessment Group List), 45 FR 53672, August 12, 1980

FIFRA—Registration Standard Development Data Collection (herbicide)

40 CFR 162.11 FIFRA-RPAR issued
Current Status: PD 1 published, 42, FR 27669 (5/31/77); comment period closed 9/9/77; NTIS# PB80 212863.
PD 2/3 completed and Notice of Determination published, 45 FR 38437 (6/9/80); comment period closed 7/9/80: NTIS# PB80 216849.
PD 4 in agency review.
Criteria Possibly Met or Exceeded: Oncogenicity and Mutagenicity

IARC—carcinogenic in animals

*RCRA 40 CFR 261
EPA hazardous waste no. U062

DIAMINE. [302-01-2]

*RCRA 40 CFR 261
EPA hazardous waste no. U133 (reactive waste, toxic waste)

p-DIAMINOANISOLE. See Phenylenediamines. [5307-02-8]

TSCA 4(e), ITC
Sixth Report of the Interagency Testing Committee to the administrator, Environmental Protection Agency, April 1980; 45 FR 35897, May 28, 1980, responded to by the EPA administrator, 47 FR 973, January 8, 1982

*TSCA 8(d), Unpublished Health and Safety Studies Reporting, 40 CFR 716, 47 FR 38800, September 2, 1982; 48 FR 13178, March 30, 1983

2,4-DIAMINOANISOLE (4-METHOXY-m-PHENYLENEDIAMINE) in hair and fur dyes.

NIOSH Current Intelligence Bulletin No. 19—cancer-related bulletin

2,4-DIAMINOANISOLE SULFATE.
See Phenylenediamines. [39156-41-7]

OSHA Candidate List of Potential Occupational Carcinogens, 45 FR 53672, August 12, 1980

NCI—carcinogenic in animals

TSCA 4(e), ITC
Sixth Report of the Interagency Testing Committee to the administrator, Environmental Protection Agency, April 1980; 45 FR 35897, May 28, 1980, responded to by the EPA administrator, 47 FR 973, January 8, 1982

*TSCA 8(d), Unpublished Health and Safety Studies Reporting, 40 CFR 716, 47 FR 38800, September 2, 1982; 48 FR 13178, March 30, 1983

2,4-DIAMINOAZOBENZENE.
[495-54-5]

TSCA-CHIP available

m-DIAMINOBENZENE. See Phenylenediamines. [108-45-2]

TSCA 4(e), ITC
Sixth Report of the Interagency Testing Committee to the administrator, Environmental Protection Agency, April 1980; 45 FR 35897, May 28, 1980, responded to by the EPA administrator, 47 FR 973, January 8, 1982

*TSCA 8(d), Unpublished Health and Safety Studies Reporting, 40 CFR 716, 47 FR 38800, September 2, 1982; 48 FR 13178, March 30, 1983

o-DIAMINOBENZENE. See Phenylenediamines. [95-54-5]

TSCA 4(e), ITC
Sixth Report of the Interagency Testing Committee to the administrator, Environmental Protection Agency, April 1980; 45 FR 35897, May 28, 1980, responded to by

the EPA administrator, 47 FR 973, January 8, 1982

*TSCA 8(d), Unpublished Health and Safety Studies Reporting, 40 CFR 716, 47 FR 38800, September 2, 1982; 48 FR 13178, March 30, 1983

p-DIAMINOBENZENE. *See* Phenylenediamines. [106-50-3]

TSCA 4(e), ITC

Sixth Report of the Interagency Testing Committee to the administrator, Environmental Protection Agency, April 1980; 45 FR 35897, May 28, 1980, responded to by the EPA administrator, 47 FR 973, January 8, 1982

*TSCA 8(d), Unpublished Health and Safety Studies Reporting, 40 CFR 716, 47 FR 38800, September 2, 1982; 48 FR 13178, March 30, 1983

3,3'-DIAMINOBENZIDINE TETRAHYDROCHLORIDE. *See* 3,3',4,4'-Biphenyltetramine tetrahydrochloride. [7411-49-6]

DIAMINOBIPHENYL ETHER.
[101-80-4]

IARC—carcinogenic in animals.

TSCA-CHIP available

4,6-DIAMINO-o-CRESOL. *See* Phenylenediamines. [15872-73-8]

TSCA 4(e), ITC

Sixth Report of the Interagency Testing Committee to the administrator, Environmental Protection Agency, April 1980; 45 FR 35897, May 28, 1980, responded to by the EPA administrator, 47 FR 973, January 8, 1982

*TSCA 8(d), Unpublished Health and Safety Studies Reporting, 40 CFR 716, 47 FR 38800, September 2, 1982; 48 FR 13178, March 30, 1983

2,2'-DIAMINODIETHYLAMINE. *See* Diethylene triamine. [111-40-0]

4,4'-DIAMINODIPHENYL ETHER.
See Diaminobiphenyl ether.
[101-80-4]

4,4'-DIAMINODIPHENYLMETHANE (DDM). [101-77-9]

NIOSH Current Intelligence Bulletin No. 8

p,p'-DIAMINODIPHENYL SULFIDE.
See Aniline, 4,4'-thiodi-. [139-65-1]

1,2-DIAMINOETHANE. *See* Ethylenediamine. [107-15-3]

1,6-DIAMINOHEXANE. [124-09-4]

TSCA-CHIP available

1,2-DIAMINO-3-METHYLBENZENE. *See* Phenylenediamines. [2687-25-4]

TSCA 4(e), ITC

Sixth Report of the Interagency Testing Committee to the administrator, Environmental Protection Agency, April 1980; 45 FR 35897, May 28, 1980, responded to by the EPA administrator, 47 FR 973, January 8, 1982

*TSCA 8(d), Unpublished Health and Safety Studies Reporting, 40 CFR 716, 47 FR 38800, September 2, 1982; 48 FR 13178, March 30, 1983

1,2-DIAMINO-4-METHYLBENZENE.
See Phenylenediamines. [496-72-0]

TSCA 4(e), ITC

Sixth Report of the Interagency Testing Committee to the administrator, Environmental Protection Agency, April 1980; 45 FR 35897, May 28, 1980, responded to by the EPA administrator, 47 FR 973, January 8, 1982

*TSCA 8(d), Unpublished Health and Safety Studies Reporting, 40 CFR 716, 47 FR 38800, September 2, 1982; 48 FR 13178, March 30, 1983

1,3-DIAMINO-4-METHYLBENZENE.
See Phenylenediamines. [95-80-7]

TSCA 4(e), ITC

Sixth Report of the Interagency Testing Committee to the administrator, Environmental Protection Agency, April 1980; 45 FR 35897, May 28, 1980, responded to by the EPA administrator, 47 FR 973, January 8, 1982

*TSCA 8(d), Unpublished Health and Safety Studies Reporting, 40 CFR 716, 47 FR 38800, September 2, 1982; 48 FR 13178, March 30, 1983

2,6-DIAMINO-1-METHYLBENZENE. See Phenylenediamines. [823-40-5]

TSCA 4(e), ITC
Sixth Report of the Interagency Testing Committee to the administrator, Environmental Protection Agency, April 1980; 45 FR 35897, May 28, 1980, responded to by the EPA administrator, 47 FR 973, January 8, 1982

*TSCA 8(d), Unpublished Health and Safety Studies Reporting, 40 CFR 716, 47 FR 38800, September 2, 1982; 48 FR 13178, March 30, 1983

4,6-DIAMINO-2-METHYLPHENYL. See Phenylenediamines. [65879-44-9]

TSCA 4(e), ITC
Sixth Report of the Interagency Testing Committee to the administrator, Environmental Protection Agency, April 1980; 45 FR 35897, May 28, 1980, responded to by the EPA administrator, 47 FR 973, January 8, 1982

*TSCA 8(d), Unpublished Health and Safety Studies Reporting, 40 CFR 716, 47 FR 38800, September 2, 1982; 48 FR 13178, March 30, 1983

1,5-DIAMINONAPHTHALENE. [2243-62-1]

NCI—carcinogenic in animals

1,4-DIAMINOPHENOL DIHYDROCHLORIDE. See Phenylenediamines. [137-09-7]

TSCA 4(e), ITC
Sixth Report of the Interagency Testing Committee to the administrator, Environmental Protection Agency, April 1980; 45 FR 35897, May 28, 1980, responded to by the EPA administrator, 47 FR 973, January 8, 1982

*TSCA 8(d), Unpublished Health and Safety Studies Reporting, 40 CFR 716, 47 FR 38800, September 2, 1982; 48 FR 13178, March 30, 1983

2,6-DIAMINO-3-(PHENYLAZO)-PYRIDINE (HYDROCHLORIDE). [136-40-3]

NCI—carcinogenic in animals

DIAMINOTOLUENE. See Phenylenediamines. [25376-45-8]

TSCA 4(e), ITC
Sixth Report of the Interagency Testing Committee to the administrator, Environmental Protection Agency, April 1980; 45 FR 35897, May 28, 1980, responded to by the EPA administrator, 47 FR 973, January 8, 1982

*TSCA 8(d), Unpublished Health and Safety Studies Reporting, 40 CFR 716, 47 FR 38800, September 2, 1982; 48 FR 13178, March 30, 1983

*RCRA 49 CER 172.101
EPA hazardous waste no. U221

2,4-DIAMINOTOLUENE. [95-80-7]

IARC—carcinogenic in animals

NCI—carcinogenic in animals

2,5-DIAMINOTOLUENE. See Phenylenediamines. [95-70-5]

TSCA 4(e), ITC
Sixth Report of the Interagency Testing Committee to the administrator, Environmental Protection Agency, April 1980; 45 FR 35897, May 28, 1980, responded to by the EPA administrator, 47 FR 973, January 8, 1982

*TSCA 8(d), Unpublished Health and Safety Studies Reporting, 40 CFR 716, 47 FR 38800, September 2, 1982; 48 FR 13178, March 30, 1983

3,5-DIAMINOTOLUENE. See Phenylenediamines. [108-71-4]

TSCA 4(e), ITC
Sixth Report of the Interagency Testing Committee to the administrator, Environmental Protection Agency, April 1980; 45 FR 35897, May 28, 1980, responded to by the EPA administrator, 47 FR 973, January 8, 1982

*TSCA 8(d), Unpublished Health and Safety Studies Reporting, 40 CFR 716, 47 FR

38800, September 2, 1982; 48 FR 13178, March 30, 1983

2,5-DIAMINOTOLUENE. SULFATE(1:1).
See Phenylenediamines. [615-50-9]

TSCA 4(e), ITC
Sixth Report of the Interagency Testing Committee to the administrator, Environmental Protection Agency, April 1980; 45 FR 35897, May 28, 1980, responded to by the EPA administrator, 47 FR 973, January 8, 1982

*TSCA 8(d), Unpublished Health and Safety Studies Reporting, 40 CFR 716, 47 FR 38800, September 2, 1982; 48 FR 13178, March 30, 1983

DIAMYLAMINE. [2050-92-2]
NFPA—hazardous chemical
health—3, flammability—2, reactivity—0

o-DIANISIDINE (3,3'-DIMETHOXYBENZIDINE-BASED DYES. [119-90-4]
TSCA 4(e), ITC
Fifth Report of the Interagency Testing Committee to the administrator, Environmental Protection Agency, November 1979; 44 FR 70664, December 7, 1979, EPA responded to the committee's recommendations for testing, 46 FR 55005, November 5, 1981

1,4-DIAZABICYCLO(2,2,2)OCTANE. [280-57-9]
TSCA-CHIP available

DIAZINON. [333-41-5]
ACGIH-TLV
Skin
TWA: 0.1 mg/m^3
STEL: 0.3 mg/m^3
*CWA 311(b)(2)(A), 40 CFR 116, 117
Discharge RQ = 1 pound (0.454 kilogram)

DIAZOMETHANE. [334-88-3]
*OSHA 29 CFR 1910.1000 Table Z-1
PEL(TWA): 0.2 ppm, 0.4 mg/m^3

ACGIH-TLV
TWA: 0.2 ppm, 0.4 mg/m^3
IARC—carcinogenic in animals

DIBENZ[a,h]ACRIDINE. [226-36-8]
OSHA Candidate List of Potential Occupational Carcinogens (EPA Carcinogen Assessment Group List), 45 FR 53672, August 12, 1980

IARC—carcinogenic in animals

DIBENZ[a,j]ACRIDINE. [224-42-0]
OSHA Candidate List of Potential Occupational Carcinogens (EPA Carcinogen Assessment Group List), 45 FR 53672, August 12, 1980

IARC—carcinogenic in animals

DIBENZ[a,h]ANTHRACENE. Syn: Dibenzo[a,h]anthracene. [53-70-3]
OSHA Candidate List of Potential Occupational Carcinogens (EPA Carcinogen Assessment Group List), 45 FR 53672, August 12, 1980

IARC—carcinogenic in animals
CWA 307(a)(1), 40 CFR 125, Priority Pollutant
*RCRA 40 CFR 261
EPA hazardous waste no. U063

1,2:5,6-DIBENZANTHRACENE. See Dibenz[a,h]anthracene. [53-70-3]

DIBENZO[a,h]ANTHRACENE. See Dibenz[a,h]anthracene. [53-70-3]

DIBENZO[c,g]CARBAZOLE. See 7H-Dibenzo[c,g]carbazole. [194-59-2]

7H-DIBENZO[c,g]CARBAZOLE. [194-59-2]
OSHA Candidate List of Potential Occupational Carcinogens (EPA Carcinogen Assessment Group List), 45 FR 53672, August 12, 1980

IARC—carcinogenic in animals

DIBENZO[h,rst]PENTAPHENE. [192-47-2]
IARC—carcinogenic in animals

DIBENZO[a,c]PYRENE.
IARC—carcinogenic in animals

DIBENZO[a,e]PYRENE. [192-65-4]
OSHA Candidate List of Potential Occupational Carcinogens (EPA Carcinogen Assessment Group List), 45 FR 53672, August 12, 1980

DIBENZO[a,i]PYRENE. [189-55-9]
IARC—carcinogenic in animals

DIBENZO[a,h]PYRENE. [198-64-0]
OSHA Candidate List of Potential Occupational Carcinogens (EPA Carcinogen Assessment Group List), 45 FR 53672, August 12, 1980

IARC—carcinogenic in animals

*RCRA 40 CFR 261
EPA hazardous waste no. U064

DIBENZO[a,j]PYRENE.
OSHA Candidate List of Potential Occupational Carcinogens (EPA Carcinogen Assessment Group List), 45 FR 53672, August 12, 1980

IARC—carcinogenic in animals

1,2:7,8-DIBENZOPYRENE. See
Dibenzo[a,i]pyrene. [189-55-9]

DIBENZOYL CHLORIDE. See Benzoyl
chloride. [98-88-4]

DIBENZOYL PEROXIDE. See Benzoyl
peroxide. [94-36-0]

DIBENZ[a,i]PYRENE. See Dibenzo[a,i]
pyrene. [189-55-9]

DIBORANE. [19287-45-7]
*OSHA 29 CFR 1910.1000 Table Z-1
PEL(TWA): 0.1 ppm, 0.1 mg/m^3

ACGIH-TLV
TWA: 0.1 ppm, 0.1 mg/m^3

NFPA—hazardous chemical
health—3, flammability—4, reactivity—3,
W—water may be hazardous in fire fighting

DIBROMOCHLOROPROPANE(DBCP).
[96-12-8]
*OSHA Standard 29 CFR 1910.1044
Cancer Hazard
Permissible exposure limit.(1) *Inhalation.*
The employer shall assure that no employee is exposed to an airborne concentration of DBCP in excess of 1 part DBCP per billion parts of air (ppb) as an 8-hour time-weighted average. (2) *Dermal and eye exposure.* The employer shall assure than no employee is exposed to eye or skin contact with SBCP.

Shall not apply to: (i) application and use solely as a pesticide; (ii) exposure to DBCP that results solely from the application and use of SBCP as a pesticide; or (iii) the storage, transportation, distribution or sale of DBCP in intact containers sealed in such a manner as to prevent exposure to DBCP vapors or liquid.

NIOSH Criteria Document (Pub. No. 78-115), NTIS Stock No. PB81 228728, September 2, 1977

"Occupational exposure to dibromochloropropane" refers to any workplace situation in which DBCP is manufactured, formulated, or stored. All sections of the standard shall apply where there is occupational exposure to DBCP.

Dibromochloropropane shall be controlled in the workplace so that no employee is exposed to airborne dibromochloropropane at a concentration greater than 10 ppb (approximately 0.1 mg/m^3) determined as a time-weighted average (TWA) concentration for up to a 10-hour work shift, 40-hour workweek.

The standard is designed to protect the health and provide for the safety of employees for up to a 10-hour work shift, 40-hour workweek, over a working lifetime.

The possible effects on the health of employees chronically exposed to DBCP may include sterility, diminished renal function, and degeneration and cirrhosis of the liver, In addition, ingestion of daily doses of DBCP by mice and rats has been found to result in the appearance of gastric cancers in both sexes of both species and in mammary cancers in female rats. Although an increased risk for cancer has not been seen with inhalation exposures, these results are no definitive: therefore the risk of cancer due to occupational exposure to DBCP remains a continuing concern. There are indications from *in vitro* experiments that mutagenic effects may occur also, but there has been no study yet of this possibility with mammalian subjects. Employees should be told of these possible effects and informed that some 20–25 years of experience in the manufacture and formulation of DBCP has not yet called such effects in employees of the pesticide industry to the notice of physicians and epidemiologists.

OSHA Candidate List of Potential Occupational Carcinogens (EPA Carcinogen Assessment Group List), 45 FR 53672, August 12, 1980

IARC—carcinogenic in animals

NCI—carcinogenic in animals

40 CFR 162.11, FIFRA-RPAR Completed Current Status: PD 1 published, 42 FR 48026 (9/22/77); comment period closed 11/7/77. Suspension Order and Notice of Intent to Cancel published, 42 FR 57543 (11/3/77). Amended Notice of Intent to Cancel and PD 2/3/4 published, 43 FR 40911 (9/13/78): NTIS# PB80 213853. Notice of Intent to Suspend all products issued, 44 FR 41783 (7/18/79). Suspension Order (affecting all uses except pineapple) and Notice of Intent to Cancel All Uses published, 44 FR 65135 (10/29/79). Agreement reached with registrants that all uses, except on pineapples, are cancelled.

Criteria Possibly Met or Exceeded: Oncogenicity and Reproductive Effects

*RCRA 40 CFR 261
EPA hazardous waste no. U066

DIBROMOETHANE. *See* Ethylene dibromide. [106-93-4]

1,2-DIBROMOETHANE. *See* Ethylene dibromide. [106-93-4]

1,6-DIBROMO-4-NITROANILINE.
See Aniline and chloro-, bromo-, and/or nitro- anilines. [827-94-1]

DIBUTYLAMINE. [111-92-2]

NFPA—hazardous chemical
health—3, flammability—2, reactivity—0

NFPA—hazardous chemical
health—3, flammability—2, reactivity—0

2-n-DIBUTYLAMINOETHANOL.
[102-81-8]

ACGIH-TLV

Skin

TWA: 2 ppm, 14 mg/m^3
STEL: 4 ppm, 28 mg/m^3

DIBUTYL ETHER (NORMAL).
[142-96-1]

NFPA—hazardous chemical
health—2, flammability—3, reactivity—0

DIBUTYLNITROSAMINE. [924-16-3]

IARC—carcinogenic in animals (used as positive carcinogenic control by NCI)

DIBUTYL PEROXIDE (TERTIARY).
[110-05-4]

NFPA—hazardous chemical
health—3, flammability—2, reactivity—4, oxy—oxidizing chemical

DIBUTYL PHOSPHATE.

*OSHA 29 CFR 1910.1000 Table Z-1
PEL(TWA): 0.1 ppm, 5 mg/m^3

ACGIH-TLV
TWA: 1 ppm, 5 mg/m^3
STEL: 2 ppm, 10 mg/m^3

DIBUTYLPHTHALATE. [84-74-2]

*OSHA 29 CFR 1910.1000 Table Z-1
PEL(TWA): 0.1 ppm, 5 mg/m^3

ACGIH-TLV
TWA: 1 ppm, 5 mg/m^3
STEL: 2 ppm, 10 mg/m^3

NFPA—hazardous chemical (flammable chemical)
health—0, flammability—1, reactivity—0
*RCRA 40 CFR 261
EPA hazardous waste no. U069 (hazardous substance)

DIBUTYLTIN BIS(ISOOCTYL MALEATE). [25168-21-2]
TSCA 4(e), ITC
Eleventh Report of the Interagency Testing Committee to the administrator, Environmental Protection Agency, November 1982; 47 FR 54626, December 3, 1982
*TSCA 8(d), Unpublished Health and Safety Studies Reporting 40 CFR 716, 47 FR 54624, December 3, 1982

DIBUTYLTIN BIS(ISOOCTYL MERCAPTOACETATE).
[25168-24-5]
TSCA 4(e), ITC
Eleventh Report of the Interagency Testing Committee to the administrator, Environmental Protection Agency, November 1982; 47 FR 54626, December 3, 1982
*TSCA 8(d), Unpublished Health and Safety Studies Reporting 40 CFR 716, 47 FR 54624, December 3, 1982

DIBUTYLTIN BIS(LAURYL MERCAPTIDE).
[1185-81-5]
TSCA 4(e), ITC
Eleventh Report of the Interagency Testing Committee to the administrator, Environmental Protection Agency, November 1982; 47 FR 54626, December 3, 1982
*TSCA 8(d), Unpublished Health and Safety Studies Reporting 40 CFR 716, 47 FR 54624, December 3, 1982

DIBUTYLTIN DILAURATE. [77-58-7]
TSCA 4(e), ITC
Eleventh Report of the Interagency Testing Committee to the administrator, Environmental Protection Agency, November 1982; 47 FR 54626, December 3, 1982

*TSCA 8(d), Unpublished Health and Safety Studies Reporting 40 CFR 716, 47 FR 54624, December 3, 1982

DICAMBA. [1918-00-9]
FIFRA—Data Call In, 45 FR 75488, November 14, 1980
Status: letter being drafted
FIFRA—Registration Standard Development—Data Collection (herbicide)
*CWA 311(b)(2)(A), 40 CFR 116, 117
Discharge RQ = 1000 pounds (454 kilograms)

DICHLOBENIL. [1194-65-6]
*CWA 311(b)(2)(A), 40 CFR 116, 117
Discharge RQ = 1000 pounds (454 kilograms)

DICHLOFOP METHYL.
FIFRA—Data Call In, 45 FR 75488, November 14, 1980
Status: letter being drafted

DICHLONE. [117-80-6]
FIFRA—Registration Standard—Issued March 1981, NTIS# 540/RS 81-0001 80 (fungicide)
*CWA 311(b)(2)(A), 40 CFR 116, 117
Discharge RQ = 1000 pounds (454 kilograms)

2,2-DICHLOROACETALDEHYDE.
[79-02-7]
TSCA-CHIP available

DICHLOROACETYLENE. [7572-29-4]
ACGIH-TLV
TWA: ceiling limit, 0.1 ppm, 0.4 mg/m^3

S-(2,3-DICHLOROALLYL) DIISOPROPYLTHIOCARBAMATE.
[2303-16-4]
*RCRA 40 CFR 261
EPA hazardous waste no. U062

3,4-DICHLOROANILINE. See Aniline and chloro-, bromo-, and/or nitro-, anilines. [95-76-1]

NFPA—hazardous chemical
health—3, flammability—1, reactivity—0

2,3-DICHLOROBENZAMINE. See Aniline and chloro-, bromo-, and/or nitro-, anilines [608-27-5]

*TSCA 8(d), Unpublished Health and Safety Studies Reporting, 40 CFR 716, 47 FR 387, September 2, 1982

2,4-DICHLOROBENZENAMINE. See Aniline and chloro-, bromo-, and/or nitro-, anilines. [554-00-7]

2,5-DICHLOROBENZENAMINE. See Aniline and chloro-, bromo-, and/or nitro-, anilines. [95-82-9]

3,4-DICHLOROBENZENAMINE. See 3,4-Dichloroaniline. [95-76-1]

3,5-DICHLOROBENZENAMINE. See Aniline and chloro-, bromo-, and/or nitro-, anilines. [626-43-7]

m-DICHLOROBENZENE. [541-73-1]
TSCA 4(e), ITC
Initial Report to the administrator, Environmental Protection Agency, TSCA Interagency Testing Committee, October 1, 1977; 42 FR 55026, October 12, 1977, responded to by the EPA administrator, 43 FR 50134, October 26, 1978 and 45 FR 48524, July 18, 1980
CWA 307(a)(1), 40 CFR 125, Priority Pollutant
*RCRA 40 CFR 261
EPA hazardous waste no. U071 (hazardous substance)

o-DICHLOROBENZENE. [95-50-1]
*OSHA 29 CFR 1910.1000 Table Z-1
PEL(Ceiling): 50 ppm, 300 mg/m^3
ACGIH-TLV
TWA: ceiling limit, 50 ppm, 300 mg/m^3
STEL: 2 ppm, 10 mg/m^3
TSCA 4(e), ITC
Initial Report to the administrator, Environmental Protection Agency, TSCA Interagency Testing Committee, October 1, 1977;
42 FR 55026, October 12, 1977, responded to by the EPA administrator, 43 FR 50134, October 26, 1978 and 45 FR 48524, July 18, 1980
CWA 307(a)(1), 40 CFR 125, Priority Pollutant
NFPA—hazardous chemical
health—2, flammability—2, reactivity—0
*RCRA 40 CFR 261
EPA hazardous waste no. U070 (hazardous substance)

p-DICHLOROBENZENE. [106-46-7]
*OSHA 29 CFR 1910.1000 Table Z-1
PEL(TWA): 75 ppm, 450 mg/m^3
ACGIH-TLV
TWA: 75 ppm, 450 mg/m^3
STEL: 110 ppm, 675 mg/m^3
TSCA 4(e), ITC
Initial Report to the administrator, Environmental Protection Agency, TSCA Interagency Testing Committee, October 1, 1977; 42 FR 55026, October 12, 1977, responded to by the EPA administrator, 43 FR 50134, October 26, 1978 and 45 FR 48524, July 18, 1980
CWA 307(a)(1), Priority Pollutant
CAA 112
Chemical proposed to be assessed as a Hazardous Air Pollutant by a House of Representatives bill to amend the CAA
*RCRA 40 CFR 261
EPA hazardous waste no. U072 (hazardous substance)

1,2-DICHLOROBENZENE.
See o-Dichlorobenzene. [95-50-1]

1,3-DICHLOROBENZENE.
See m-Dichlorobenzene. [541-73-1]

1,4-DICHLOROBENZENE.
See p-Dichlorobenzene. [106-46-7]

DICHLOROBENZENES. [25321-22-6]
CWA 304(a)(1), 45 FR 79318, November 28, 1980, Water Quality Criteria

Freshwater Aquatic Life

The available data for dichlorobenzenes indicate that acute and chronic toxicity to freshwater aquatic life occurs at concentrations as low as 1120 and 763 µg/liter, respectively, and would occur at lower concentrations among species that are more sensitive than those tested.

Saltwater Aquatic Life

The available data for dichlorobenzenes indicate that acute toxicity to saltwater aquatic life occurs at concentrations as low as 1970 µg/liter and would occur at lower concentrations among species that are more sensitive than those tested. No data are available concerning the chronic toxicity of dichlorobenzenes to sensitive saltwater aquatic life.

Human Health

For the protection of human health from the toxic properties of dichlorobenzenes (all isomers) ingested through water and contaminated aquatic organisms, the ambient water criterion is determined to be 400 µg/liter.

For the protection of human health from the toxic properties of dichlorobenzenes (all isomers) ingested through contaminated aquatic organisms alone, the ambient water criterion is determined to be 2.6 mg/liter.

*CWA 311(b)(2)(A), 40 CFR 116, 117
Discharge RQ = 100 pounds (45.4 kilograms)

3,3'-DICHLOROBENZIDINE. [91-94-1]

ACGIH-TLV

Skin

TWA: Industrial substance suspect of carcinogenic potential for humans—which is suspect of inducing cancer, based on either (1) limited epidemiologic evidence, exclusive of clinical reports of single cases, or (2) demonstration of carcinogenesis in one or more animal species by appropriate methods. Worker exposure by all routes should be carefully controlled to levels consistent with the animal and human experience data.

See below.

3,3'-DICHLOROBENZIDINE. (and its salts). [91-94-1]

*OSHA Standard 29 CFR 1910.1007
Cancer-Suspect Agent
Shall not apply to solid or liquid mixtures containing less than 1% by weight or volume of 3,3'-dichlorobenzidine (or its salts)

OSHA Candidate List of Potential Occupational Carcinogens (EPA Carcinogen Assessment Group List), 45 FR 53672, August 12, 1980

IARC—carcinogenic in animals
CWA 307(a)(1), Priority Pollutant
*RCRA 40 CFR 261
EPA hazardous waste no. U073

DICHLOROBENZIDINES.

CWA 304(a)(1), 45 FR 79318, November 28, 1980, Water Quality Criteria

Freshwater Aquatic Life

The data base available for dichlorobenzidines and freshwater organisms is limited to one test on bioconcentration of 3,3'-dichlorobenzidine and no statement can be made concerning acute or chronic toxicity.

Saltwater Aquatic Life

No saltwater organisms have been tested with any dichlorobenzidine and no statement can be made concerning acute or chronic toxicity.

Human Health

For the maximum protection of human health from the potential carcinogenic effects due to exposure of dichlorobenzidine through ingestion of contaminated water and contaminated aquatic organisms, the ambient water concetration should be zero based on the nonthreshold assumption for this chemical. However, zero level may not be attainable at the present time. Therefore, the levels that may result in incremental increase of cancer risk over the lifetime are estimated at 10^{-5}, 10^{-6}, and 10^{-7}. The corresponding criteria are 0.103 µg/liter, 0.0103 µg/liter, and 0.00103 µg/liter, respectively. If the above estimates are made for consumption of aquatic organisms only, excluding consumption of water, the levels are 0.204 µg/liter, 0.0204 µg/liter, and 0.00204 µg/liter, respectively. Other concentrations representing different risk levels may be calculated by use of the guidelines. The risk estimate range is presented for information purposes and does not represent an agency judgment on an "acceptable" risk level.

DICHLOROBROMOMETHANE.
[75-27-4]

CWA 307(a)(1), Priority Pollutant

1,4-DICHLOROBUTANE.

NFPA—hazardous chemical
health—2, flammability—2, reactivity—0

1,4-DICHLORO-2-BUTENE. [764-41-0]

*RCRA 40 CFR 261
EPA hazardous waste no. U074 (ignitable waste, toxic waste)

3,3'-DICHLORO-4,4'-DIAMINODIPHENYL ETHER.
[28434-86-8]

IARC—carcinogenic in animals

2,7-DICHLORODIBENZO-p-DIOXIN(DCDD). [33857-26-0]

NCI—may be carcinogenic in animals

2,2'-DICHLORODIETHYL ETHER.
See Dichloroethyl ether. [11-44-4]

DICHLORODIFLUOROMETHANE.
[75-71-8]

*OSHA 29 CFR 1910.1000 Table Z-1
PEL(TWA): 1000 ppm, 4950 mg/m^3
ACGIH-TLV
TWA: 1000 ppm, 4950 mg/m^3
STEL: 1250 ppm, 6200 mg/m^3

1,3-DICHLORO-5,5-DIMETHYL HYDANTHION. [118-52-5]

*OSHA 29 CFR 1910.1000 Table Z-1
PEL(TWA): 0.2 mg/m^3
ACGIH-TLV
TWA: 0.2 mg/m^3
STEL: 0.4 mg/m^3

3,5-DICHLORO-N-(1,1-DIMETHYL-2-PROPYNYL)BENZAMIDE.
[23950-58-5]

*RCRA 40 CFR 261
EPA hazardous waste no. U192

DICHLORODIPHENYLDICHLOROETHANE. See DDE. [72-54-8]

DICHLORODIPHENYLTRICHLOROETHANE. See DDT. [50-29-3]

1,1-DICHLOROETHANE. [75-34-3]

*OSHA 29 CFR 1910.1000 Table Z-1
PEL(TWA): 100 ppm, 400 mg/m^3
ACGIH-TLV
TWA: 200 ppm, 810 mg/m^3
STEL: 250 ppm, 1010 mg/m^3
*RCRA 40 CFR 261
EPA hazardous waste no. U076

1,2-DICHLOROETHANE. See
Ethylene dichloride. [107-06-2]

1,1-DICHLOROETHENE. See
1,1-Dichloroethylene. [75-35-47]

1,2-DICHLOROETHENE. See
1,2-Dichloroethylene. [540-59-0]

1,1-DICHLOROETHYLENE. [75-35-4]

ACGIH-TLV
TWA: 10 ppm, 40 mg/m^3
STEL: 20 ppm, 80 mg/m^3
Intended changes for 1982 (1982 revision or addition):
TWA: 5 ppm, 20 mg/m^3
STEL: 20 ppm, 80 mg/m^3
OSHA Candidate List of Potential Occupational Carcinogens (EPA Carcinogen Assessment Group List), 45 FR 53672, August 12, 1980

NFPA—hazardous chemical
health—2, flammability—4, reactivity—2,

CWA 307(a)(1), Priority Pollutant
*CWA 311(b)(2)(A), 40 CFR 116, 117
Discharge RQ = 5000 pounds (2270 kilograms)

CAA 112
Chemical proposed assessed as a Hazardous

Air Pollutant by a House of Representatives bill to amend the CAA

*RCRA 40 CFR 261
EPA hazardous waste no. U078 (hazardous substance)

See Dichloroethylenes.

1,2-DICHLOROETHYLENE. [540-59-0]

*OSHA 29 CFR 1910.1000 Table Z-1
PEL(TWA): 200 ppm, 790 mg/m^3

ACGIH-TLV
TWA: 200 ppm, 790 mg/m^3
STEL: 250 ppm, 1000 mg/m^3

NFPA—hazardous chemical
health—2, flammability—3, reactivity—2

*RCRA 40 CFR 261
EPA hazardous waste no. U079

See Dichloroethylenes.

1,2-*trans*-DICHLOROETHYLENE. [156-60-5]

CWA 307(a)(1), 40 CFR 125, Priority Pollutant

DICHLOROETHYLENES.

CWA 304(a)(1), 45 FR 79318, November 28, 1980, Water Quality Criteria

Freshwater Aquatic Life

The available data for dichloroethylenes indicate that acute toxicity to freshwater aquatic life occurs at concentrations as low as 11,600 µg/liter and would occur at lower concentrations among species that are more sensitive than those tested. No definitive data are available concerning the chronic toxicity of dichloroethylenes to sensitive freshwater aquatic life.

Saltwater Aquatic Life

The available data for dichloroethylenes indicate that acute toxicity to saltwater aquatic life occurs at concentrations as low as 224,000 µg/liter and would occur at lower concentrations among species that are more sensitive than those tested. No data are available concerning the chronic toxicity of dichloroethylenes to sensitive saltwater aquatic life.

Human Health

For the maximum protection of human health from the potential carcinogenic effects due to exposure of 1,1-dichloroethylene through ingestion of contaminated water and contaminated aquatic organisms, the ambient water concentration should be zero based on the nonthreshold assumption for this chemical. However, zero level may not be attainable at the present time. Therefore, the levels that may result in incremental increase of cancer risk over the lifetime are estimated at 10^{-5}, 10^{-6}, and 10^{-7}. The corresponding criteria are 0.33 µg/liter, 0.033 µg/liter, and 0.0033 µg/liter, respectively. If the above estimates are made for consumption of aquatic organisms only, excluding consumption of water, the levels are 18.5 µg/liter, 1.85 µg/liter, and 0.185 µg/liter, respectively. Other concentrations representing different risk levels may be calculated by use of the guidelines. The risk estimate range is presented for information purposes and does not represent an agency judgment on an "acceptable" risk level.

Using the present guidelines, a satisfactory criterion cannot be derived at this time due to the insufficiency in the available data for 1,2-dichloroethylene.

DICHLOROETHYL ETHER. [111-44-4]

*OSHA 29 CFR 1910.1000 Table Z-1

Skin

*OSHA 29 CFR 1910.1000 Table Z-1
PEL(Ceiling): 15 ppm, 90 mg/m^3

ACGIH-TLV

Skin

TWA: 5 ppm, 30 mg/m^3
STEL: 10 ppm, 60 mg/m^3

NFPA—hazardous chemical
health—3, flammability—2, reactivity—0

*RCRA 40 CFR 261
EPA hazardous waste no. U025

DICHLOROFLUOROMETHANE. [75-43-4]

ACGIH-TLV
TWA: 10 ppm, 40 mg/m^3

CWA 307(a)(1), 40 CFR 125, Priority Pollutant
*RCRA 40 CFR 261
EPA hazardous waste no. U075

DICHLOROISOCYANURIC ACID. See Dichloro-s-Triazinetrione. [2782-57-2]

DICHLOROMETHANE. See Methylene chloride. [75-09-2]

DICHLOROMONOFLUORO-METHANE. [74-43-4]
*OSHA 29 CFR 1910.1000 Table Z-1
PEL(TWA): 1000 ppm, 4200 mg/m^3

2,6-DICHLORO-4-NITRO-ANILINE. See Aniline and chloro-, bromo-, and/or nitro- anilines. [99-30-9]

2,6-DICHLORO-4-NITROBENZENAMINE. See Aniline and chloro-, bromo-, and/or nitro- anilines. [99-30-9]

1,1-DICHLORO-1-NITROETHANE. [594-72-9]
*OSHA 29 CFR 1910.1000 Table Z-1
PEL(Ceiling): 10 ppm, 60 mg/m^3
ACGIH-TLV
TWA: 2 ppm, 10 mg/m^3
STEL: 10 ppm, 60 mg/m^3

2,4-DICHLORO-1-(4-NITROPHENOXY)BENZENE. See Ether, 2,4-dichlorophenyl p-nitrophenyl. [1836-75-5]

2,4-DICHLOROPHENOL. [120-83-2]
CWA 304(a)(1), 45 FR 79318, November 28, 1980, Water Quality Criteria

Freshwater Aquatic Life
The available data for 2,4-dichlorophenol indicate that acute and chronic toxicity to freshwater aquatic life occurs at concentrations as low as 2020 and 365 µg/liter, respectively, and would occur at lower concentrations among species that are more sensitive than those tested. Mortality to early life stages of one species of fish occurs at concentrations as low as 70 µg/liter.

Saltwater Aquatic Life
Only one test has been conducted with saltwater organisms on 2,4-dichlorophenol and no statement can be made concerning acute or chronic toxicity.

Human Health
For comparison purposes, two approaches were used to derive criterion levels for 2,4-dichlorophenol. Based on available toxicity data, for the protection of public health, the derived level is 3.09 mg/liter. Using available organoleptic data, for controlling undesirable taste and odor quality of ambient water, the estimated level is 0.3 µg/liter. It should be recognized that organoleptic data as a basis for establishing a water quality criterion have limitations and have no demonstrated relationship to potential adverse human health effects.

CWA 307(a)(1), Priority Polutant
*RCRA 40 CFR 261
EPA hazardous waste no. U081

2,6-DICHLOROPHENOL.
*RCRA 40 CFR 261
EPA hazardous waste no. U082

DICHLOROPHENYLARSINE. [696-28-6]
*RCRA 40 CFR 261
EPA hazardous waste no. P036

DICHLOROPROPANE. [26638-19-7]
*CWA 311(b)(2)(A), 40 CFR 116, 117
Discharge RQ = 5000 pounds (2270 kilograms)
See Dichloropropanes/dichloropropenes.

1,2-DICHLOROPROPANE. See Propylene dichloride. [78-87-5]

1,3-DICHLOROPROPANE. [142-28-9]
TSCA-CHIP available
*RCRA 40 CFR 261
EPA hazardous waste no. U084 (hazardous substance)

DICHLOROPROPANES/ DICHLOROPROOPENES. [8003-19-8]
CWA 304(a)(1), 45 FR 79318, November 28, 1980, Water Quality Criteria

Freshwater Aquatic Life

The available data for dichloropropanes indicate that acute and chronic toxicity to freshwater aquatic life occurs at concentrations as low as 23,000 and 5700 µg/liter, respectively, and would occur at lower concentrations among species that are more sensitive than those tested.

The available data for dichloropropenes indicate that acute and chronic toxicity to freshwater aquatic life occurs at concentrations as low as 6060 and 244 µg/liter, respectively, and would occur at lower concentrations among species that are more sensitive than those tested.

Saltwater Aquatic Life

The available data for dichloropropanes indicate that acute and chronic toxicity to saltwater aquatic life occurs at concentrations as low as 10,300 and 3040 µg/liter, respectively, and would occur at lower concentrations among species that are more sensitive than those tested.

The available data for dichloropropenes indicate that acute toxicity to saltwater aquatic life occurs at concentrations as low as 790 µg/liter, and would occur at lower concentrations among species that are more sensitive than those tested. No data are available concerning the chronic toxicity of dichloropropenes to sensitive saltwater aquatic life.

Human Health

Using the present guidelines, a satisfactory criterion cannot be derived at this time due to the insufficiency in the available data for dichloropropanes.

For the protection of human health from the toxic properties of dichloropropenes ingested through water and contaminated aquatic organisms, the ambient water criterion is determined to be 87 µg/liter.

For the protection of human health from the toxic properties of dichloropropenes ingested through contaminated aquatic organimsm alone, the ambient water criterion is determined to be 14.1 mg/liter/

*CWA 311(b)(2)(A), 40 CFR 116, 117
Discharge RQ = 5000 pounds (2270 kilograms)

DICHLOROPROPENE. [26952-23-8]
ACGIH-TLV

Skin
TWA: 1 ppm, 5 mg/m^3
STEL: 10 ppm, 50 mg/m^3
*CWA 311(b)(2)(A), 40 CFR 116, 117
Discharge RQ = 5000 pounds (2270 kilograms)
See Dichloropropanes/dichloropropenes.

1,3-DICHLOROPROPENE (cis and trans). [10061-01-5, 10061-02-6]
NFPA—hazardous chemical
health—3, flammability—3, reactivity—0
CWA 307(a)(1), Priority Pollutant
See Dichloropropanes/dichloropropenes.

2,2-DICHLOROPROPIONIC ACID. [75-99-0]
ACGIH-TLV
TWA: 1 ppm, 6 mg/m^3
*CWA 311(b)(2)(A), 40 CFR 116, 117
Discharge RQ = 5000 pounds (2270 kilograms)

1,3-DICHLOROPROPYLENE. See 1,3-Dichloropropene. [542-75-6]

DICHLOROTETRAFLUOROETHANE. [76-14-2]
*OSHA 29 CFR 1910.1000 Table Z-1
PEL(TWA): 1000 ppm, 7000 mg/m^3
ACGIH-TLV
TWA: 1000 ppm, 7000 mg/m^3
STEL: 1250 ppm, 8750 mg/m^3

DICHLORO-s-TRIAZINETRIONE. [2782-57-2]
NFPA—hazardous chemical
health—3, flammability—0, reactivity—0, oxy—oxidizing chemical

DICHLORVOS (DDVP, 2,2-DICHLOROVINYL DIMETHYL PHOSPHATE). [62-73-7]

ACGIH-TLV

Skin

TWA: 0.1 ppm, 1 mg/m^3
STEL: 0.3 ppm, 3 mg/m^3

40 CFR 162.11, FIFRA Pre-RPAR Review
Current Status: Pre-RPAR evaluation underway
Criteria Possibly Met or Exceeded: Oncogenicity, Mutagenicity, Reproductive Effects, Fetotoxicity, and Neurotoxicity

*CWA 311(b)(2)(A), 40 CFR 116, 117
Discharge RQ = 10 pounds (4.54 kilograms)

DICHROMATES (except AMMONIUM DICHROMATE).

NFPA-hazardous chemical
health—1, flammability—0, reactivity—1, oxy—oxidizing chemical

DICOFOL. [115-32-2]

NCI—carcinogenic in animals

DICROTOPHOS (BIDRIN). [141-66-2]

ACGIH-TLV

Skin

TWA: 0.25 mg/m^3

FIFRA—Registration Standard Development—Data Evaluation and Development of Regulatory Position (insecticide)

DICYCLOHEXYLMETHANE, 4,4'-DIISOCYANATE.

See Diisocyanates, NIOSH Criteria Document

DICYCLOPENTADIENE. [77-73-6]

ACGIH-TLV
TWA: 5 ppm, 30 mg/m^3

DICYCLOPENTADIENYL IRON. [102-54-5]

ACGIH-TLV
TWA: 10 mg/m^3
STEL: 20 mg/m^3

DIELDRIN.* [60-57-1]

*OSHA 29 CFR 1910.1000 Table Z-1

Skin

PEL(TWA): 0.25 mg/m^3

ACGIH-TLV

Skin

TWA: 0.25 mg/m^3
STEL: 0.75 mg/m^3

OSHA Candidate List of Potential Occupational Carcinogens (EPA Carcinogen Assessment Group List), 45 FR 53672, August 12, 1980

IARC—carcinogenic in animals

NCI—carcinogenic in animals

CWA 304(a)(1), 45 FR 79318, November 28, 1980, Water Quality Criteria

Freshwater Aquatic Life

For dieldrin the criterion to protect freshwater aquatic life as derived using the guidelines is 0.0019 µg/liter as a 24-hour average and the concentration should not exceed 2.5 µg/liter at any time.

Saltwater Aquatic Life

For dieldrin the criterion to protect saltwater aquatic life as derived using the guidelines is 0.0019 µg/liter as a 24-hour average and the concentration should not exceed 0.71 µg/liter at any time.

Human Health

For the maximum protection of human health from the potential carcinogenic effects due to exposure of dieldrin through ingestion of contaminated water and contaminated aquatic organisms, the ambient water

---

*Generic name for an insecticidal product containing not less than 85% of 1,2,3,4,10,10-hexachloro-6,7-epoxy-1,4,4a,5,6,7,8,8a-octahydro-1,4-endo,exo-5,8-dimethanonaphthalene, and not less than 15% active related compounds.

concentration should be zero based on the nonthreshold assumption for this chemical. However, zero level may not be attainable at the present time. Therefore, the levels that may result in incremental increase of cancer risk over the lifetime are estimated at 10^{-5}, 10^{-6}, and 10^{-7}. The corresponding criteria are 0.71 ng/liter, 0.071 ng/liter, and 0.0071 ng/liter, respectively. If the above estimates are made for consumption of aquatic organisms only, excluding consumption of water, the levels are 0.76 ng/liter, 0.076 ng/liter, and 0.0076 ng/liter, respectively. Other concentrations representing different risk levels may be calculated by use of the guidelines. The risk estimate range is presented for information purposes and does not represent an agency judgment on an "acceptable" risk level.

*CWA 307(a)(1), 40 CFR 125, Priority Pollutant

*CWA 311(b)(2)(A), 40 CFR 116, 117
Discharge RQ = 1 pound (0.454 kilogram)

*RCRA 40 CFR 261
EPA hazardous waste no. P037 (hazardous substance)

DIEPOXYBUTANE. [1464-53-5]

OSHA Candidate List of Potential Occupational Carcinogens (EPA Carcinogen Assessment Group List), 45 FR 53672, August 12, 1980

IARC—carcinogenic in animals
*RCRA 40 CFR 261
EPA hazardous waste no. U085 (ignitable waste, toxic waste)

1,2:3,4-DIEPOXYBUTANE.

See Diepoxybutane. [1464-53-5]

DI(2,3-EPOXYPROPYL)ETHER.

[2238-07-5]
See Glycidyl ethers, NIOSH Criteria Document.

DIESEL FUEL OIL NO. 1-D.

NFPA—hazardous chemical (flammable chemical)
health—0, flammability—2, reactivity—0

DIESEL FUEL OIL NO. 2-D.

NFPA—hazardous chemical (flammable chemical)
health—0, flammability—2, reactivity—0

DIESEL FUEL OIL NO. 4-D.

NFPA—hazardous chemical (flammable chemical)
health—0, flammability—2, reactivity—0

DIETHANOLAMINE. [111-42-2]

ACGIH-TLV
TWA: 3 ppm, 15 mg/m^3
NFPA—hazardous chemical (flammable chemical)
health—1, flammability—1, reactivity—0

DIETHYLALUMINUM CHLORIDE.

See Chlorodiethylaluminum.

DIETHYLAMINE. [109-89-7]

*OSHA 29 CFR 1910.1000 Table Z-1
PEL(TWA): 25 ppm, 75 mg/m^3
ACGIH-TLV
TWA: 10 ppm, 30 mg/m^3
STEL: 25 ppm, 75 mg/m^3
NFPA—hazardous chemical (flammable chemical)
health—2, flammability—3, reactivity—0
*CWA 311(b)(2)(A), 40 CFR 116, 117
Discharge RQ = 1000 pounds (454 kilograms)

DIETHYLAMINE, 2,2'-DICHLORO-N-METHYL-, HYDROCHLORIDE.

Syn: Mechlorethamine hydrochloride. [55-86-7]

OSHA Candidate List of Potential Occupational Carcinogens, 45 FR 53672, August 12, 1980

DIETHYLAMINO ETHANOL.

[100-37-8]
*OSHA 29 CFR 1910.1000 Table Z-1

Skin

PEL(TWA): 10 ppm, 50 mg/m^3
ACGIH-TLV

Skin
TWA: 10 ppm, 50 mg/m³

DIETHYLARSINE.
*RCRA 40 CFR 261
EPA hazardous waste no. P038

DIETHYLCARBAMOYL CHLORIDE (DECC).
NIOSH Current Intelligence Bulletin No. 12

DIETHYLCHLOROSILANE.
See Chlorodiethylsilane.

DIETHYLCHLOROTHIOPHOSPHATE.
[2524-04-1]
TSCA-CHIP available

1,4-DIETHYLENE DIOXIDE.
See Dioxane. [123-91-1]

DIETHYLENE GLYCOL. [111-46-6]
NFPA—hazardous chemical (flammable chemical)
health—1, flammability—1, reactivity—0
TSCA-CHIP available

DIETHYLENETRIAMINE. [111-40-0]
ACGIH-TLV

Skin
TWA: 1 ppm, 4 mg/m³
NFPA—hazardous chemical (flammable chemical)
health—3, flammability—1, reactivity—0
TSCA 4(e), ITC
Eighth Report of the TSCA Interagency Testing Committee to the administrator, Environmental Protection Agency, April 1981; 46 FR 28138, May 22, 1981, EPA responded to the committee's recommendations for testing, 47 FR 18386, April 29, 1982
*TSCA 8(d), Unpublished Health and Safety Studies Reporting, 40 CFR 716, 47 FR 38800. September 2, 1982; 48 FR 13178, March 30, 1983

DIETHYL ETHER. See Ethyl ether.
[60-29-7]

See Waste anesthetic gases and vapors, NIOSH Criteria Document
NFPA—hazardous chemical
health—2, flammability—4, reactivity—1

O,O-DIETHYL S-[2-(ETHYLTHIO)ETHYL]PHOSPHORODITHIOATE.
[298-04-4]
*RCRA 40 CFR 261
EPA hazardous waste no. P039 (hazardous substance)

DI(2-ETHYLHEXYL)ADIPATE.
[103-23-1]
NCI—carcinogenic in animals

DI(2-ETHYLHEXYL)PHTHALATE.
See Dioctyl phthalate. [117-81-7]

1,2-DIETHYLHYDRAZINE. [1615-80-1]
OSHA Candidate List of Potential Occupational Carcinogens, (EPA Carcinogenic Assessment Group List) 45 FR 53672, August 12, 1980
IARC—carcinogenic in animals
*RCRA 40 CFR 261
EPA hazardous waste no. U086

DIETHYL KETONE. [96-22-0]
ACGIH-TLV
TWA: 200 ppm, 705 mg/m³

O,O-DIETHYL-S-METHYLDITHIOPHOSPHATE.
*RCRA 40 CFR 261
EPA hazardous waste no. U087

DIETHYL p-NITROPHENYL PHOSPHATE. [311-45-5]
*RCRA 40 CFR 261
EPA hazardous waste no. P041

O,O-DIETHYL O-p-NITROPHENYL PHOSPHOROTHIOATE.
See Parathion. [56-38-2]

O,O-DIETHYL-para-NITROPHENYLTHIOPHOSPHATE.
See Parathion [56-38-2]

DIETHYLNITROSAMINE. [55-18-5]
IARC—carcinogenic in animals (used as positive carcinogenic control by NCI)

DI(p-ETHYLPHENYL) DICHLOROETHANE(p,p-ETHYL-DDD: PERTHANE). [72-56-0]
NCI—may be carcinogenic in animals

DIETHYL PHTHALATE. [84-66-2]
ACGIH-TLV
TWA: 5 mg/m^3
STEL: 10 mg/m^3
NFPA—hazardous chemical (flammable chemical)
health—0, flammability—1, reactivity—0
*RCRA 40 CFR 261
EPA hazardous waste no. U088

O,O-DIETHYL O-PYRAZINYL PHOSPHOROTHIOATE. [297-97-2]
*RCRA 40 CFR 261
EPA hazardous waste no. P040

DIETHYLSTILBESTROL. [56-53-1]
OSHA Candidate List of Potential Occupational Carcinogens (EPA Carcinogen Assessment Group List), 45 FR 53672, August 12, 1980
IARC—carcinogenic in humans
*RCRA 40 CFR 261
EPA hazardous waste no. U089

DIETHYLSTILBESTROL DIPROPIONATE.
See 4,4'-Stilbenediol, alpha,alpha'-diethyl-, dipropionate, (E)-. [130-80-3]

DIETHYL SULFATE. [64-67-5]
IARC—carcinogenic in animals
NFPA—hazardous chemical
health—3, flammability—1, reactivity—1

1,3-DIETHYL-2-THIOUREA.
See Urea, 1,3-diethyl-2-thio. [105-55-5]

N,N-DIETHYLTHIOUREA.
See Urea, 1,3-diethyl-2-thio. [105-55-5]

DIETHYLZINC. [557-20-2]
NFPA—hazardous chemical
health—0, flammability—3, reactivity—3,
W—water may be hazardous in fire fighting

DIFLUBENZURON DIMILIN. [35367-38-5]
40 CFR 162.11, FIFRA-RPAR, Notice of Intent to Register Issued
Current Status: RPAR action completed in April 1979. Registrant consented to restricted uses.
Criteria Possibly Met or Exceeded: Oncogenicity and Hazard to Wildlife and Other Chronic Effects

DIFLUORODIBROMOMETHANE. [75-61-6]
*OSHA 29 CFR 1910.1000 Table Z-1
PEL(TWA): 100 ppm, 860 mg/m^3
ACGIH-TLV
TWA: 100 ppm, 860 mg/m^3
STEL: 150 ppm, 1290 mg/m^3

DIGLYCIDYL ETHER (DGE). [2238-07-5]
*OSHA 29 CFR 1910.1000 Table Z-1
PEL(Ceiling): 0.5 ppm, 2.8 mg/m^3
ACGIH-TLV
TWA: 0.1 ppm, 0.5 mg/m^3

DIGLYCIDYL ETHER OF BISPHENOL A. [1675-54-3]
TSCA 4(e), ITC
Third Report of the TSCA Interagency Testing Committee to the administrator, Environmental Protection Agency, October 1978; 43 FR 50630, October 30, 1978

1,2-DIHYDRO-5-NITROACENAPHTHYLENE.
See Acenaphthene, 5-nitro-. [602-87-9]

1,2-DIHYDRO-3,6-PYRADIZINEDIONE. [123-33-1]
*RCRA 40 CFR 261
EPA hazardous waste no. U148

DIHYDROSAFROLE. [94-58-6]
OSHA Candidate List of Potential Occupational Carcinogens (EPA Carcinogen Assessment Group List), 45 FR 53672, August 12, 1980

IARC—carcinogenic in animals

*RCRA 40 CFR 261

EPA hazardous waste no. U090

DIHYDROXYBENZENE.
See Hydroquinone. [123-31-9]

DIISOBUTYL KETONE. [108-83-8]
*OSHA 29 CFR 1910.1000 Table Z-1

PEL(TWA): 50 ppm, 290 mg/m^3

ACGIH-TLV

TWA: 25 ppm, 150 mg/m^3

See Ketones, NIOSH Criteria Document.

DIISOCYANATES.
NIOSH Criteria Document (Pub. No. 78-215), NTIS Stock No. PB81-226615, September 27, 1978

The present recommended standard includes all diisocyanates, but not their polymerized forms.

"Occupational exposure to diisocyanates" is defined as exposure to airborne diisocyanates at concentrations above one-half the recommended time-weighted average (TWA) occupational exposure limit or above the recommended ceiling limit. Adherence to all provisions of the standard is required if employees are occupationally exposed to airborne diisocyanates. If employees are exposed to airborne diisocyanates at concentrations of one-half the recommended TWA workplace environmental limit or less, the employer shall comply with all sections of the recommended standard except Sections 2(b), 4(c), 8(b), and the monitoring provisions of 8(c).

Exposure to diisocyanates shall be controlled so that no employee is exposed to concentrations greater than the limits specified below. These limits expressed in micrograms per cubic meter are equivalent to a vapor concentration of 5 ppb as a TWA concentration for up to a 10-hour work shift, 40-hour workweek, and 20 ppb as a ceiling concentration for any 10-minute sampling period. The microgram equivalents for selected diisocyanates are as follows:

	TWA	Ceiling
Toluene diisocyanate (TDI)	35 µg/m^3	140 µg/m^3
Diphenylmethane diisocyanate (MDI)	50 µg/m^3	200 µg/m^3
Hexamethylene diisocyanate (HDI)	35 µg/m^3	140 µg/m^3
Naphthalene diisocyanate (NDI)	40 µg/m^3	170 µg/m^3
Isophorone diisocyanate (IPDI)	45 µg/m^3	180 µg/m^3
Dicyclohexylmethane, 4,4'-diisocyanate (hydrogenated MDI)	55 µg/m^3	210 µg/m^3

If other diisocyanates are used, employers should observe environmental limits equivalent to a ceiling concentration of 20 ppb and a TWA concentration of 5 ppb.

The standard is designed to protect the health and provide for the safety of employees for up to a 10-hour work shift, 40-hour workweek, over a working lifetime.

4,4'-DIISOCYANATO-3,3'-DIMETHOXYBIPHENYL. [91-93-0]
NCI—carcinogenic in animals

DIISOPROPYLAMINE. [108-18-9]
*OSHA 29 CFR 1910.1000 Table Z-1

Skin

PEL(TWA): 5 ppm, 20 mg/m^3

ACGIH-TLV

Skin

TWA: 5 ppm, 20 mg/m^3

NFPA—hazardous chemical

health—3, flammability—3, reactivity—0

DIISOPROPYL ETHER. [108-20-3]
*OSHA 29 CFR 1910.1000 Table Z-1
PEL(TWA): 500 ppm, 2100 mg/m^3
ACGIH-TLV
TWA: 250 ppm, 1050 mg/m^3
STEL: 310 ppm, 1320 mg/m^3
NFPA—hazardous chemical
health—2, flammability—3, reactivity—1

DIISOPROPYL FLUOROPHOSPHATE. [55-91-4]
*RCRA 40 CFR 261
EPA hazardous waste no. P043

DIISOPROPYL PEROXYDICARBONATE. [105-64-6]
NFPA—hazardous chemical
health—0, flammability—4, reactivity—4, oxy—oxidizing chemical

DIKETENE. [674-82-8]
NFPA—hazardous chemical
health—2, flammability—2, reactivity—2

DILAUROYL PEROXIDE. [105-74-8]
NFPA—hazardous chemical
health—0, flammability—2, reactivity—3, oxy—oxidizing chemical

DIMETHOATE. [60-51-5]
FIFRA—Data Call In, 45 FR 75488, November 14, 1980
Status: letter issued 11/27/81
40 CFR 162.11, FIFRA-RPAR Completed
Current Status: PD 1 published, 42 FR 45806 (9/22/77); comment period closed 1/6/78.
PD 2/3 completed and Notice of Determination published, 44 FR 66558 (11/19/79); comment period closed 12/19/79: NTIS# PB80 213846.
PD 4 published, 46 FR 5334 (1/19/81)
Criteria Possibly Met or Exceeded: Oncogenicity, Mutagenicity, Fetotoxicity, and Reproductive Effects
*RCRA 40 CFR 261
EPA hazardous waste no. P044

DIMETHOXANE. [828-00-2]
IARC—carcinogenic in animals
TSCA-CHIP available

3,3'-DIMETHOXYBENZIDINE. [119-90-4]
*RCRA 40 CFR 261
EPA hazardous waste no. U091
OSHA Candidate List of Potential Occupational Carcinogens (EPA Carcinogen Assessment Group List), 45 FR 53672, August 12, 1980
IARC—carcinogenic in animals

3,3'-DIMETHOXYBENZIDINE-4,4'-DIISOCYANATE. [91-93-0]
NCI—carcinogenic in animals

DIMETHOXYMETHANE.
See Methylal. [109-87-5]

DIMETHYL ACETAMIDE. [127-19-5]
*OSHA 29 CFR 1910.1000 Table Z-1

Skin

PEL(TWA): 10 ppm, 35 mg/m^3
ACGIH-TLV

Skin

TWA: 10 ppm, 35 mg/m^3
STEL: 15 ppm, 50 mg/m^3

DIMETHYLAMINE.
See Methylamines. [124-40-3]
*OSHA 29 CFR 1910.1000 Table Z-1
PEL(TWA): 10 ppm, 18 mg/m^3
ACGIH-TLV
TWA: 10 ppm, 18 mg/m^3
*CWA 311(b)(2)(A), 40 CFR 116, 117
Discharge RQ = 1000 pounds (454 kilograms)
*RCRA 40 CFR 261
EPA hazardous waste no. U092 (hazardous substance, ignitable waste)

p-DIMETHYLAMINOAZOBENZENE. [60-11-7]
*OSHA Standard 29 CFR 1910.1015

Cancer-Suspect Agent
Shall not apply to solid or liquid mixtures containing less than 1.0% by weight or volume of 4-Dimethylaminoazobenzene.
OSHA Candidate List of Potential Occupational Carcinogens (EPA Carcinogen Assessment Group List), 45 FR 53672, August 12, 1980
IARC—carcinogenic in animals

4-DIMETHYLAMINOAZOBENZENE.
See p-Dimethylaminoazobenzene.
[60-11-7]

DIMETHYLAMINOBENZENE.
See Xylidene. [1300-73-8]

trans-2-[(DIMETHYLAMINO) METHYLIMINO]-5-[2-(5-NITRO-2-FURYL)VINYL]-1,3,4-OXADIAZOLE.
[55738-54-0]
IARC—carcinogenic in animals

DIMETHYLANILINE (n-DIMETHYLANILINE).
[121-69-7]
*OSHA 29 CFR 1910.1000 Table Z-1

Skin
PEL(TWA): 5 ppm, 25 mg/m^3
ACGIH-TLV

Skin
TWA: 5 ppm, 25 mg/m^3
STEL: 10 ppm, 50 mg/m^3

7,1,2-DIMETHYLBENZ[a] ANTHRACENE. [57-97-6]
OSHA Candidate List of Potential Occupational Carcinogens (EPA Carcinogen Assessment Group List), 45 FR 53672, August 12, 1980
*RCRA 40 CFR 261
EPA hazardous waste no. U094

DIMETHYLBENZENE. See Xylene.
[95-47-6, 108-38-3, 106-42-3]

m-DIMETHYLBENZENE.
See m-Xylene. [108-38-3]

o-DIMETHYLBENZENE.
See o-Xylene. [95-47-6]

p-DIMETHYLBENZENE. See p-Xylene.
[106-42-3]

3,3'-DIMETHYLBENZIDINE.
[119-93-7]
See σ-Tolidine, NIOSH Criteria Document.
OSHA Candidate List of Potential Occupational Carcinogens (EPA Carcinogen Assessment Group List), 45 FR 53672, August 12, 1980
IARC—carcinogenic in animals
*RCRA 40 CFR 261
EPA hazardous waste no. U095

α,α-DIMETHYLBENZYLHYDROPEROXIDE. [80-15-9]
*RCRA 40 CFR 261
EPA hazardous waste no. U096 (reactive waste)

DIMETHYLCARBAMOYL CHLORIDE (DMCC). [79-44-7]
NIOSH Current Intelligence Bulletin No. 11
OSHA Candidate List of Potential Occupational Carcinogens (EPA Carcinogen Assessment Group List), 45 FR 53672, August 12, 1980
IARC—carcinogenic in animals
*RCRA 40 CFR 261
EPA hazardous waste no. U097

DIMETHYL CARBAMYL CHLORIDE.
[79-44-7]
ACGIH-TLV
TWA: Industrial substance suspect of carcinogenic potential for humans—which is suspect of inducing cancer, based on either (1) limited epidemiologic evidence, exclusive of clinical reports of single cases, or (2) demonstration of carcinogenesis in one or more animal species by appropriate methods.
Worker exposure by all routes should be

carefully controlled to levels consistent with the animal and human experience data.

DIMETHYL-1, 2-DIBROMO-2,2-DICHLOROETHYL PHOSPHATE.
See naled. [300-76-5]

O,O-DIMETHYL-S-(1,2-DICARBOETHOXYETHYL)DITHIO-PHOSPHATE. [121-75-5]

See Malathion, NIOSH Criteria Document.

DIMETHYL ETHER. [115-10-6]

NFPA—hazardous chemical
health—2, flammability—4, reactivity—1

DIMETHYLFORMAMIDE. [68-12-2]
*OSHA 29 CFR 1910.1000 Table Z-1

Skin

PEL(TWA): 10 ppm, 30 mg/m³
ACGIH-TLV

Skin

TWA: 10 ppm, 30 mg/m³
STEL: 20 ppm, 60 mg/m³
TSCA-CHIP available

2,6-DIMETHYLHEPTANONE. See diisobutyl ketone. [108-83-8]

2,6-DIMETHYL-4-HEPTANONE. See diisobutyl ketone. [108-83-8]

1,1-DIMETHYLHYDRAZINE. [57-14-7]
*OSHA 29 CFR 1910.1000 Table Z-1

Skin

PEL(TWA): 0.5 ppm, 1 mg/m³
ACGIH-TLV

Skin

TWA: 0.5 ppm, 1 mg/m³
STEL: 1 ppm, 2 mg/m³
Industrial substance suspect of carcinogenic potential for humans—which is suspect of inducing cancer, based on either (1) limited epidemiologic evidence, exclusive of clinical reports of single cases, or (2) demonstration of carcinogenesis in one or more animal species by appropriate methods. Worker exposure by all routes should be carefully controlled to levels consistent with the animal and human experience data.

See Hydrazine, NIOSH Criteria Document.

OSHA Candidate List of Potential Occupational Carcinogens (EPA Carcinogen Assessment Group List), 45 FR 53672, August 12, 1980
IARC—carcinogenic in animals
*RCRA 40 CFR 261
EPA hazardous waste no. U098

1,2-DIMETHYLHYDRAZINE. Syn: dimethylhydrazine dihydrochloride. See hydrazine, 1,2-dimethyl-, dihydrochloride. [540-73-8]

OSHA Candidate List of Potential Occupational Carcinogens (EPA Carcinogen Assessment Group List), 45 FR 53672, August 12, 1980
IARC—carcinogenic in animals
*RCRA 40 CFR 261
EPA hazardous waste no. U099

N,N-DIMETHYL-4-((3-METHYLPHENYL)AZO)BENZENAMINE.
See aniline, N,N-dimethyl-p-(m-tolylazo)-.
[55-80-1]

3,3-DIMETHYL-1-(METHLYTHIO)-2-BUTANONE, O-[(METHYLAMINO)CARBONYL] OXIME. [1063-55-4]
*RCRA 40 CFR 261
EPA hazardous waste no. P045

O,O-DIMETHYL O-p-NITROPHENYL PHOSPHOROTHIOSATE. [298-00-0]

See Methyl parathion, NIOSH Criteria Document
*RCRA 40 CFR 261
EPA hazardous waste no. P071 (hazardous substance)

DIMETHYLNITROSAMINE. [62-75-9]

IARC—carcinogenic in animals (used as positive carcinogenic control by NCI)

CAA 112
Chemical proposed to be assessed as a Hazardous Air Pollutant by a House of Representatives bill to amend the CAA

*RCRA 40 CFR 261
EPA hazardous waste no. P082

3,3-DIMETHYL-2-OXETANONE. [1955-45-9]

NCI—carcinogenic in animals

α, α-DIMETHYLPHENETHYLAMINE. [122-09-8]

*RCRA 40 CFR 261
EPA hazardous waste no. P046

2,4-DIMETHYLPHENOL. [105-67-9]

CWA 304(a)(1), 45 FR 79318, November 28, 1980, Water Quality Criteria

Freshwater Aquatic Life

The available data for 2,4-dimethylphenol indicate that acute toxicity to freshwater aquatic life occurs at concentrations as low as 2120 µg/liter and would occur at lower concentrations among species that are more sensitive than those tested. No data are available concerning the chronic toxicity of dimethylphenol to sensitive freshwater acquatic life.

Saltwater Aquatic Life

No saltwater organisms have been tested with 2,4-dimethylphenol and no statement can be made concerning acute and chronic toxicity.

Human Health

Sufficient data are not available for 2,4-dimethylphenol to derive a level that would protect against the potential toxicity of this compound. Using available organoleptic data, for controlling undesirable taste and odor quality of ambient water, the estimated level is 400 µg/liter. It should be recognized that organoleptic data as a basis for establishing a water quality criterion have limitations and have no demonstrated relationship to potential adverse human health effects.

CWA 307(a)(1), Priority Pollutant

*RCRA 40 CFR 261
EPA hazardous waste no. U101 (hazardous substance)

DIMETHYLPHOSPHOROCHLORO-DITHIOATE. [2524-03-0]

TSCA-CHIP available

DIMETHYLPHTHALATE. [131-11-3]

*OSHA 29 CFR 1910.1000 Table Z-1
PEL(TWA): 5 mg/m^3
ACGIH-TLV
TWA: 5 mg/m^3
STEL: 10 mg/m^3
*RCRA 40 CFR 261
EPA hazardous waste no. U102

DIMETHYL SULFATE. [77-78-1]

*OSHA 29 CFR 1910.1000 Table Z-1

Skin

PEL(TWA): 1ppm, 5 mg/m^3
ACGIH-TLV

Skin

TWA: 0.1 ppm, 0.5 mg/m^3
Industrial substance suspect of carcinogenic potential for humans—which is suspect of inducing cancer, based on either (1) limited epidemiologic evidence, exclusive of clinical reports of single cases, or (2) demonstration of carcinogenesis in one or more animal species by appropriate methods. Worker exposure by all routes should be carefully controlled to levels consistent with the animal and human experience data.

OSHA Candidate List of Potential Occupational Carcinogens (EPA Carcinogen Assessment Group List), 45 FR 53672, August 12, 1980

IARC—carcinogenic in animals

NFPA—hazardous chemical
health—4, flammability—2, reactivity—0

*RCRA 40 CFR 261
EPA hazardous waste no. U103

DIMETHYL SULFIDE. [75-18-3]
NFPA—hazardous chemical
health—2, flammability—4, reactivity—0

DIMETHYLTHIOUREA. [534-13-4]
TSCA-CHIP available

DIMETHYLTIN BIS(ISOOCTYL MERCAPTOACETATE). [26636-01-1]
TSCA 4(e), ITC
Eleventh Report of the Interagency Testing Committee to the administrator, Environmental Protection Agency, November 1982; 47 FR 54626, December 3, 1982

*TSCA 8(d), Unpublished Health and Safety Studies Reporting, 40 CFR 716, 47 FR 54624, December 23, 1982

DIMILIN. [35367-38-5]
FIFRA—Data Call In, 45 FR 75488, November 14, 1980
Status: letter being drafted

DINASEB.
*RCRA 40 CFR 261
EPA hazardous waste no. P020

DINITOLMIDE. [148-01-6]
ACGIH-TLV
TWA: 5 mg/m^3
STEL: 10 mg/m^3

DINITRAMINE. [29091-05-2]
FIFRA—Data Call In, 45 FR 75488, November 14, 1980
Status: letter issued 8/28/81

2,4-DINITROANILINE. See Aniline and chloro-, bromo-, and/or nitro-anilines. [97-02-9]
NFPA—hazardous chemical
health—3, flammability—1, reactivity—3

2,4-DINITROBENZENAMINE. See 2,4-Dinitroaniline. [97-02-9]

DINITROBENZENE (all isomers).-[25154-54-5]
*OSHA 29 CFR 1910.1000 Table Z-1

Skin
PEL(TWA): 1 mg/m^3
ACGIH-TLV

Skin
TWA: 0.15 ppm, 1 mg/m^3
STEL: 0.5 ppm, 3 mg/m^3
*CWA 311(b)(2)(A), 40 CFR 116, 117
Discharge RQ = 1000 pounds (454 kilograms)

o-DINITROBENZENE. [528-29-0]
NFPA—hazardous chemical
health—3, flammability—1, reactivity—4

2,4-DINITROCHLOROBENZENE.
See 1-chloro-2,4-dinitrobenzene
[97-00-7]

DINITRO-o-CRESOL. [534-52-1]
*OSHA 29 CFR 1910.1000 Table Z-1

Skin
PEL(TWA): 0.2 mg/m^3
ACGIH-TLV

Skin
TWA: 0.2 mg/m^3
STEL: 0.6 mg/m^3
NIOSH Criteria Document (Pub. No. 78-131), NTIS Stock No. PB80 175870, February 1978
The criteria and recommendations for dinitro-*ortho*-cresol (DNOC) apply to exposure of employees to any DNOC ($C_7H_6N_2O_5$) isomer or to any of the salts of DNOC.
"Occupational exposure to DNOC," because of systemic effects, absorption through the skin on contact, and possible dermal irritation, is defined as work in any area where DNOC is manufactured, formulated, processed, stored, or otherwise used. The "action level" is defined as one-half the recommended time-weighted average (TWA) environmental limit. Adherence to all provisions of the standard is required if any employee is exposed to airborne DNOC at concentrations above the action level. If any employee is occupationally exposed at con-

centrations equal to or below the action level, then all sections of the recommended standard except Sections 4(c)(2) and 8(a) shall be complied with because adverse effects can be produced by skin and eye contact.

When skin exposure is prevented, occupational exposure to DNOC shall be controlled so that no employee is exposed to DNOC at a concentration greater than 0.2 mg/m^3 of air, determined as a time-weighted average (TWA) concentration for up to a 10-hour work shift and 40-hour workweek.

The standard is designed to protect the health and provide for the safety of employees for up to a 10-hour work shift, 40-hour workweek, over a working lifetime.

CWA 307 (a) (1), Priority Pollutant

4,6-DINITRO-o-CRESOL AND SALTS. [534-52-1]

*RCRA 40 CFR 261
EPA hazardous waste no. P047

4,6-DINITRO-o-CYCLOHEXYLPHENOL. [131-89-5]

*RCRA 40 CFR 261
EPA hazardous waste no. P034

1,2-DINITROETHANEDIOL. See Ethylene glyol dinitrate and/or nitroglycerin. [628-96-6]

2,4-DINITROPHENOL. [51-28-5]

TSCA-CHIP available

CWA 307(a)(1), Priority Pollutant

*CWA 311(b)(2)(A), 40 CFR 116, 117
Discharge RQ = 1000 pounds (454 kilograms)

*RCRA 40 CFR 261
EPA hazardous waste no. P048 (hazardous substance)

DINITROSOPENTAMETHYLENE-TETRAMINE. [101-25-7]

TSCA-CHIP available

N-N'-DINITROSOPIPERAZINE.

See piperazine, 1,4-dinitroso-. [140-79-4]

3,5-DINITRO-o-TOLUAMIDE.

See Dinitolmide. [148-01-6]

2,4-DINITROTOLUENE. [121-14-2]

*OSHA 29 CFR 1910.1000 Table Z-1

Skin

PEL(TWA): 1.5 mg/m^3
ACGIH-TLV

Skin

TWA: 1.5 mg/m^3
STEL: 5 mg/m^3

OSHA Candidate List of Potential Occupational Carcinogens (EPA Carcinogen Assessment Group List), 45 FR 53672, August 12, 1980

NFPA—hazardous chemical
health—3, flammability—1, reactivity—3

TSCA-CHIP available

CWA 304(a)(1), 45 FR 79318, November 28, 1980, Water Quality Criteria

Freshwater Aquatic Life

The available data for 2,4-dinitrotoluene indicate that acute and chronic toxicity to freshwater aquatic life occurs at concentrations as low as 330 and 230 µg/liter, respectively, and would occur at lower concentrations among species that are more sensitive than those tested.

Saltwater Aquatic Life

The available data for 2,4-dinitrotoluenes indicate that acute toxicity to saltwater aquatic life occurs at concentrations as low as 590 µg/liter and would occur at lower concentrations among species that are more sensitive than those tested. No data are available concerning the chronic toxicity of 2,4-dinitrotoluenes to sensitive saltwater aquatic life, but a decrease in algal cell numbers occurs at concentrations as low as 370 µg/liter.

Human Health

For the maximum protection of human health from the potential carcinogenic effects due to exposure of 2,4-dinitrotoluene through ingestion of contaminated water and contaminated aquatic organisms, the ambient water concentration should be zero based on the nonthreshold assumption for this chemical. However, zero level may not be attainable at the present time. Therefore, the levels that may result in incremental increase of cancer risk over the lifetime are estimated at 10^{-5}, 10^{-6}, and 10^{-7}. The corresponding criteria are 1.1 µg/liter, 0.11 µg/liter, and 0.011 µg/liter, respectively. If the above estimates are made for consumption of aquatic organisms only, excluding consumption of water, the levels are 91 µg/liter, 9.1 µg/liter, and 0.91 µg/liter, respectively. Other concentrations representing different risk levels may be calculated by use of the guidelines. The risk estimate range is presented for information purposes and does not represent an agency judgment on an "acceptable" risk level.

CWA 307(a)(1), Priority Pollutant
*CWA 311(b)(2)(A), 40 CFR 116, 117
Discharge RQ = 1000 pounds (454 kilograms)
*RCRA 40 CFR 261
EPA hazardous waste no. U105 (hazardous substance)

2,6-DINITROTOLUENE. [606-20-2]

CWA 307(a)(1), Priority Pollutant
*RCRA 49 CFR 261
EPA hazardous waste no. U106 (hazardous substance)

DINOCAP. [6119-92-2]

FIFRA—Data Call In, 45 FR 75488, November 14, 1980
Status: letter issued 9/22/81

DIOCTYL PHTHALATE.
Syn: Di-2-ethylhexylphthalate. [117-81-7]

*OSHA 29 CFR 1910.1000 Table Z-1
PEL(TWA): 5 mg/m^3
ACGIH-TLV
TWA: 5 mg/m^3
STEL: 10 mg/m^3
IARC—carcinogenic in animals
NFPA—hazardous chemical (flammable chemical)
health—0, flammability—1, reactivity—0
*RCRA 40 CFR 261
EPA hazardous waste no. U107

DIOXANE (DIETHYLENE DIOXIDE, 1,4-DIOXANE). [123-91-1]

*OSHA 29 CFR 1910.1000 Table Z-1

Skin

PEL(TWA): 100 ppm, 360 mg/m^3
ACGIH-TLV

Skin

TWA: 25 ppm, 90 mg/m^3
STEL: 100 ppm, 360 mg/m^3
TSCA-CHIP available

NIOSH Criteria Document (Pub. No. 77-226), NTIS Stock No. PB 274810, September 1, 1977
Synonyms for dioxane include p-dioxane, 1,4-dioxane, diethylene dioxide, and glycol ethylene ether.

Because of the carcinogenic action of dioxane and its ability to penetrate the skin, "occupational exposure to dioxane" is defined as any work in workplaces where dioxane is handled, manufactured, or otherwise used, except where it is present as an unintentional contaminant in other chemical substances at less than 1% by weight or where it is only stored in leak-proof containers.

Occupational exposure to dioxane shall be controlled so that employees are not exposed at airborne concentrations greater than 1 ppm, (3.6 mg/m^3) based on a 30-minute sampling period.

The standard is designed to protect the health and provide for the safety of employees for up to a 10-hour workday, 40-hour workweek, during their working lifetime.

OSHA Candidate List of Potential Occupational Carcinogens (EPA Carcinogen Assessment Group List), 45 FR 53672, August 12, 1980
IARC—carcinogenic in animals
NCI—carcinogenic in animals
NFPA—hazardous chemical
health—2, flammability—3, reactivity—1

*RCRA 40 CFR 261
EPA hazardous waste no. U108

m-DIOXAN-4-OL, 2,6-DIMETHYL-, ACETATE.   Syn: Acetomethoxan.
[828-00-2]

OSHA Candidate List of Potential Occupational Carcinogens, 45 FR 53672, August 12, 1980

DIOXANTHION.   [78-34-2]

ACGIH-TLV

Skin

TWA: 0.2 mg/m^3

FIFRA—Registration Standard Development—Data Collection (insecticide)

DIOXIN.
See 2,3,7,8-Tetrachlorodibenzo-p-dioxin.
[1746-01-6]

1,3-DIOXOLONE.   [646-06-0]

TSCA 5(e), ITC

Eleventh Report of the Interagency Testing Committee to the administrator, Environmental Protection Agency, November 1982; 47 FR 54626, December 3, 1982

*TSCA 8(d), Unpublished Health and Safety Studies Reporting, 40 CFR 716, 47 FR 54624, December 3, 1982

DIOXOTHION.   [78-34-2]

FIFRA—Data Call In, 45 FR 75488, November 14, 1980
Status: agency decision reached

DIPHENYL.   See Biphenyl.   [92-52-4]

DIPHENYLAMINE.   [122-39-4]

ACGIH-TLV
TWA: 10 mg/m^3
STEL: 20 mg/m^3

DIPHENYLAMINE, N-NITROSO-,
Syn: Diphenylnitrosoamine.   [80-30-6]

OSHA Candidate List of Potential Occupational Carcinogens, 45 FR 53672, August 12, 1980

DIPHENYLAMINE, 4-NITROSO-.
Syn: p-Nitroso-N-phenylaniline.
[156-10-5]

OSHA Candidate List of Potential Occupational Carcinogens, 45 FR 53672, August 12, 1980

1,2-DIPHENYLHYDRAZINE.
[122-66-7]

OSHA Candidate List of Potential Occupational Carcinogens, (EPA Carcinogen Assessment Group List), 45 FR 53672, August 12, 1980

NCI—carcinogenic in animals

CWA 304(a)(1), 45 FR 79319, November 28, 1980, Water Quality Criteria

Freshwater Aquatic Life

The available data for 1,2-diphenylhydrazine indicate that acute toxicity to freshwater aquatic life occurs at concentrations as low as 270 µg/liter and would occur at lower concentrations among species that are more sensitive than those tested. No data are available concerning the chronic toxicity of 1,2-diphenylhydrazine to sensitive freshwater life.

Saltwater Aquatic Life

No saltwater organisms have been tested with 1,2-diphenylhydrazine and no statement can be made concerning acute and chronic toxicity.

Human Health

For the maximum protection of human health from the potential carcinogenic effects due to exposure of 1,2-diphenylhydrazine through ingestion of contaminated water and contaminated aquatic organisms, the ambient water concentration should be zero based on the nonthreshold assumption for this chemical. However, zero level may not be attainable at the present time. Therefore, the levels that may result in incremental increase of cancer risk over the lifetime are estimated at 10^{-5}, 10^{-6}, and 10^{-7}. The corresponding criteria are 422 ng/liter, 42 ng/liter, and 4 ng/liter, respectively. If the above estimates are made for consumption of aquatic organisms only, excluding consumption of water, the levels are 5.6 µg/liter, 0.56 µg/liter, 0.056 µg/liter, respectively. Other concentrations representing different risk levels may be calculated by use of the guidelines. The risk estimate range is presented for information purposes and does not represent an agency judgment on an "acceptable" risk level.

CWA 307(a)(1), Priority Pollutant
*RCRA 40 CFR 261
EPA hazardous waste no. U109

DIPHENYLMETHANE DIISOCYANATE.
See Methylene biphenyl isocyanate (MDI).
[101-68-8]
See Diisocyanates, NIOSH Criteria Document.

DIPHENYLNITROSAMINE.
See Diphenylamine, N-nitroso-.
[86-30-6]

DIPHOSPHORAMIDE, OCTAMETHYL-.
*RCRA 40 CFR 261
EPA hazardous waste no. P085

DIPROPETRYN. [4147-51-7]
FIFRA—Data Call In, 45 FR 75488, November 14, 1980
Status: letter being drafted

DIPROPYLAMINE. [142-84-7]
*RCRA 40 CFR 261
EPA hazardous waste no. U110 (ignitable waste)

DIPROPYLENE GLYCOL. [110-98-5]
NFPA—hazardous chemical (flammable chemical)
health—0, flammability—1, reactivity—0

DIPROPYLENE GLYCOL METHYL ETHER. [34590-94-8]
*OSHA 29 CFR 1910.1000 Table Z-1

Skin

PEL(TWA): 100 ppm, 600 mg/m^3

ACGIH-TLV
TWA: 100 ppm, 600 mg/m^3
STEL: 150 ppm, 900 mg/m^3

DIPROPYL KETONE. [123-19-3]
ACGIH-TLV
TWA: 50 ppm, 235 mg/m^3

DI-n-PROPYLNITROSAMINE.
[621-64-7]
*RCRA 40 CFR 261
EPA hazardous waste no. U111
OSHA Candidate List of Potential Occupational Carcinogens, (EPA Carcinogen Assessment Group List), 45 FR 53672, August 12, 1980

DIQUAT. [85-00-7, 2764-72-9]
ACGIH-TLV
TWA: 0.5 mg/m^3
STEL: 1 mg/m^3
*CWA 311(b)(2)(A), 40 CFR 116, 117
Discharge RQ = 1000 pounds (454 kilograms)
*RCRA 40 CFR 261
EPA hazardous waste no. P039 (hazardous substance)

DISULFIRAM. [97-77-8]
ACGIH-TLV
TWA: 2 mg/m^3
STEL: 5 mg/m^3

DISULFOTON. [298-04-4]
ACGIH-TLV
TWA: 0.1 mg/m^3
STEL: 0.3 mg/m^3
FIFRA—Data Call In, 45 FR 75488, November 14, 1980
Status: agency decision reached
*CWA 311(b)(2)(A), 40 CFR 116, 117
Discharge RQ = 1 pound (0.454 kilogram)

2,6-DITERT-BUTYL-p-CRESOL.
[128-37-0]
ACGIH-TLV
TWA: 10 mg/m^3
STEL: 20 mg/m^3

2,4-DITHIOBIURET. [541-53-7]

*RCRA 40 CFR 261
EPA hazardous waste no. P049

DITHIOPYROPHOSPHORIC ACID, TETRAETHYL ESTER. [3689-24-5]

*RCRA 40 CFR 261
EPA hazardous waste no. P109

DITHRANOL. [480-22-8]

IARC—carcinogenic in animals

DIURON. [330-54-1]

ACGIH-TLV
TWA: 10 mg/m^3
FIFRA—Registration Standard Development—Data Evaluation and Development of Regulatory Position (herbicide)
*CWA 311(b)(2)(A), 40 CFR 116, 117
Discharge RQ = 100 pounds (45.4 kilograms)

DIVINYL BENZENE. [108-57-6]

ACGIH-TLV
TWA: 10 ppm, 50 mg/m^3
NFPA—hazardous chemical
health—2, flammability—2, reactivity—2

DIVINYL ETHER. [109-93-3]

NFPA—hazardous chemical
health—2, flammability—3, reactivity—2

DIURON. [330-54-1]

DMS. See Dimethyl sulfate. [75-18-3]

DNT. See Dinitrotoluene. [121-41-2]

1-DODECANETHIOL. [112-55-0]
See Thiols, NIOSH Criteria Document.

DODECANOYL PEROXIDE.
See Dilauroyl peroxide. [105-74-8]

DODECYLBENZENESULFONIC ACID. [27176-87-0]

*CWA 311(b)(2)(A), 40 CFR 116, 117
Discharge RQ = 1000 pounds (454 kilograms)

DODINE. [2439-10-3]

FIFRA—Data Call In, 45 FR 75488, November 14, 1980
Status: letter issued 8/12/81

DSMA. [144-21-8] See Methanearsonates.

FIFRA—Data Call In, 45 FR 75488, November 14, 1980
Status: letter being drafted

E

EBDC. *See* Ethylene bis dithiocarbamate. Syn: Maneb, mancozeb, metiram, nabam, zineb, amobam. [142-59-6, 3566-10-7]
40 CFR 162.11, FIFRA-RPAR issued
Current Status: PD 1 published, 42 FR 40618 (8/10/77); comment period closed 12/23/77.
PD 2/3 in agency review.
Criteria Possibly Met or Exceeded: Oncogenicity, Teratogenicity, and Hazard to Wildlife

EDTA. [60-00-4]
*CWA 311(b)(2)(A), 40 CFR 116, 117
Discharge RQ = 5000 pounds (2270 kilograms)

EGT. [2514-53-6]
FIFRA—Data Call In, 45 FR 75488, November 14, 1980
Status: letter being drafted

EMERY. [112-62-9]
ACGIH-TLV
TWA: Nuisance particulate, 30 mppcf, or 10 mg/m^3 of total dust < 1% quartz, or 5 mg/m^3 respirable dust
STEL: 20 mg/m^3

ENDOSULFAN. [115-29-7]
ACGIH-TLV

Skin
TWA: 0.1 mg/m^3
STEL: 0.3 mg/m^3
FIFRA—Registration Standard Development—Data Evaluation and Development of Regulatory Position (insecticide)
CWA 304(a)(1), 45 FR 79319, November 28, 1980, Water Quality Criteria

Freshwater Aquatic Life

For endosulfan the criterion to protect freshwater aquatic life as derived using the guidelines is 0.056 μg/liter as a 24-hour average and the concentration should not exceed 0.22 μg/liter at any time.

Saltwater Aquatic Life

For endosulfan the criterion to protect saltwater aquatic life as derived using the guidelines is 0.0087 μg/liter as a 24-hour average and the concentration should not exceed 0.034 μg/liter at any time.

Human Health

For the protection of human health from the toxic properties of endosulfan ingested through water and contaminated aquatic organisms, the ambient water criterion is determined to be 74 μg/liter. For the protection of human health from the toxic properties of endosulfan ingested through contaminated aquatic organisms alone, the ambient water criterion is determined to be 159 μg/liter.
*CWA 311(b)(2)(A), 40 CFR 116, 117
Discharge RQ = 1 pound (0.454 kilogram)
*RCRA 40 CFR 261
EPA hazardous waste no. P050 (hazardous substance)

ENDOSULFAN, ALPHA and BETA. Syn: 6,7,8,9,10,10-Hexachloro-1,5,5a,6,9,9a-hexahydro-6,9-methano-2,4,3,-benzodioxathiepin 3-oxide. [115-29-7]
CWA 307(a)(1), Priority Pollutant

ENDOSULFAN SULFITE.
CWA 307(a)(1), Priority Pollutant

ENDOTHALL. [145-73-3]

*RCRA 40 CFR 261
EPA hazardous waste no. P088

ENDRIN (1,2,3,4,10,10-HEXACHLORO-6,7-EPOXY-1,4,4a,5,6,7,8,8a-OCTAHYDRO-1,4-ENDO,ENDO-5,8-DIMETHANONAPHTHALENE. [72-20-8]

*OSHA 29 CFR 1910.1000 Table Z-1

Skin

PEL(TWA): 0.1 mg/m^3

ACGIH-TLV

Skin

TWA: 0.1 mg/m^3
STEL: 0.3 mg/m^3

40 CFR 162.11, FIFRA-RPAR Completed Current Status: PD 1 published, 41 FR 31316 (7/26/76); comment period closed 11/4/76: NTIS# PB81 112690.
PF 2/3 completed and Notice of Proposed Determination published, 43 FR 51132 (11/2/78).
PD 4 published (including cancellations) 44 FR 43631 (7/25/79) NTIS# PB81 10980. PD 4 follow-up notice mailed 8/28/80, cancellations began 30 days later. Some uses are conditionally registered because requirements have been met. Herb Harrison, Registration Division (557-7020), is contact point for inquiries.
Criteria Possibly Met or Exceeded: Oncogenicity, Teratogenicity, and Reduction in Endangered Species and Nontarget Species
FIFRA—Data Call In, 45 FR 75488, November 14, 1980
Status: letter issued 12/7/81, no additional chronic toxicological data required at this time.
NFPA—hazardous chemical
when dry: health—2, flammability—0, reactivity—0
when solutions: health—3, flammability—1, reactivity—0
CWA 304(a)(1), 45 FR 79318, November 28, 1980, Water Quality Criteria

Freshwater Aquatic Life

For endrin the criterion to protect freshwater aquatic life as derived using the guidelines is 0.0023 µg/liter as a 24-hour average and the concentration should not exceed 0.18 µg/liter at any time.

Saltwater Aquatic Life

For endrin the criterion to protect saltwater aquatic life as derived using the guidelines is 0.0023 µg/liter as a 24-hour average and the concentration should not exceed 0.037 µg/liter at any time.

Human Health

The ambient water quality criterion for endrin is recommended to be identical to the existing drinking water standard which is 1 µg/liter. Analysis of the toxic effects data resulted in a calculated level that is protective of human health against the ingestion of contaminated water and contaminated aquatic organisms. The calculated value is comparable to the present standard. For this reason a selective criterion based on exposure solely from consumption of 6.5 g of aquatic organisms was not derived.

CWA 307(a)(1), Priority Pollutant
*CWA 311(b)(2)(A), 40 CFR 116, 117
Discharge RQ = 1 pound (0.454 kilogram)
*National Interim Primary Drinking Water Regulations, 40 CFR 141; 40 FR 59565, December 24, 1975; amended by 41 FR 28402, July 9, 1976; 44 FR 68641, November 29, 1979; corrected by 45 FR 15542, March 11, 1980; 45 FR 57342, August 27, 1980; 47 FR 18998, March 3, 1982; corrected by 47 FR 10998, March 12, 1982
Maximum contaminant level—0.0002 mg/liter
*RCRA 40 CFR 261
EPA hazardous waste no. P051 (hazardous substance)

ENDRIN ALDEHYDE. See Endrin. [72-20-8]

ENFLURANE. [13838-16-9]

See Waste anesthetic gases and vapors, NIOSH Criteria Document.

ENTRAMIN. See Thiazole, 2-amino-5-nitro-. [121-66-4]

EPI. See Chloromethyloxirane.

EPICHLOROHYDRIN. [106-89-8]

*OSHA 29 CFR 1910.1000 Table Z-1

Skin

PEL(TWA): 5 ppm, 19 mg/m^3

ACGIH-TLV

Skin

TWA: 2 ppm, 10 mg/m^3
STEL: 5 ppm, 20 mg/m^3

NIOSH Criteria Document (Pub. No. 76-206), NTIS Stock No. PB81 227019, September 17, 1976
These criteria and the recommended standard apply to exposure of employees to solid, liquid, or gaseous 3-chloro-1,2-epoxypropane, hereafter referred to as "epichlorohydrin," either alone or in combination with other substances. Synonyms for epichlorohydrin include 1-chloro-2,3-epoxypropane, (chloromethyl)ethylene oxide, chloromethyloxirane, chloropropylene oxide, gamma-chloropropylene oxide, 3-chloro-1,2-propylene oxide, 2-chloromethyloxirane, alpha-epichlorohydrin, epichlorhydrin, and epichlorophydrin.

"Occupational exposure to epichlorohydrin" is defined as exposure to airborne epichlorohydrin at concentrations exceeding the action level. The "action level" is defined as one-half of the recommended time-weighted average (TWA) environmental concentration for epichlorohydrin. Exposure to airborne epichlorohydrin at concentrations equal to or less than one-half of the workplace environmental limit, as determined in accordance with Section 8, will require adherence to Sections 2, 3, 4(a), 5(a,b,c,d), 6, and 7.

"Occupational exposure to epichlorohydrin shall be controlled so that employees are not exposed to epichlorohydrin at concentrations greater than 2 mg/m^3 of air (approximately 0.5 parts per million parts of air by volume) determined as a TWA concentration for up to a 10-hour workday, 40-hour workweek, with a ceiling concentration of 19 mg/m^3 of air (approximately 5 parts per million parts of air by volume) as determined by a sampling time of 15 minutes. The standard is designed to protect the health and safety of employees for up to a 10-hour workday, 40-hour workweek, over a working lifetime.

NIOSH Current Intelligence Bulletin No. 30—cancer-related bulletin

OSHA Candidate List of Potential Occupational Carcinogens (EPA Carcinogen Assessment Group List), 45 FR 53672, August 12, 1980

IARC—carcinogenic in animals

NFPA—hazardous chemical
health—3, flammability—3, reactivity—2

*TSCA 8(a), 40 CFR 712, 47 FR 26992, June 22, 1982, Chemical Information Rule

*CWA 311(b)(2)(A), 40 CFR 116, 117
Discharge RQ = 1000 pounds (454 kilograms)

CAA 112
Chemical proposed to be assessed as a Hazardous Air Pollutant by a House of Representatives bill to amend the CAA

*RCRA 40 CFR 261
EPA hazardous waste no. U041

EPINEPHRINE. [51-43-4]

*RCRA 40 CFR 261
EPA hazardous waste no. P042

EPN (ETHYL p-NITRO-PHENYL THIO-NOBENZENEPHOSPHONATE). [2014-64-5]

*OSHA 29 CFR 1910.1000 Table Z-1

Skin

PEL(TWA): 0.5 mg/m^3

ACGIH-TLV

Skin

TWA: 0.5 mg/m^3
STEL: 2 mg/m^3

40 CFR 162.11, FIFRA-RPAR issued
Current Status: PD 1 published, 44 FR 54384 (9/19/79); comment period closed 12/29/79: NTIS# PB80 216815.

PD 2/3 in agency review.
Criteria Possibly Met or Exceeded: Neurotoxicity

1-EPOXYETHYL-3,4-EPOXYCYCLOHEXANE. [106-87-6]

IARC—carcinogenic in animals

3,4-EPOXY-6-METHYLCYCLOHEXYLMETHYL-3,4-EPOXY-6-METHYL CYCLOHEXANE CARBOXYLATE. [141-37-7]

IARC—carcinogenic in animals

1,2-EPOXYPROPANE. See Propylene oxide. [75-56-9]

2,3-EPOXY-1-PROPANOL. See Glycidol. [556-52-5]

EPTC. [759-94-4]

FIFRA—Data Call In, 45 FR 75488, November 14, 1980
Status: letter issued 10/29/81
FIFRA—Registration Standard Development—Data Collection (herbicide)

ERBON. [50-28-2]

40 CFR 162.11, FIFRA-RPAR, Final Action, Voluntary Cancellation
Current Status: 45 FR 58770 (10/4/80)
Criteria Possibly Met or Exceeded: Oncogenicity, Teratogenicity, and Fetotoxicity

ESTRADIOL MUSTARD. [22966-79-6]

IARC—carcinogenic in animals
NCI—carcinogenic in animals

EHTANAL. See Acetaldehyde. [75-07-0]

ETHANAMINE, 2,2-DIMETHOXY-N-METHYL-. [122-07-6]

*TSCA 8(a), 40 CFR 712, 47 FR 26992, June 22, 1982, Chemical Information Rule

ETHANAMINE, 1,1-DIMETHYL-2-PHENYL-.

*RCRA 40 CFR 261
EPA hazardous waste no. P046

EHTANAMINE, N-ETHYL-N-NITROSO-. [55-18-5]

*RCRA 40 CFR 261
EPA hazardous waste no. U174

ETHANAMINE, N-METHYL-N-NITROSO-. [10595-95-6]

*RCRA 40 CFR 261
EPA hazardous waste no. P084

ETHANE. [74-84-0]

ACGIH-TLV
TWA: Asphyxiant—inert gas or vapors, when present in high concentrations in air, act primarily as simple asphyxiants without other significant physiologic effects

ETHANE, 1,2-BIS(CHLOROMETHOXY)-. [13483-18-6]

OSHA Candidate List of Potential Occupational Carcinogens, 45 FR 53672, August 12, 1980

1,2-ETHANEDIAMINE, N-(2-AMINOETHYL)-. [111-40-0]

*TSCA 8(a), 40 CFR 712, 47 FR 26992, June 22, 1982, Chemical Information Rule

ETHANE, 1,2-DIBROMO-. [106-93-4]

*RCRA 40 CFR 261
EPA hazardous waste no. U067 (hazardous substance)

ETHANE, 1,1-DICHLORO-. [1300-21-6]

*RCRA 40 CFR 261
EPA hazardous waste no. U076

ETHANE, 1,2-DICHLORO-. [107-06-2]

*RCRA 40 CFR 261
EPA hazardous waste no. U077 (hazardous substance)

ETHANE, 1,1-DICHLORO-2,2-BIS(p-CHLOROPHENYL)-. Syn: TDE. See DDT and metabolites. [72-54-8]

OSHA Candidate List of Potential Occupational Carcinogens, 45 FR 53672, August 12, 1980

*CWA 311(b)(2)(A), 40 CFR 116, 117
Discharge RQ = 1 pound (0.454 kilogram)

ETHANE, 1,1-DICHLORO-2,2-BIS(p-ETHYLPHENYL)-. See DDE. [72-56-0]

ETHANE, 1,1-DIMETHOXY-. [5341-51-6]

*TSCA 8(a), 40 CFR 712, 47 FR 26992, June 22, 1982, Chemical Information Rule

ETHANEDIOIC ACID. See Oxalic acid. [144-62-7]

1,2-ETHANEDIYLBISCARBAMODITHIOIC ACID.

*RCRA 40 CFR 261
EPA hazardous waste no. U114

ETHANE, 1,1,1,2,2,2-HEXACHLORO- [67-72-1]

*RCRA 40 CFR 261
EPA hazardous waste no. U131

ETHANE, 1,1'-[METHYLENEBIS(OXY)]BIS[2-CHLORO-]. [111-91-1]

*RCRA 40 CFR 261
EPA hazardous waste no. U024

ETHANENITRILE. [75-05-8]

*RCRA 40 CFR 261
EPA hazardous waste no. U003 (hazardous substance, ignitable waste, toxic waste)

ETHANE, 1,1'-OXYBIS-. [60-29-7]

*RCRA 40 CFR 261
EPA hazardous waste no. U117 (ignitable waste)

ETHANE, 1,1'-OXYBIS(2-CHLORO). [111-44-4]

*RCRA 40 CFR 261
EPA hazardous waste no. U025

ETHANE, PENTACHLORO-. [76-01-7]

*RCRA 40 CFR 261
EPA hazardous waste no. U184

ETHANE, 1,1,1,2-TETRACHLORO-. [630-20-6]

*RCRA 40 CFR 261
EPA hazardous waste no. U208

ETHANE, 1,1,2,2-TETRACHLORO-. [79-34-5]

*RCRA 40 CFR 261
EPA hazardous waste no. U209

ETHANETHIOAMIDE. [62-55-5]

*RCRA 40 CFR 261
EPA hazardous waste no. U218

ETHANETHIOL. See Ethyl mercaptan. [75-08-1]

See Thiols, NIOSH Criteria Document.

ETHANE, 1,1,1-TRICHLORO-. [71-55-6]

*TSCA 8(a), 40 CFR 712, 47 FR 26992, June 22, 1982, Chemical Information Rule

ETHANE, 1,1,2-TRICHLORO-. [79-00-5]

*RCRA 40 CFR 261
EPA hazardous waste no. U227

ETHANOLAMINE. [102-71-6, 111-42-2, 141-43-5]

*OSHA 29 CFR 1910.1000 Table Z-1
PEL(TWA): 3 ppm, 6 mg/m^3

NFPA—hazardous chemical
health—2, flammability—2, reactivity—0

ACGIH-TLV
TWA: 3 ppm, 8 mg/m^3
STEL: 6 ppm, 15 mg/m^3
TSCA-CHIP available

ETHANOL, 2-CHLORO-, PHOSPHITE(3:1). [140-08-9]

*TSCA 8(a), 40 CFR 712, 47 FR 26992, June 22, 1982, Chemical Information Rule

ETHANOL, 2-(2,4-DIAMINO-PHENOXY)-DIHYDROCHLORIDE. See Phenylenediamines. [66422-95-5]

TSCA 4(e), ITC
Sixth Report of the Interagency Testing Committee to the administrator, Environmental Protection Agency, April 1980; 45 FR 35897, May 28, 1980, responded to by the EPA administrator 47 FR 973, January 8, 1982

*TSCA 8(d) Unpublished Health and Safety Studies Reporting, 40 CFR 716, 47 FR 38800, September 2, 1982; 48 FR 13178, March 30, 1983

ETHANOL, 1,2-DIBROMO-, ACETATE. [24442-57-7]

*TSCA 8(a), 40 CFR 712, 47 FR 26992, June 22, 1982, Chemical Information Rule

ETHANOL, 2-HYDRAZINO-. Syn: beta-Hydroxyethylhydrazine. [109-84-2]

OSHA Candidate List of Potential Occupational Carcinogens, 45 FR 53672, August 12, 1980

ETHANOL, 2,2'-(NITROSOIMINO)BIS-. [1116-54-7]

*RCRA 40 CFR 261
EPA hazardous waste no. U173

ETHANONE, 1-PHENYL-. [98-86-2]

*RCRA 40 CFR 261
EPA hazardous waste no. U004

ETHANOYL CHLORIDE. [75-36-5]
*RCRA 40 CFR 261
EPA hazardous waste no. U006 (hazardous substance, corrosive waste, reactive waste, toxic waste)

ETHENE. [74-85-1]
NFPA—hazardous chemical
health—1, flammability—4, reactivity—2

ETHENE, BROMO-. [593-60-2]
*TSCA 8(a), 40 CFR 712, 47 FR 26992, June 22, 1982, Chemical Information Rule

ETHENE, CHLORO-. [75-01-4]
*RCRA 40 CFR 261
EPA hazardous waste no. U043

ETHENE, 2-CHLOROETHOXY-.
*RCRA 40 CFR 261
EPA hazardous waste no. U042

ETHENE, 1,1-DICHLORO-. [75-35-4]
*RCRA 40 CFR 261
EPA hazardous waste no. U078 (hazardous substance)

ETHENE, *trans*-1,2-DICHLORO-. [156-60-5]
*RCRA 40 CFR 261
EPA hazardous waste no. U079

ETHENE, 1,1-DIFLUORO-. [75-38-7]
*TSCA 8(a), 40 CFR 712, 47 FR 26992, June 22, 1982, Chemical Information Rule

ETHENE, 1,1,2,2-TETRACHLORO-. [127-18-4]
*RCRA 40 CFR 261
EPA hazardous waste no. U210

ETHER. *See* Diethyl ether. [60-29-7]

ETHER, 2,4-DICHLOROPHENYL *p*-NITROPHENYL. Syn: 2,4-Dichloro-1-(4-nitrophenoxy)benzene. [1836-75-5]
OSHA Candidate List of Potential Occupational Carcinogens, 45 FR 53672, August 12, 1980

ETHINYLESTRADIOL. [57-63-6]
IARC—carcinogenic in animals

ETHION. [563-12-2]
ACGIH-TLV

Skin

TWA: 0.4 mg/m^3
FIFRA—Registration Standard Development—Data Collection
FIFRA—Data Call In, 45 FR 75488, November 14, 1980
Status: letter issued 6/18/81
*CWA 311(b)(2)(A), 40 CFR 116, 117
Discharge RQ = 10 pounds (4.54 kilograms)

ETHIONAMIDE.
IARC—carcinogenic in animals
NCI—carcinogenic in animals

ETHIONINE. *See* Butyric acid, 2-amino-4-(ethylthio)-, L-. [13073-35-3]

DL-**ETHIONINE.** *See* Butyric acid, 2-amino-4-(ethylthio)-, DL-. [67-21-0]

ETHOPROP. [13194-48-4]
FIFRA—Data Call In, 45 FR 75488, November 14, 1980
Status: agency decision reached

2-ETHOXYETHANOL. [110-80-5]
*OSHA 29 CFR 1910.1000 Table Z-1

Skin

PEL(TWA): 200 ppm, 740 mg/m^3

ACGIH-TLV

Skin

TWA: 50 ppm, 185 mg/m^3
STEL: 100 ppm, 370 mg/m^3
Intended changes for 1982 (1982 Revision or Addition):

Skin

TWA: 5 ppm, 19 mg/m^3

NIOSH Current Intelligence Bulletin No. 39—cancer-related bulletin

TSCA-CHIP available

2-ETHOXYETHYLACETATE (CELLOSOLVE ACETATE). [111-15-9]
*OSHA 29 CFR 1910.1000 Table Z-1

Skin

PEL(TWA): 100 ppm, 540 mg/m^3

ACGIH-TLV

Skin

TWA: 50 ppm, 270 mg/m^3
STEL: 100 ppm, 540 mg/m^3
Intended changes for 1982 (1982 Revision or Addition):

Skin

TWA: 5 ppm, 27 mg/m^3

TSCA-CHIP available

ETHOXYQUIN. [91-53-2]
FIFRA—Registration Standard—Issued March 1981, NTIS# 540/RS 81-0002 (growth regulator)

ETHYL ACETATE. [141-78-6]
*OSHA 29 CFR 1910.1000 Table Z-1
PEL(TWA): 400 ppm, 1400 mg/m^3

ACGIH-TLV

TWA: 400 ppm, 1400 mg/m^3

NFPA—hazardous chemical (flammable chemical)
health—1, flammability—3, reactivity—0

*RCRA 40 CFR 261
EPA hazardous waste no. U112 (ignitable waste)

ETHYL ACRYLATE. [140-88-5]
*OSHA 29 CFR 1910.1000 Table Z-1

Skin

PEL(TWA): 25 ppm, 100 mg/m^3

ACGIH-TLV

Skin

TWA: 5 ppm, 20 mg/m^3
STEL: 25 ppm, 100 mg/m^3

NFPA—hazardous chemical
health—2, flammability—3, reactivity—2

TSCA-CHIP available

*RCRA 40 CFR 261
EPA hazardous waste no. U113 (ignitable waste)

ETHYL ALCOHOL (ETHANOL). [46-17-5]
*OSHA 29 CFR 1910.1000 Table Z-1
PEL(TWA): 1000 ppm, 1900 mg/m^3

ACGIH-TLV

TWA: 1000 ppm, 1900 mg/m^3

NFPA—hazardous chemical (flammable chemical)
health—0, flammability—3, reactivity—0

ETHYLAMINES. [75-04-7, 109-89-7, 121-44-8]
*OSHA 29 CFR 1910.1000 Table Z-1
PEL(TWA): 10 ppm, 18 mg/m^3

ACGIH-TLV

TWA: 10 ppm, 18 mg/m^3

NFPA—hazardous chemical
health—3, flammability—4, reactivity—0

TSCA-CHIP available

ETHYL-sec-AMYL KETONE (5-METHYL-3-HEPTANONE). [41-85-5]
*OSHA 29 CFR 1910.1000 Table Z-1
PEL(TWA): 25 ppm, 130 mg/m^3

ACGIH-TLV

TWA: 25 ppm, 130 mg/m^3

N-ETHYLANILINE. [103-69-5]

NFPA—hazardous chemical
health—3, flammability—2, reactivity—0

ETHYLBENZENE. [100-41-4]

*OSHA 29 CFR 1910.1000 Table Z-1
PEL(TWA): 100 ppm, 435 mg/m^3

ACGIH-TLV
TWA: 100 ppm, 435 mg/m^3
STEL: 125 ppm, 545 mg/m^3

NFPA—hazardous chemical
health—2, flammability—3, reactivity—0
CWA 304 (a)(1), 45 FR 79318, November 28, 1980 Water Quality Criteria

Freshwater Aquatic Life

The available data for ethylbenzene indicate that acute toxicity to freshwater aquatic life occurs at concentrations as low as 32,000 µg/liter and would occur at lower concentrations among species that are more sensitive than those tested. No definitive data are available concerning the chronic toxicity of ethylbenzene to sensitive freshwater aquatic life.

Saltwater Aquatic Life

The available data for ethylbenzene indicate that acute toxicity to saltwater aquatic life occurs at concentrations as low as 430 µg/liter and would occur at lower concentrations among species that are more sensitive than those tested. No data are available concerning the chronic toxicity of ethylbenzene to sensitive saltwater aquatic life.

Human Health

For the protection of human health from the toxic properties of ethylbenzene ingested through water and contaminated aquatic organisms, the ambient water criterion is determined to be 1.4 mg/liter.

For the protection of human health from the toxic properties of ethylbenzene ingested through contaminated aquatic organisms alone, the ambient water criterion is determined to be 3.28 mg/liter.

CWA 307(a)(1), Priority Pollutant
*CWA 311(b)(2)(A), 40 CFR 116, 117
Discharge RQ = 1000 pounds (454 kilograms)

ETHYL BENZOATE. [93-89-0]

NFPA—hazardous chemical (flammable chemical)
health—1, flammability—1, reactivity—0

ETHYL BROMIDE. [74-96-4]

*OSHA 29 CFR 1910.1000 Table Z-1
PEL(TWA): 200 ppm, 890 mg/m^3

ACGIH-TLV
TWA: 200 ppm, 890 mg/m^3
STEL: 250 ppm, 1,110 mg/m^3

2-ETHYLBUTANOL. [97-95-0]

NFPA—hazardous chemical (flammable chemical)
health—1, flammability—2, reactivity—0

ETHYLBUTYL ACETATE. [123-66-0]

NFPA—hazardous chemical (flammable chemical)
health—1, flammability—2, reactivity—0

ETHYL BUTYL KETONE (3-HEPTANONE). [106-35-4]

*OSHA 29 CFR 1910.1000 Table Z-1
PEL(TWA): 50 ppm, 230 mg/m^3

ACGIH-TLV
TWA: 50 ppm, 230 mg/m^3
STEL: 75 ppm, 345 mg/m^3

NFPA—hazardous chemical (flammable chemical)
health—1, flammability—2, reactivity—0

ETHYL CARBAMATE (URETHAN). [51-79-6]

*RCRA 40 CFR 261
EPA hazardous waste no. U238

ETHYL CHLORIDE. [75-00-3]

*OSHA 29 CFR 1910.1000 Table Z-1
PEL(TWA): 1000 ppm, 2600 mg/m^3

ACGIH-TLV
TWA: 1000 ppm, 2600 mg/m^3
STEL: 1250 ppm, 3250 mg/m^3

NIOSH Current Intelligence Bulletin No. 27—cancer-related bulletin

CWA 307(a)(1), 40 CFR 125, Priority Pollutant

TSCA-CHIP available

NFPA—hazardous chemical
health—2, flammability—4, reactivity—0

ETHYL CYANIDE. [107-12-0]

*RCRA 40 CFR 261
EPA hazardous waste no. P101

p,p'-ETHYL-DDD.
See DDE. Ethane, 1,1-dichloro-2, 2-bis(p-ethylphenyl)-. [72-56-0]

ETHYL 4,4'-DICHLOROBENZILATE.
[510-15-6]

*RCRA 40 CFR 261
EPA hazardous waste no. U038

ETHYLENE. See Ethene. [74-85-1]
ACGIH-TLV
TWA: Asphyxiant—inert gas or vapors, when present in high concentrations in air, act primarily as simple asphyxiant without other significant physiologic effects.

ETHYLENE BIS DITHIOCARBAMATE.
Syn: EBDC. [142-59-6, 3566-10-7]

OSHA Candidate List of Potential Occupational Carcinogens (EPA Carcinogen Assessment Group List), 45 FR 53672, August 12, 1980

*RCRA 40 CFR 261
EPA hazardous waste no. U114

ETHYLENE BROMIDE. See Ethylene dibromide. [106-93-4]

ETHYLENE, BROMO-. Syn: Vinyl bromide. [593-60-2]

OSHA Candidate List of Potential Occupational Carcinogens, 45 FR 53672, August 12, 1980
ACGIH-TLV
TWA: 5 ppm, 20 mg/m^3
Industrial substance suspect of carcinogenic potential for humans—which is suspect of inducing cancer, based on either (1) limited epidemiologic evidence, exclusive of clinical reports of single cases, or (2) demonstration of carcinogenesis in one or more animal species by appropriate methods. Worker exposure by all routes should be carefully controlled to levels consistent with the animal and human experience data.
TSCA-CHIP available

ETHYLENE CHLORIDE. See Ethylene dichloride. [107-06-3]

ETHYLENE CHLOROHYDRIN.
[107-07-3]

*OSHA 29 CFR 1910.1000 Table Z-1

Skin

PEL(TWA): 5 ppm, 16 mg/m^3
ACGIH-TLV

Skin

TWA: ceiling limit, 1 ppm, 3 mg/m^3

ETHYLENE CYANOHYDRIN. See 3-Hydroxypropanenitrile. [109-78-4]

ETHYLENEDIAMINE. [107-15-3]

*OSHA 29 CFR 1910.1000 Table Z-1
PEL(TWA): 10 ppm, 25 mg/m^3
ACGIH-TLV
TWA: 10 ppm, 25 mg/m^3
NFPA—hazardous chemical
health—3, flammability—2, reactivity—0
TSCA-CHIP available
*CWA 311(b)(2)(A), 40 CFR 116, 117
Discharge RQ = 1000 pounds (454 kilograms)

ETHYLENE DIBROMIDE (EDB).
Syn: 1,2-Dibromoethane. [106-93-4]

*OSHA 29 CFR 1910.1000 Table Z-1
20 ppm, 8-hour TWA
30 ppm, acceptable ceiling concentration
50 ppm/5 minutes acceptable maximum peak
ACGIH-TLV (1982 Addition)

Skin

TWA: Industrial substance suspect of carcinogenic potential for humans—which is suspect of inducing cancer, based on either (1) limited epidemiologic evidence, exclusive of clinical reports of single cases, or (2) demonstration of carcinogenesis in one or more animal species by appropriate methods. Worker exposure by all routes should be carefully controlled to levels consistent with the animal and human experience data.

NIOSH Criteria Document (Pub. No. 77-221), NTIS Stock No. PB 276621, August 26, 1977

Synonyms for ethylene dibromide include ethylene bromide, dibromoethane, sym-dibromoethane, 1,2-dibromoethane, glycol dibromide, and EDB. The major uses of ethylene dibromide are as an additive to leaded gasoline and as a component of fumigants.

"Occupational exposure to ethylene dibromide" is defined as work in any establishment where ethylene dibromide is manufactured, blended, stored, used, handled, or otherwise present. The "action level" is defined as one-half of the recommended workplace exposure concentration designated as a ceiling limit for ethylene dibromide. Exposure to airborne ethylene dibromide at concentrations less than the action level, as determined in accordance with Section 8, will not require adherence to Sections 2, 3, 4(a), or 8(c,d).

The employer shall control workplace concentrations of ethylene dibromide so that no employee is exposed in the workplace to concentrations greater than 1.0 mg/m^3 (0.13 ppm) as a ceiling limit, as determined by sampling period of 15 minutes. The standard is designed to protect the health and provide for the safety of employees for up to a 10-hour workday, 40-hour workweek, over a working lifetime.

NIOSH Current Intelligence Bulletin No. 3—cancer-related bulletin

NIOSH Current Intelligence Bulletin No. 37—revised cancer-related bulletin

OSHA Candidate List of Potential Occupational Carcinogens (EPA Carcinogen Assessment Group List), 45 FR 53672, August 12, 1980

IARC—carcinogenic in animals

NCI—carcinogenic in animals

40 CFR 162.11, FIFRA-RPAR issued
Current Status: PD 1 published, 42 FR 63134 (12/14/77); comment period closed 4/3/78: NTIS# PB80 109456.

PD 2/3 completed and Notice of Determination published, 45 FR 81586 (12/10/81); comment period closed 3/9/81.

PD 4 being developed. Criteria Possibly Met or Exceeded: Oncogenicity, Mutagenicity, and Reproductive Effects

NFPA—hazardous chemical
health—3, flammability—0, reactivity—0

TSCA-CHIP available

*CWA 311(b)(A)(2), 40 CFR 116, 117
Discharge RQ = 1000 pounds (454 kilograms)

*RCRA 40 CFR 261
EPA hazardous waste no. U067 (hazardous substance)

ETHYLENE DIBROMIDE and DISULFIRAM TOXIC INTERACTION.
[106-93-4, 97-77-8]

NIOSH Current Intelligence Bulletin No. 23—cancer-related bulletin

ETHYLENE DICHLORIDE. Syn:
1,2-Dichloroethane. [107-06-2]

*OSHA 1910.1000 Table Z-2
50 ppm, 8-hour TWA
100 ppm, acceptable ceiling concentration
200 ppm/5 minutes in any 3 hours, acceptable maximum peak

ACGIH-TLV
TWA: 10 ppm, 40 mg/m^3
STEL: 15 ppm, 60 mg/m^3

NIOSH Criteria Document, revised (Pub. No. 78-211), NTIS Stock No. PB80 176092, September 1978

The synonym for ethylene dichloride is 1,2-dichloroethane. The term "ethylene dichloride" refers to all physical forms of the compound and its solutions. "Occupational exposure to ethylene dichloride" is defined as work in any area in which ethylene dichloride is producd, stored, used, packaged, or distributed. If ethylene dichloride is handled or stored only in intact, sealed containers (e.g., during shipment), adherence to Sections 3, 5(a), 6(g), and 8(a) is required.

Occupational exposure to ethylene dichloride shall be controlled so that employees are not exposed at a concentration in excess of 1 ppm (4 mg/m^3) in air as determined by a personal air sample collected at a rate of flow between 25 and 200 ml/minute. A ceiling concentration of 2 ppm (8 mg/m^3), as determined over a 15-minute sampling period, is also recommended. The recommended standard is designed to protect the health and provide for the safety of employees for up to a 10-hour work shift, 40-hour workweek, over a working lifetime.

NIOSH Current Intelligence Bulletin No. 25—cancer-related bulletin

OSHA Candidate List of Potential Occupational Carcinogens (EPA Carcinogen Assessment Group List), 45 FR 53672, August 12, 1980

NCI—carcinogenic in animals

FIFRA—Data Call In, 45 FR 75488, November 14, 1980
Status: letter issued 7/2/81

NFPA—hazardous chemical
health—2, flammability—3, reactivity—0

TSCA-CHIP available

CWA 307(a)(1), 40 CFR 125, Priority Pollutant

*CWA 311(b)(2)(A), 40 CFR 116, 117
Discharge RQ = 5000 pounds (2270 kilograms)

CAA 112
Chemical proposed to be assessed as a Hazardous Air Pollutant by a House of Representatives bill to amend the CAA

*RCRA 40 CFR 261
EPA hazardous waste no. U077 (hazardous substance)

ETHYLENE, 1,1-DICHLORO-2,2-BIS (p-CHLOROPHENYL)-. See DDE. [72-55-9]

ETHYLENE GLYCOL. [107-21-1]
ACGIH-TLV
TWA: ceiling limit, 50 ppm, 125 mg/m^3

NFPA—hazardous chemical (flammable chemical)
health—1, flammability—1, reactivity—0

ETHYLENE GLYCOL, PARTICULATE. [107-21-1]
ACGIH-TLV
TWA: 10 mg/m^3
STEL: 20 mg/m^3
Intended changes for 1982 (1982 Revision or Addition): particulate DELETE

ETHYLENE GLYCOL BIS(CHLOROMETHYL) ETHER. See Ethane, 1,2-bis(chloromethoxy)-. [13483-18-6]

ETHYLENE GLYCOL DINITRATE and/or NITROGLYCERIN. [628-96-6]
*OSHA 29 CFR 1910.1000 Table Z-1

Skin

PEL(ceiling): 0.2 ppm,* 1 mg/m^3
ACGIH-TLV

Skin

TWA: 0.02 ppm, 0.1 mg/m^3
STEL: 0.05 ppm, 0.3 mg/m^3
Intended changes for 1982:

Skin

TWA: 0.05 ppm, 0.3 mg/m^3
STEL: 0.1 ppm, 0.6 mg/m^3
See Nitroglycerin and ethylene glycol dinitrate, NIOSH Criteria Document

ETHYLENE GLYCOL METHYL ETHER ACETATE. See 2-Methoxyethyl acetate. [110-49-6]

ETHYLENE GLYCOL MONOMETHYL ETHER ACETATE. See Methyl cellosolve acetate. [110-49-6]

ETHYLENEIMINE. [151-56-4]
*OSHA Standard 29 CFR 1910.1012
Cancer-Suspect Agent

---

*An atmospheric concentration of not more than 0.02 ppm, or personal protection may be necessary to avoid headache.

Shall not apply to solid or liquid mixtures containing less than 1.0% by weight or volume of ethyleneimine

*OSHA 29 CFR 1910.1000 Table Z-1

Skin

PEL(TWA): 0.5 ppm, 1 mg/m^3

ACGIH-TLV

Skin

TWA: 0.5 ppm, 1 mg/m^3

OSHA Candidate List of Potential Occupational Carcinogens (EPA Carcinogen Assessment Group List), 45 FR 53672, August 12, 1980

NFPA—hazardous chemical
health—3, flammability—3, reactivity—2

*RCRA 40 CFR 261

EPA hazardous waste no. P054

ETHYLENE OXIDE (EO). [75-21-8]

*OSHA 29 CFR 1910 (proposed), 48 FR 17284, April 21, 1983

OSHA is proposing to amend its existing occupational standard that regulates employee exposure to ethylene oxide (EtO). The basis for this action is a determination by the assistant secretary, based on animal and human data, that exposure to EtO at OSHA's current permissible exposure limit(PEL) of 50 parts EtO per million parts of air (50 ppm) as an 8-hour time-weighted average (TWA) is inadequate for employee health protection. OSHA proposes to reduce the PEL for EtO to a TWA of 1 ppm. An "action level" of 0.5 ppm as a TWA is included in the proposal as a mechanism for exempting an employer from the obligation to comply with certain requirements, such as employee exposure monitoring and medical surveillance, in instances where the employer can demonstrate that the employees' exposures are at very low levels. The proposal would provide for, among other requirements, certain methods of exposure control, personal protective equipment, measurement of employee exposures, training, medical surveillance, signs and labels, regulated areas, emergency procedures, and record keeping.

*OSHA 29 CFR 1910.1000 Table Z-1
PEL(TWA): 50 ppm, 90 mg/m^3

ACGIH-TLV
TWA: 10 ppm, 20 mg/m^3
Intended changes for 1982 (1982 Revision or Addition):
TWA: 1 ppm, 2 mg/m^3
Industrial substance suspect of carcinogenic potential for humans—which is suspect of inducing cancer, based on either (1) limited epidemiologic evidence, exclusive of clinical reports of single cases, or (2) demonstration of carcinogenesis in one or more animal species by appropriate methods. Worker exposure by all routes should be carefully controlled to levels consistent with the animal and human experience data.

NIOSH Current Intelligence Bulletin No. 35—cancer-related bulletin

OSHA Candidate List of Potential Occupational Carcinogens (EPA Carcinogen Assessment Group List), 45 FR 53672, August 12, 1980

40 CFR 162.11, FIFRA-RPAR issued
Current Status: PD 1 published, 43 FR 3801 (1/27/78); comment period closed 5/15/78: NTIS# PB80 213903.
PD 2/3 being developed. Criteria Possibly Met or Exceeded: Mutagenicity and Testicular Effects

NFPA—hazardous chemical
health—2, flammability—4, reactivity—3

*TSCA 8(a), 40 CFR 712, 47 FR 26992, June 22, 1982, Chemical Information Rule

*TSCA 8(d) Unpublished Health and Safety Studies Reporting, 40 CFR 716, 47 FR 387, September 2, 1982

TSCA-CHIP available

CAA 112
Chemical proposed to be assessed as a Hazardous Air Pollutant by a House of Representatives bill to amend the CAA

*RCRA 40 CFR 261
EPA hazardous waste no. U115 (ignitable waste, toxic waste)

ETHYLENE SULFIDE. [420-12-2]

IARC—carcinogenic in animals

ETHYLENE THIOUREA (ETU).
[96-45-7]

NIOSH Current Intelligence Bulletin No. 22—cancer-related bulletin

OSHA Candidate List of Potential Occupational Carcinogens (EPA Carcinogen Assessment Group List), 45 FR 53672, August 12, 1980

IARC—carcinogenic in animals

*RCRA 40 CFR 261
EPA hazardous waste no. U116

ETHYL ETHER. See Diethyl ether.
[60-29-7]

*OSHA 29 CFR 1910.1000 Table Z-1
PEL(TWA): 400 ppm, 1200 mg/m^3

ACGIH-TLV
TWA: 400 ppm, 1200 mg/m^3
STEL: 500 ppm, 1500 mg/m^3

*RCRA 40 CFR 261
EPA hazardous waste no. U117 (ignitable waste)

ETHYL FORMATE. [109-94-4]

*OSHA 29 CFR 1910.1000 Table Z-1
PEL(TWA): 100 ppm, 300 mg/m^3

ACGIH—TLV
TWA: 100 ppm, 300 mg/m^3
STEL: 150 ppm, 450 mg/m^3

2-ETHYL-1,3-HEXANEDIOL (6-12).
[94-96-2]

FIFRA—Registration Standard—Issued June 1981, NTIS# PB81 234098 (insecticide)

2-ETHYLHEXYL ACRYLATE.
[103-11-7]

NFPA—hazardous chemical
health—2, flammability—2, reactivity—2

TSCA—CHIP available

ETHYLIDENE CHLORIDE.
See 1,1-Dichloroethane. [75-34-3]

ETHYLIDINE DICHLORIDE.
See 1,1-Dichloroethane. [75-34-3]

ETHYLIDINE NORBORNENE.
[16219-75-3]

ACGIH-TLV
TWA: ceiling limit, 5 ppm, 25 mg/m^3

ETHYL MERCAPTAN. [75-08-1]

*OSHA 29 CFR 1910.1000 Table Z-1
PEL(Ceiling): 10 ppm, 25 mg/m^3

ACGIH-TLV
TWA: 0.5 ppm, 1 mg/m^3
STEL: 2 ppm, 3 mg/m^3

ETHYL METHACRYLATE. [97-63-2]

*RCRA 40 CFR 261
EPA hazardous waste no. U118

ETHYL METHANESULFONATE.
[62-50-0]

OSHA Candidate List of Potential Occupational Carcinogens (EPA Carcinogen Assessment Group List), 45 FR 53672, August 12, 1980

IARC—carcinogenic in animals

*RCRA 40 CFR 261
EPA hazardous waste no. U119

ETHYL METHYL ETHER. [540-67-0]

NFPA—hazardous chemical
health—2, flammability—4, reactivity—1

ETHYL METHYL KETONE. [78-93-3]

NFPA—hazardous chemical (flammable chemical)
health—1, flammability—3, reactivity—0

ETHYL METHYL KETONE PEROXIDE.

NFPA—hazardous chemical
health—2, flammability—2, reactivity—4, oxy—oxidizing chemical

5-ETHYL-2-METHYLPYRIDINE.
[104-90-5]

NFPA—hazardous chemical
health—2, flammability—2, reactivity—0

N-ETHYLMORPHOLINE. [100-74-3]

*OSHA 29 CFR 1910.1000 Table Z-1

Skin

PEL(TWA): 20 ppm, 94 mg/m^3

ACGIH-TLV (1982 Addition)

Skin

TWA: 5 ppm, 40 mg/m^3
STEL: 20 ppm, 95 mg/m^3

ETHYL NITRITE. [75-05-8]

NFPA—hazardous chemical
health—2, flammability—4, reactivity—4

ETHYL OXIRANE. See Alkyl epoxides.
[106-88-7]

*TSCA 8(d) Unpublished Health and Safety Studies Reporting, 40 CFR 716, 47 FR 387, September 2, 1982

5-ETHYL-2-PICOLINE. See 5-Ethyl-2-methylpyridine. [104-90-5]

ETHYL SILICATE. [78-10-4]

*OSHA 29 CFR 1910.1000 Table Z-1
PEL(TWA): 100 ppm, 850 mg/m^3

ACGIH-TLV
TWA: 10 ppm, 85 mg/m^3
STEL: 30 ppm, 255 mg/m^3

ETHYL SULFATE. See Sulfuric acid, diethyl ester; or Diethyl sulfate. [64-67-5]

ETHYL TELLURAC. See Tellurium, tetrabis(diethyldithiocarbamato)-.
[20941-65-5, 30145-38-1]

NCI—carcinogenic in animals

ETHYLTOLUENE (mixed isomers of ortho, meta, and para).
[25550-14-5]

TSCA 4(e), ITC
Tenth Report of the Interagency Testing Committee to the administrator, Environmental Protection Agency, May 1982; 47 FR 22585, May 25, 1982

*TSCA 8(d) Unpublished Health and Safety Studies Reporting, 40 CFR 716, 47 FR 38800, September 2, 1982; 48 FR 13178, March 30, 1983

ETHYNODIOL DIACETATE. [297-76-7]

IARC—carcinogenic in animals

EUGENOL. [97-53-0]

NCI—may be carcinogenic in animals

EVANS BLUE. [314-1-6]

IARC—carcinogenic in animals

F

FAMPHUR. [52-85-7]

*RCRA 40 CFR 261
EPA hazardous waste no. P097

FAST GREEN FCF. [569-64-2]

IARC—carcinogenic in animals

FENSULFOTHION. [115-90-2]

ACGIH-TLV
TWA: 0.1 mg/m^3

FIFRA—Data Call In, 45 FR 75488, November 14, 1980
Status: agency decision reached

FENTHION. [55-38-9]

ACGIH-TLV
TWA: 0.1 mg/m^3
STEL: 0.3 mg/m^3
Intended change for 1982:
TWA: 0.2 mg/m^3

NCI—may be carcinogenic in animals

FENVALERATE.

FIFRA—Data Call In, 45 FR 75488, November 14, 1980
Status: letter issued 9/9/81

FERBAM. [14484-64-1]

*OSHA 29 CFR 1910.1000 Table Z-1
PEL(TWA): 15 mg/m^3

ACGIH-TLV
TWA: 10 mg/m^3
STEL 20 mg/m^3
FIFRA—Data Call In, 45 FR 75488, November 14, 1980
Status: letter issued 9/14/81

FERRIC AMMONIUM CITRATE.
[1185-57-5]

*CWA 311(b)(2)(A), 40 CFR 116, 117
Discharge RQ = 1000 pounds (454 kilograms)

FERRIC AMMONIUM OXALATE.
[2944-67-4, 55488-87-4]

*CWA 311(b)(2)(A), 40 CFR 116, 117
Discharge RQ = 1000 pounds (454 kilograms)

FERRIC CHLORIDE. [7705-08-0]

*CWA 311(b)(2)(A), 40 CFR 116, 117
Discharge RQ = 1000 pounds (454 kilograms)

FERRIC DEXTRAN. [9004-66-4]

*RCRA 40 CFR 261
EPA hazardous waste no. U139

FERRIC FLUORIDE. [7783-50-8]

*CWA 311(b)(2)(A), 40 CFR 116, 117
Discharge RQ = 100 pounds (45.4 kilograms)

FERRIC NITRATE. [10421-48-4]

*CWA 311(b)(2)(A), 40 CFR 116, 117
Discharge RQ = 1000 pounds (454 kilograms)

FERRIC SULFATE. [10028-22-5]

*CWA 311(b)(2)(A), 40 CFR 116, 117
Discharge RQ = 1000 pounds (454 kilograms)

FERROUS AMMONIUM SULFATE.
[10045-89-3]

*CWA 311(b)(2)(A), 40 CFR 116, 117
Discharge RQ = 1000 pounds (454 kilograms)

FERROUS CHLORIDE. [7758-94-3]

*CWA 311(b)(2)(A), 40 CFR 116, 117
Discharge RQ = 100 pounds (45.4 kilograms)

FERROUS SULFATE. [7720-78-7, 7782-63-0]

*CWA 311(b)(2)(A), 40 CFR 116, 117
Discharge RQ = 1000 pounds (454 kilograms)

FERROVANADIUM DUST.
[12604-58-9]
*OSHA 29 CFR 1910.1000 Table Z-1
PEL(TWA): 1 mg/m^3
ACGIH-TLV
TWA: 1 mg/m^3
STEL: 0.3 mg/m^3

FIBROUS GLASS DUST. [14808-60-7]
ACGIH-TLV (<7 μm in diameter)
TWA: 10 mg/m^3
NIOSH Criteria Document (Pub. No. 77-152), NTIS Stock No. PB 274195, April 15, 1977

Fibrous glass is the name for a manufactured fiber in which the fiber-forming substance is glass. Glasses are a class of materials made from mixtures of silicon dioxide with oxides of various metals and other elements, that solidify from the molten state without crystallization.

Synonyms for fibrous glass include fiberglass and glass fibers. A fiber is considered to be a particle with a length-to-diameter ratio of 3 to 1 or greater. An "action level" is defined as half the recommended time-weighted average (TWA) environmental limit. "Occupational exposure" is defined as exposure to airborne fibrous glass above the action level. In addition, because workers may be exposed to fibrous glass by dermal or eye contact, occupational exposure includes contact with the skin and eyes to fibrous glass where it is manufactured, used, handled, or stored. When environmental concentrations are at or below the action level, adherence to Sections 1, 2(b), 4(c), and 8(b) is not required.

In addition, although this criteria document addresses occupational exposure to fibrous glass, NIOSH considers that until more information is available, the recommended standard can also be applied to other man-made mineral fibers.

Occupational exposure to fibrous glass shall be controlled so that no worker is exposed at an airborne concentration greater than 3 million fibers per cubic meter of air (3 fibers per cubic centimeter of air) having a diameter equal to or less than 3.5 μm and a length equal to or greater than 10 μm determined as a time-weighted average (TWA) concentration for up to a 10-hour work shift in a 40-hour workweek; airborne concentrations determined as total fibrous glass shall be limited to a TWA concentration of 5 mg/m^3 of air. The standard is designed to protect the health and safety of workers for up to a 10-hour workday, 40-hour workweek, over a normal working lifetime.

FLUCHLORALIN. [33245-39-5]
FIFRA—Data Call In, 45 FR 75488, November 14, 1980
Status: letter being drafted

FLUOMETURON. [2164-17-2]
NCI—may be carcinogenic in animals

FLUORANTHENE. [206-44-0]
CWA 304(a)(1), 45 FR 79318, November 28, 1980 Water Quality Criteria

Freshwater Aquatic Life

The available data for fluoranthene indicate that acute toxicity to freshwater aquatic life occurs at concentrations as low as 3980 μg/liter and would occur at lower concentrations among species that are more sensitive than those tested. No data are available concerning the chronic toxicity of fluoranthene to sensitive freshwater aquatic life.

Saltwater Aquatic Life

The available data for fluoranthene indicate that acute and chronic toxicity to saltwater aquatic life occur at concentrations as low as 40 and 16 μg/liter, respectively, and would occur at lower concentrations among species that are more sensitive than those tested.

Human Health

For the protection of human health from the toxic properties of fluoranthene ingested through water and contaminated aquatic organisms, the ambient water criterion is determined to be 42 μg/liter. For the protection of human health from the toxic properties of fluoranthene ingested through contaminated aquatic organisms alone, the ambient water criterion is determined to be 54 µg/liter.

CWA 307(a)(1), Priority Pollutant

*RCRA 40 CFR 261

EPA hazardous waste no. U120

FLUORENE.

CWA 307(a)(1), 40 CFR 125, Priority Pollutant

FLUORENE, 2-NITRO-. Syn: 2-Nitro-9h-fluorene. [607-57-8]

OSHA Candidate List of Potential Occupational Carcinogens, 45 FR 53672, August 12, 1980

FLUOREN-9-ONE, 2,4,7-TRINITRO-. Syn: 2,4,7-Trinitro-9h-fluoren-9-one. [129-79-3]

OSHA Candidate List of Potential Occupational Carcinogens, 45 FR 53672, August 12, 1980

TSCA-CHIP available

FLUORESCEIN SODIUM. *See* C.I. acid yellow 73.

FLUORIDE.

*OSHA 29 CFR 1910.1000 Table Z-1
PEL(TWA): 2.5 mg/m^3 (as F)

*OSHA 1910.1000 Table Z-2
2.5 mg/m^3, 8-hour TWA (as dust)

ACGIH-TLV
TWA: 2.5 mg/m^3

NIOSH Criteria Document (Pub. No. 76-103), NTIS Stock No. PB 246692, June 30, 1975

Fluorides are defined as those compounds of fluoride that are inorganic solids at normal workroom temperature (20°C), compounds without radioactive elements, compounds having components that do not have exposure limits more restrictive than that proposed here for fluoride, and any gaseous fluorides that are emitted simultaneously with particulate fluorides as defined above.

"Occupational exposure to fluorides" is defined as exposure above one-half the recommended workroom limit. Occupational exposure shall be controlled so that no worker is exposed to a concentration of fluorides greater than 2.5 mg of F (combined ionic fluoride, atomic weight 19) per cubic meter of air determined as a time-weighted average (TWA) exposure for up to a 10-hour workday, 40-hour workweek. The standard is designed to protect the health and safety of workers for up to a 10-hour workday, 40-hour workweek, over a working lifetime.

*National Interim Primary Drinking Water Regulations, 40 CFR 141; 40 FR 59565, December 24, 1975; amended by 41 FR 28402, July 9, 1976; 44 FR 68641, November 29, 1979; corrected by 45 FR 15542, March 11, 1980; 45 FR 57342, August 27, 1980; 47 FR 18998, March 3, 1982; corrected by 47 FR 10998, March 12, 1982

Fluoride at optimum levels in drinking water has ben shown to have beneficial effects in reducing the occurrence of tooth decay. When the annual average of the maximum daily air temperature for the location in which the community water system is situated is the following, the maximum contaminant levels for fluoride are:

Temperature, degrees Fahrenheit	Degrees Celsius	Level, mg/liter
53.7 and below	12.0 and below	2.4
53.8–58.3	12.1–14.6	2.2
58.4–63.8	14.7–17.6	2.0
63.9–70.6	17.7–21.4	1.8
70.7–79.2	21.5–26.2	1.6
79.3–90.5	26.3–32.5	1.4

FLUORINE. [7782-41-4]

*OSHA 29 CFR 1910.1000 Table Z-1
PEL(TWA): 0.1 ppm, 0.2 mg/m^3

ACGIH-TLV
TWA: 1 ppm, 2 mg/m^3
STEL: 2 ppm, 4 mg/m^3

NFPA—hazardous chemical
health—4, flammability—0, reactivity—3, W̶ —water may be hazardous in fire fighting, oxy—oxidizing chemical reacts explosively

with water or steam to produce toxic and corrosive fumes

*RCRA 40 CFR 261
EPA hazardous waste no. P056

FLUOROACETAMIDE. [640-19-7]
*RCRA 40 CFR 261
EPA hazardous waste no. P057

FLUOROACETIC ACID, SODIUM SALT. [62-74-8]
*RCRA 40 CFR 261
EPA hazardous waste no. P058

FLUOROALKENES. See individual fluoroalkenes.

This category is defined as fluoroalkenes of general formula: $C_nH_{2n-x}F_x$, where n equals 2 or 3 and x equals 1-6. Six fluoroalkenes, meeting the category definition, were identified in the public portion of the TSCA Chemical Substance Inventory. This category includes those six fluoroalkenes (shown below) but is not limited to them:

Chemical	CAS No.
Tetrafluoroethene	[116-14-3]
Trifluoroethene	[359-11-5]
Vinylidine fluoride (VDF)	[75-38-7]
Vinyl fluoride (VF)	[75-02-5]
Hexafluoropropene	[116-15-4]
Trifluoromethylethene	[677-21-4]

TSCA 4(e), ITC
Seventh Report of the Interagency Testing Committee to the administrator, Environmental Protection Agency, October 1980; 45 FR 78432, November 25, 1980, EPA responded to the committee's recommendations for testing 46 FR 53704, October 30, 1981
*TSCA 8(d) Unpublished Health and Safety Studies Reporting, 40 CFR 716, 47 FR 38800, September 2, 1982; 48 FR 13178, March 30, 1983

FLUOROCARBON POLYMERS, DECOMPOSITION PRODUCTS OF.

NIOSH Criteria Document (Pub. No. 77-193), NTIS Stock No. PB 274727, September 22, 1977

Fluorocarbon polymers include polymers of substituted polyethylene monomers of the general formula $(XCX-XCF)_n$, where X can be F, H, Cl, CF_3, or CF_3-CF_2-CF_2-O. The decomposition products of fluorocarbon polymers are defined as substances that are thermally generated from fluorocarbon polymers. Adherence to all provisions of the recommended standard is required in work areas in which fluorocarbon polymers are used, regardless of the concentration of airborne decomposition products of fluorocarbon polymers. Occupational exposure is defined as any exposure to fluorocarbon polymers that may involve the production of decomposition products. Because there is insufficient information on which to establish a safe workplace environmental concentration, none is recommended. Both employers and employees should take all steps possible to keep occupational exposure to the decomposition products of fluorocarbon polymers as near to zero as possible.

Since the decomposition of fluorocarbon polymers can lead to occupational exposure to inorganic fluorides, including hydrogen fluoride, workroom air shall be monitored for inorganic fluorides and hydrogen fluoride in accordance with the requirements of the recommended standards for these substances. Recommended methods for sampling and analysis of workroom air for inorganic fluorides are given in *Criteria for a Recommended Standard... Occupational Exposure to Inorganic Fluorides*. Recommended methods for sampling and analysis of workroom air for hydrogen fluoride are given in *Criteria for a Recommended Standard... Occupational Exposure to Hydrogen Fluoride*. Adherence to the recommended standards for inorganic fluorides and hydrogen fluoride may not protect the worker from adverse effects caused by other decomposition products of fluorocarbon polymers. The standard is designed to protect the health and provide for the safety of employees for up to a 10-hour work shift, 40-hour workweek, over a working lifetime.

FLUOROTRICHLOROMETHANE. See Trichlorofluoromethane. [75-69-4]

FOAMING AGENTS.

*National Secondary Drinking Water Regulations, 40 CFR 143; 44 FR 42198, July 19, 1979, effective January 19, 1981
Maximum contaminant level—0.5 mg/liter

FOLPET. [133-07-3]
FIFRA—Data Call In, 45 FR 75488, November 14, 1980
Status: letter issued 9/4/81

FONOFOS. [944-22-9]
ACGIH-TLV

Skin

TWA: 0.1 mg/m^3
FIFRA—Data Call In, 45 FR 75488, November 14, 1980
Status: agency decision reached

FORMALDEHYDE. [50-00-0]
*OSHA 29 CFR 1910.1000 Table Z-2
3 ppm, 8-hour TWA
5 ppm, acceptable ceiling concentration
10 ppm/30 minutes, acceptable maximum peak
ACGIH-TLV
TWA: ceiling limit, 2 ppm, 3 mg/m^3
Intended changes for 1982 (1982 Revision or Addition):
TWA: ceiling limit, 1 ppm, 1.5 mg/m^3
Industrial substance suspect of carcinogenic potential for humans—which is suspect of inducing cancer, based on either (1) limited epidemiologic evidence, exclusive of clinical reports of single cases, or (2) demonstration of carcinogenesis in one or more animal species by appropriate methods. Worker exposure by all routes should be carefully controlled to levels consistent with the animal and human experience data.

NIOSH Criteria Document (Pub. No. 77-126), NTIS Stock No. PB 273805, December 30, 1976
"Occupational exposure to formaldehyde" is defined as exposure to formaldehyde in air at a concentration in excess of 0.6 mg/m^3 (0.5 ppm), based on a 30-minute sampling period, or by contact with formaldehyde in liquid or solid form. Adherence to all provisions of Sections 3–6 is required in occupational environments where formaldehyde is used regardless of the concentration of airborne formaldehyde. Medical surveillance and environmental monitoring are required as specified in Sections 2 and 8, respectively.

Exposure to formaldehyde shall be controlled so that no employee is exposed to formaldehyde at a concentration greater than 1.2 mg/m^3 of air (1 ppm) for any 30-minute sampling period. The standard is designed to protect the health and to provide for the safety of employees for up to a 10-hour workday, 40-hour workweek, over a working lifetime.
NIOSH Current Intelligence Bulletin No. 34—cancer related bulletin
OSHA Candidate List of Potential Occupational Carcinogens (EPA Carcinogen Assessment Group List), 45 FR 53672, August 12, 1980
IARC—carcinogenic in animals
NFPA—hazardous chemical
when water solution: health—2, flammability—2, reactivity—0
when gas: health—2, flammability—4, reactivity—0
TSCA-CHIP available
*CWA 311(b)(2)(A), 40 CFR 116, 117
Discharge RQ = 1000 pounds (454 kilograms)
CAA 112
Chemical proposed to be assessed as a Hazardous Air Pollutant by a House of Representatives bill to amend the CAA
*RCRA 40 CFR 261
EPA hazardous waste no. U122 (hazardous substance)

FORMALIN. See Formaldehyde. [50-00-0]

FORMAMIDE. [75-12-7]
ACGIH-TLV
TWA: 20 ppm, 30 mg/m^3
STEL: 30 ppm, 45 mg/m^3
TSCA 4(e), ITC
Tenth Report of the Interagency Testing Committee to the administrator, Environ-

mental Protection Agency, May 1982; 47 FR 22585, May 25, 1982

*TSCA 8(a), 40 CFR 712, 47 FR 26992, June 22, 1982, Chemical Information Rule

*TSCA 8(d) Unpublished Health and Safety Studies Reporting, 40 CFR 716, 47 FR 38800, September 2, 1982; 48 FR 13178, March 30, 1983

TSCA-CHIP available

FORMETANATE HYDROCHLORIDE.
[23422-53-9]

FIFRA—Data Call In, 45 FR 75488, November 14, 1980
Status: letter issued 5/15/81

FIFRA—Registration Standard Development—Data Collection (insecticide)

FORMIC ACID. [64-18-6]

*OSHA 29 CFR 1910.1000 Table Z-1
PEL(TWA): 5 ppm, 9 mg/m^3
ACGIH-TLV
TWA: 5 ppm, 9 mg/m^3
NFPA—hazardous chemical
health—3, flammability—2, reactivity—0

*CWA 311(b)(2)(A), 40 CFR 116, 117
Discharge RQ = 5000 pounds (2270 kilograms)

*RCRA 40 CFR 261
EPA hazardous waste no. U123 (hazardous substance, corrosive waste, toxic waste)

2-(2-FORMYLHYDRAZINO)-4-(5-NITRO-2-FURYL) THIAZOLE.
[3570-75-0]

IARC—carcinogenic in animals

FUEL OIL NO. 1 (RANGE OIL, KEROSENE).

NFPA—hazardous chemical (flammable chemical)
health—0, flammability—2, reactivity—0

FUEL OIL NO. 2.

NFPA—hazardous chemical (flammable chemical)
health—0, flammability—2, reactivity—0

FUEL OIL NO. 4.

NFPA—hazardous chemical (flammable chemical)
health—0, flammability—2, reactivity—0

FUEL OIL NO. 5.

NFPA—hazardous chemical (flammable chemical)
health—0, flammability—2, reactivity—0

FUEL OIL NO. 6.

NFPA—hazardous chemical (flammable chemical)
health—0, flammability—2, reactivity—0

FULMINIC ACID, MERCURY(II) SALT.
[628-86-4]

*RCRA 40 CFR 261
EPA hazardous waste no. P065 (reactive waste, toxic waste)

FUMARIC ACID. [110-17-8]

*CWA 311(b)(2)(A), 40 CFR 116, 117
Discharge = 5000 pounds (2270 kilograms)

FUMARIN® (COUMAFURYL).
[117-52-2]

FIFRA—Registration Standard—Issued September 1980. NTIS# PB81 123812 (rodent toxicant and anticoagulant)

FUMING LIQUID ARSENIC. See Arsenic chloride. [7784-34-1]

2-FURALDEHYDE. See Furfural.
[98-01-1]

2-FURALDEHYDE, 5-NITRO-, SEMICARBAZONE. Syn: Nitrofurazone.
[59-87-0]

OSHA Candidate List of Potential Occupational Carcinogens, 45 FR 53672, August 12, 1980

FURAN. [110-00-9]

*RCRA 40 CFR 261
EPA hazardous waste no. U124 (ignitable waste)

FURAN, TETRAHYDRO-. [109-99-9]

*RCRA 40 CFR 261
EPA hazardous waste no. U213 (ignitable waste)

2-FURANCARBOXALDEHYDE.
[98-01-1]

*RCRA 40 CFR 261
EPA hazardous waste no. U125 (hazardous substance, ignitable waste)

2,5-FURANDIONE. [108-31-6]

*TSCA 8(a), 40 CFR 712, 47 FR 26992, June 22, 1982, Chemical Information Rule
*RCRA 40 CFR 261
EPA hazardous waste no. U147 (hazardous substance)

FURFURAL. [98-01-1]

*OSHA 29 CFR 1910.1000 Table Z-1

Skin

PEl(TWA): 5 ppm, 20 mg/m^3
ACGIH-TLV

Skin

TWA: 2 ppm, 8 mg/m^3
STEL: 10 ppm, 40 mg/m^3
NFPA—hazardous chemical
health—2, flammability—2, reactivity—0
*CWA 311(b)(2)(A), 40 CFR 116, 117
Discharge RQ = pounds (454 kilograms)
*RCRA 40 CFR 261
EPA hazardous waste no. U125 (hazardous substance, ignitable waste)

FURFURAN. See Furan. [110-00-9]

FURFURYL ALCOHOL. [98-00-0]

*OSHA 29 CFR 1910.1000 Table Z-1
PEL(TWA): 50 ppm, 200 mg/m^3
ACGIH-TLV (1982 Addition)

Skin

TWA: 10 ppm, 40 mg/m^3
STEL: 15 ppm, 60 mg/m^3
NIOSH Criteria Document (Pub. No. 79-133), NTIS Stock No. PB80 176050, 1979
Synonyms for furfuryl alcohol include: 2-furylmethanol, 2-furylcarbinol, 2-furan methanol, furfural alcohol, and 2-(hydroxymethyl) furan.

An "action level" is defined as one-half the recommended time weighted average (TWA) Environmental limit. "Occupational exposure to furfuryl alcohol" is defined as exposure to airborne furfuryl alcohol above the action level. Exposures at lower concentrations will require adherence to Sections 2(a, e), 3(a), 4(a, c), 5, 6(a, d, e, f), 7, and 8(a).

Occupational exposure to furfuryl alcohol shall be controlled so that employees are not exposed to furfuryl alcohol at a concentration greater than 200 mg/m^3 of air, equivalent to 50 parts per million parts of air by volume, determined as a TWA concentration for up to a 10-hour work shift and 40-hour workweek.

The recommended standard is designed to protect the health and provide for the safety of employees for up to a 10-hour work shift, 40-hour workweek, over a working lifetime.

FYROL FR-2

TSCA-CHIP available

G

GASOLINE. [8006-61-9]
ACGIH-TLV (1982 Addition)
TWA: 300 ppm, 900 mg/m³
STEL: 500 ppm, 1500 mg/m³
NFPA—hazardous chemical (flammable chemical)
health—1, flammability—3, reactivity—0

GENTIAN VIOLET. [548-62-9]
TSCA-CHIP available

GERMANIUM TETRAHYDRIDE. [7782-65-2]
ACGIH-TLV
TWA: 0.2 ppm, 0.6 mg/m³
STEL: 0.6 ppm, 1.8 mg/m³

GLASS, FIBROUS OR DUST. See Fibrous glass dust. [14808-60-7]

GLEAN.
FIFRA—Registration Standard Development—Data Evaluation and Development of Regulatory Position (herbicide)

D-GLUCOPYRANOSE,2-DEOXY-2(3-METHYL-3-NITROSOUREIDO)-. [18883-66-4]
*RCRA 40 CFR 261
EPA hazardous waste no. U206

GLUTARALDEHYDE. [111-30-8]
ACGIH-TLV
TWA: ceiling limit, 0.2 ppm, 0.7 mg/m³

GLYCERIN MIST. [56-81-5]
ACGIH-TLV
TWA: nuisance particulate—30 mppcf or 10 mg/m³ of total dust <1% quartz, or 5 mg/m³ respirable dust

NFPA—hazardous chemical (flammable chemical)
health—1, flammabiity—1, reactivity—0

GLYCIDALDEHYDE. [765-34-4]
IARC—carcinogenic in animals

GLYCIDOL (2,3-EPOXY-1-PROPANOL) AND ITS DERIVATIVES. [556-52-5]

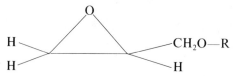

R = H; alkyl, alkenyl or alkynyl; aryl; acyl where R = alkyl, alkenyl, alkynyl, aryl, or acyl; any substituents or functional groups may be present with the alkyl, etc., groups. [77-83-8, 101-90-6, 106-90-1, 106-91-2, 106-92-3, 121-39-1, 122-60-1, 556-52-5, 930-37-0, 2238-07-5, 2425-79-8, 2426-08-6, 2461-18-9, 4016-11-9, 4016-14-2, 13236-02-7, 13561-08-5, 25085-99-8, 26447-14-3]
*OSHA 29 CFR 1910.1000 Table Z-1
PEL(TWA): 50 ppm, 150 mg/m³
ACGIH-TLV
TWA: 25 ppm, 75 mg/m³
STEL: 100 ppm, 300 mg/m³
TSCA 4(e), ITC
Third Report of the TSCA Interagency Testing Committee to the Administrator, Environmental Protection Agency, October 1978; 43 FR 50630, October 30, 1978
TSCA 4(e), ITC
This category consists of glycidol (CAS No. 556-52-5) and its derivatives. Example chemicals in this category are glycidyl acrylate (CAS No. 106-90-1), glycidyl methacrylate (CAS No. 106-91-2), allyl glycidyl ether

(CAS No. 106-92-3), *n*-butyl glycidyl ether (CAS No. 2426-08-6), *para*-cresyl glycidyl ether (CAS No. 2186-24-5), phenyl glycidyl ether (CAS No. 122-60-1), and the diglycidyl ether of bisphenol A (CAS No. 1675-54-3). Third Report of the Interagency Testing Committee to the Administrator, Environmental Protection Agency, October 1978; 43 FR 50630, October 30, 1978

*TSCA 8(d) Unpublished Health and Safety Studies Reporting, 40 CFR 716, 47 FR 387, September 2, 1982

GLYCIDYL ACRYLATE. [106-90-1]

TSCA 4(e), ITC
Third Report of the Interagency Testing Committee to the administrator, Environmental Protection Agency, October 1978; 43 FR 50630, October 30, 1978

GLYCIDYL ALDEHYDE. [765-34-4]

OSHA Candidate List of Potential Occupational Carcinogens (EPA Carcinogen Assessment Group List), 45 FR 53672, August 12, 1980

*RCRA 40 CFR 261
EPA hazardous waste no. U126

GLYCIDYL ETHERS.

NIOSH Criteria Document (Pub. No. 78-166), NTIS Stock No. PB81 229700, June 30, 1978

This recommended standard applies to monoglycidyl ethers and diglycidyl ethers that contain an alkyl group, an aromatic group, or a moiety of the structure -$(RO)_n$-R. It does not include any halogenated compounds or polymerized forms.

"Occupational exposure to glycidyl ethers" is defined as work in any area where these substances are manufactured, stored, used, or handled.

"Occupational exposure to glycidyl ether shall be controlled so that concentrations do not exceed the following ceiling concentration limits, listed in milligrams per cubic meter of air and converted to parts per million, as determined during a 15-minute sampling period:

Allyl glycidyl ether (AGE)	45 mg/m^3	(9.6 ppm)
Isopropyl glycidyl ether (IGE)	240 mg/m^3	(50 ppm)
Phenyl glycidyl ether (PGE)	5 mg/m^3	(1 ppm)
n-Butyl glycidyl ether (BGE)	30 mg/m^3	(4.4 ppm)
Di(2,3-epoxypropyl) ether (DGE)	1 mg/m^3	(0.2 ppm)

The recommended standard is designed to protect the health and provide for the safety of the employees for up to a 10-hour workshift, 40-hour workweek, over a working lifetime.
NIOSH Current Intelligence Bulletin No. 29

GLYCIDYL METHACRYLATE. [106-91-2]

TSCA 4(e), ITC
Third Report of the TSCA Interagency Testing Committee to the administrator, Environmental Protection Agency, October 1978; 43 FR 50630, October 30, 1978

GLYCOL CYANOHYDRIN. See 3-Hydroxypropanenitrile. [109-78-4]

GLYCOL MONOETHYL ETHER.
See 2-Ethoxyethanol. [110-80-5]

GLYCOLONITRILE. [107-16-4]
See Nitriles, NIOSH Criteria Document.

GLYPHOSPHATE ISOPROPYLAMINE SALT.

FIFRA—Data Call In, 45 FR 75488, November 14, 1980
Status: letter being drafted

GOAL.

40 CFR 162.11, FIFRA-RPAR issued
Current Status: PD 12/3 completed.
PD 4 in agency review.
Criteria Possibly Met of Exceeded: Oncogenicity

GOSSYPLURE®. [50933-33-0]

FIFRA—Data Call In, 45 FR, November 14, 1980

Status: letter issued 6/2/81, no additional chronic toxicological data required at this time

GRANDLURE®.
FIFRA—Data Call In, 45 FR 75488, November 14, 1980
Status: letter issued 6/1/81, no additional chronic toxicological data required at this time

GRAPHITE (NATURAL).
See Mineral dusts. [7782-42-5]
ACGIH-TLV
15 mppcf

GRAPHITE (SYNTHETIC).
ACGIH-TLV
TWA: nuisance particulate—30 mppcf or 10 mg/m^3 of total dust <1% quartz, or 5 mg/m^3 respirable dust

GUANIDINE, N-NITROSO-N-METHYL-N'-NITRO-. [70-25-7]
*RCRA 40 CFR 261
EPA hazardous waste no. U163

GUINEA GREEN B. [4680-78-8]
IARC—carcinogenic in animals

GUNCOTTON. See Cellulose nitrate. [9004-70-0]

GUTHION®. See Azinphos-methyl. [865-0-0]
*CWA 311(b)(2)(A), 40 CFR 116, 117
Discharge RQ = 1 pound (0.454 kilogram)

GYPSUM. [1010-14-4]
ACGIH-TLV
TWA: nuisance particulate—30 mppcf or 10 mg/m^3 of total dust <1% quartz, or 5 mg/m^3 respirable dust
STEL: 20 mg/m^3

H

HAFNIUM. [7440-58-6]
*OSHA 29 CFR 1910.1000 Table Z-1
PEL(TWA): 0.5 mg/m^3
ACGIH-TLV
TWA: 0.5 mg/m^3
STEL: 1.5 mg/m^3

HALOETHERS.

CWA 304(a)(1), 45 FR 79318, November 28, 1980 Water Quality Criteria

Freshwater Aquatic Life

The available data for haloethers indicate that acute and chronic toxicity to freshwater aquatic life occur at concentrations as low as 360 and 122 µg/liter, respectively, and would occur at lower concentrations among species that are more sensitive than those tested.

Saltwater Aquatic Life

No saltwater organisms have been tested with any haloether and no statement can be made concerning acute or chronic toxicity.

Human Health

Using the present guidelines, a satisfactory criterion cannot be derived at this time due to the insufficiency in the available data for haloethers.

HALOGENATED ALKYL EPOXIDES.

Halogenated noncyclic aliphatic hydrocarbons with one or more epoxy functional groups.

$$\begin{array}{c} O \\ R1 \diagdown \diagup R3 \\ R2 \diagup \diagdown R4 \end{array}$$

R1 = X or $C_nH_{2n+1-y^xy}$ (y = 1 to $2n$ + 1), R2 = H or X or $C_nH_{2n+1-y^xy}$ (y = 0 to $2n$ + 1), R3 = H or X or $C_nH_{2n+1-y^xy}$ (y = 0 to $2n$ + 1), R4 = H or X or $C_nH_{2n+1-y^xy}$ (y = 0 to $2n$ + 1), X = halogen, groups R1 — R4 may contain one or more epoxide functions. [106-89-8, 428-59-1, 3083-25-8, 3132-64-7]

TSCA 4(e), ITC
This category consists of halogenated noncyclic aliphatic hydrocarbons with one or more epoxy functional groups. At the time of designation, epichlorohydrin was the only widely used halogenated epoxide. Second Report of the TSCA Interagency Testing Committee to the administrator, Environmental Protection Agency, April 1978, 43 FR 16684, April 19, 1978, 44 FR 28095, responded to by the EPA administrator 44 FR 28095, May 14, 1979

*TSCA 8(d) Unpublished Health and Safety Studies Reporting, 40 CFR 716, 47 FR 387, September 2, 1982

HALOMETHANES.

CWA 304(a)(1), 45 FR 79318, November 28, 1980 Water Qualtiy Criteria

Freshwater Aquatic Life

The available data for halomethanes indicate that acute toxicity to freshwater aquatic life occurs at concentrations as low as 11,000 µg/liter and would occur at lower concentrations among species that are more sensitive than those tested. No data are available concerning the toxicity of halomethanes to sensitive freshwater aquatic life.

Saltwater Aquatic Life

The available data for halomethanes indicate that acute and chronic toxicity to saltwater aquatic life occur at concentrations as low as

12,000 and 6400 µg/liter, respectively, and would occur at lower concentrations among species that are more sensitive than those tested. A decrease in algal cell numbers occurs at concentrations as low as 11,500 µg/liter.

Human Health

For the maximum protection of human health from the potential carcinogenic effects due to exposure of chloromethane, bromomethane, dichloromethane, bromodichloromethane, tribromomethane, dichlorodifluoromethane, trichlorofluoromethane, or combinations of these chemicals through ingestion of contaminated water and contaminated aquatic organisms, the ambient water concentration should be zero based on the nonthreshold assumption for this chemical. However, zero level may not be attainable at the present time. Therefore, the levels that may result in incremental increase of cancer risk over the lifetimes are estimated at 10^{-5}, 10^{-6}, and 10^{-7}. The corresponding criteria are 1.9 µg/liter, 0.19 µg/liter, and 0.019 µg/liter, respectively. If the above estimates are made for consumption of aquatic organisms only, excluding consumption of water, the levels are 157 µg/liter, 15.7 µg/liter, and 1.57 µg/liter, respectively. Other concentrations representing different risk levels may be calculated by use of guidelines. The risk estimate range is presented for information purposes and does not represent an agency judgment on an "acceptable" risk level.

HALOTHANE. [151-67-7]

See Waste anesthetic gases and vapors, NIOSH Criteria Document.

HELIOTHIS VIRUS.

FIFRA—Data Call In, 45 FR 75488, November 14, 1980
Status: letter issued 5/27/81, no additional chronic toxicological data required at this time

HELIUM. [7440-59-7]

ACGIH-TLV
TWA: asphyxiant—inert gas or vapors, when present in high concentrations in air, act primarily as simple asphyxiant without other significant physiologic effects

HEMATITE MINING (RANDON).

IARC—carcinogenic in humans

HEPTACHLOR. Syn: 1,4,5,6,7,8,8-Heptachloro-3a,4,7,7a-tetrahydro-4,7-methanoindene. [76-44-8]

*OSHA 29 CFR 1910.1000 Table Z-1

Skin

PEL(TWA): 0.5 mg/m^3
ACGIH-TLV

Skin

TWA: 0.5 mg/m^3
STEL: 2 mg/m^3

OSHA Candidate List of Potential Occupational Carcinogens (EPA Carcinogen Assessment Group List), 45 FR 53672, August 12, 1980

NCI—carcinogenic in animals
CWA 304(a)(1), 45 FR 79318, November 28, 1980, Water Quality Criteria

Freshwater Aquatic Life

For heptachlor the criterion to protect freshwater aquatic life as derived using the guidelines is 0.0038 µg/liter as a 24-hour average and the concentration should not exceed 0.52 µg/liter at any time.

Saltwater Aquatic Life

For heptachlor the criterion to protect saltwater aquatic life as derived using the guidelines is 0.0036 µg/liter as a 24-hour average and the concentration should not exceed 0.053 µg/liter at any time.

Human Health

For the maximum protection of human health from the potential carcinogenic effects due to exposure of heptachlor through ingestion of contaminated water and contaminated aquatic organisms, the ambient water concentration should be zero based on the nonthreshold assumption for this chemical. However, zero level may not be attainable at the present time. Therefore, the levels that may result in incremental increase of cancer risk over the lifetimes are estimated at 10^{-5}, 10^{-6}, and 10^{-7}. The corresponding criteria are 2.78 ng/liter, 0.28 ng/liter, and 0.028 ng/liter, respectively. If the above estimates are made for consumption of aquatic organisms only, excluding consumption of water, the levels are 2.85 ng/liter, 0.29 ng/liter, and 0.029 ng/liter, respectively. Other concentrations representing different risk levels may be calculated by use of the guidelines. The risk estimate range is presented for information purposes and does not represent an agency judgment on an "acceptable" risk level.

CWA 307(a)(1), Priority Pollutant

*CWA 311(b)(2)(A), 40 CFR 116, 117
Discharge RQ = 1 pound (0.454 kilogram)

*RCRA 40 CFR 261
EPA hazardous waste no. P059 (hazardous substance)

HEPTACHLOR EPOXIDE. [1024-57-3]

CWA 307(a)(1), 40 CFR 125, Priority Pollutant

n-HEPTANE. [142-82-5]

*OSHA 29 CFR 1910.1000 Table Z-1
PEL(TWA): 500 ppm, 2000 mg/m^3

ACGIH-TLV
TWA: 400 ppm, 1600 mg/m^3
STEL: 500 ppm, 2000 mg/m^3

See Alkanes (C_5–C_8), NIOSH Criteria Document.

NFPA—hazardous chemical (flammable chemical)
health—1, flammability—3, reactivity—0

1-HEPTANETHIOL. [1639-09-4]

See Thiols, NIOSH Criteria Document.

2-HEPTANONE. See Ethyl n-amyl ketone. [110-43-0]

3-HEPTANONE. See Ethyl butyl ketone. [106-35-4]

HEXACHLOROBENZENE. [118-74-1]

OSHA Candidate List of Potential Occupational Carcinogens (EPA Carcinogen Assessment Group List), 45 FR 53672, August 12, 1980

*RCRA 40 CFR 261
EPA hazardous waste no. U127

HEXACHLOROBUTADIENE. [87-68-3]

ACGIH-TLV
TWA: 0.02 ppm, 0.24 mg/m^3
Industrial substance suspect of carcinogenic potential for humans—which is suspect of inducing cancer, based on either (1) limited epidemiologic evidence, exclusive of clinical reports of single cases, or (2) demonstration of carcinogenesis in one or more animal species by appropriate methods. Worker exposure by all routes should be carefully controlled to levels consistent with the animal and human experience data.

OSHA Candidate List of Potential Occupational Carcinogens (EPA Carcinogen Assessment Group List), 45 FR 53672, August 12, 1980

TSCA 4(e), ITC
Initial Report to the administrator, Environmental Protection Agency, TSCA Interagency Testing Committee, October 1, 1977; 42 FR 55026, October 12, 1977 responded to by the EPA administrator 43 FR 50134, October 26, 1978; 47 FR 58029, December 29, 1982

*TSCA 8(d) Unpublished Health and Safety Studies Reporting, 40 CFR 716, 47 FR 387, September 2, 1982

CWA 304(a)(1), 45 FR 79318, November 28, 1980, Water Quality Criteria

Freshwater Aquatic Life

The available data for hexachlorobutadiene indicate that acute and chronic toxicity to freshwater aquatic life occur at concentrations as low as 90 and 9.3 µg/liter, respectively, and would occur at lower concentrations among species that are more sensitive than those tested.

Saltwater Aquatic Life

The available data for hexachlorobutadiene indicate that acute toxicity to saltwater aquatic life occurs at concentrations as low as 32 µg/liter and would occur at lower concentra-

152 HEXACHLOROCYCLOHEXANE (ALL ISOMERS)

tions among species that are more sensitive than those tested. No data are available concerning the chronic toxicity of hexachlorobutadiene to sensitive saltwater aquatic life.

Human Health

For the maximum protection of human health from the potential carcinogenic effects due to exposure of hexachlorobutadiene through ingestion of contaminated water and contaminated aquatic organisms, the ambient water concentration should be zero based on the nonthreshold assumption for this chemical. However, zero level may not be attainable at the present time. Therefore, the levels that may result in incremental increase of cancer risk over the lifetimes are estimated at 10^{-5}, 10^{-6}, and 10^{-7}. The corresponding criteria are 4.47 µg/liter, 0.45 µg/liter, and 0.045 µg/liter, respectively. If the above estimates are made for consumption of aquatic organisms only, excluding consumption of water, the levels are 500 µg/liter, 50 µg/liter, and 5 µg/liter, respectively. Other concentrations representing different risk levels may be calculated by use of the guidelines. The risk estimate range is presented for information purposes and does not represent an agency judgment on an "acceptable" risk level.

CWA 307(a)(1), Priority Pollutant

*RCRA 40 CFR 261

EPA hazardous waste no. U128

HEXACHLOROCYCLOHEXANE (ALL ISOMERS). See Lindane.

OSHA Candidate List of Potential Occupational Carcinogens (EPA Carcinogen Assessment Group List), 45 FR 53672, August 12, 1980

CWA 304(a)(1), 45 FR 79318, November 28, 1980, Water Quality Criteria

Freshwater Aquatic Life

For lindane the criterion to protect freshwater aquatic life as derived using the guidelines is 0.080 µg/liter as a 24-hour average and the concentration should not exceed 2.0 µg/liter at any time.

Saltwater Aquatic Life

For saltwater aquatic life the concentration of lindane should not exceed 0.16 µg/liter at any time. No data are available concerning the chronic toxicity of lindane to sensitive saltwater aquatic life.

Freshwater Aquatic Life

The available data for a mixture of isomers of BHC indicate that acute toxicity ot freshwater aquatic life occurs at concentrations as low as 100 µg/liter and would occur at lower concentrations among species that are more sensitive than those tested. No data are available concerning the chronic toxicity of a mixture of isomers of BHC to sensitive freshwater aquatic life.

Saltwater Aquatic Life

The available data for a mixture of isomers of BHC indicate that acute toxicity to saltwater aquatic life occurs at concentrations as low as 0.34 µg/liter and would occur at lower concentrations among species that are more sensitive than those tested. No data are available concerning the chronic toxicity of a mixture of isomers of BHC to sensitive saltwater aquatic life.

Human Health

For the maximum protection of human health from the potential carcinogenic effects due to exposure of alpha-HCH through ingestion of contaminated water and contaminated aquatic organisms, the ambient water concentration should be zero based on the nonthreshold assumption for this chemical. However, zero level may not be attainable at the present time. Therefore, the levels that may result in incremental increase of cancer risk over the lifetimes are estimated at 10^{-5}, 10^{-6}, and 10^{-7}. The corresponding criteria are 92 ng/liter, 9.2 ng/liter, and 0.92 ng/liter, respectively. If the above estimates are made for consumption of aquatic organisms only, excluding consumption of water, the levels are 310 ng/liter, 31.0 ng/liter, and 3.1 ng/liter, respectively. Other concentrations representing different risk levels may be calculated by use of the guidelines. The risk estimate range is presented for information purposes and does not represent an agency judgment on an "acceptable" risk level.

For the maximum protection of human health from the potential carcinogenic effects due to exposure of beta-HCH through ingestion of contaminated water and contaminated aquatic organisms, the ambient water concentraion should be zero based on the nonthreshold assumption for this chemical. However, zero level may not be attainable at the present time. Therefore, the levels that may result in incremental increase of cancer risk over the lifetimes are estimated at 10^{-5}, 10^{-6}, and 10^{-7}. The corresponding criteria are 163 ng/liter, 16.3 ng/liter, and 1.63 ng/liter, respectively. If the above estimates are made for aquatic organisms only, excluding consumption of water, the levels are 547 ng/liter, 54.7 ng/liter, and 5.47 ng/liter, respectively. Other concentrations representing different risk levels may be calculated by use of the guidelines. The risk estimate range is presented for information purposes and does not represent an agency judgment on an "acceptable" risk level.

For the maximum protection of human health from the potential carcinogenic effects due to exposure of tech-HCH through ingestion of contaminated water and contaminated aquatic organisms, the ambient water concentration should be zero based on the nonthreshold assumption for this chemical. However, zero level may not be attainable at the present time. Therefore, the levels that may result in incremental increase of cancer risk over the lifetimes are estimated at 10^{-5}, 10^{-6}, and 10^{-7}. The corresponding criteria are 123 ng/liter, 12.3 ng/liter, and 1.23 ng/liter, respectively. If the above estimates are made for consumption of aquatic organisms only, excluding consumption of water, the levels are 414 ng/liter, 41.4 ng/liter, and 4.14 ng/liter, respectively. Other concentrations representing different risk levels may be calculated by use of the guidelines. The risk estimate range is presented for information purposes and does not represent an agency judgment on an "acceptable" risk level.

For the maximum protection of human health from the potential carcinogenic effects due to exposure of gamma-HCH through ingestion of contaminated water and contaminated aquatic organisms, the ambient water concentrations should be zero based on the nonthreshold assumption for this chemical. However, zero level may not be attainable at the present time. Therefore, the levels that may result in incremental increase of cancer risk over the lifetimes are estimated at 10^{-5}, 10^{-6}, and 10^{-7}. The corresponding criteria are 186 ng/liter, 18.6 ng/liter, and 1.86 ng/liter, respectively. If the above estimates are made for consumption of aquatic organisms only, excluding consumption of water, the levels are 625 ng/liter, 62.5 ng/liter, and 6.25 ng/liter, respectively. Other concentrations representing different risk levels may be calculated by use of the guidelines. The risk estimate range is presented for information purposes and does not represent an agency judgment on an "acceptable" risk level.

Using the present guidelines, a satisfactory criterion cannot be derived at this time due to the insufficiency in the available data for delta-HCH.

Using the present guidelines, a satisfactory criterion cannot be derived at this time due to the insufficiency in the available data for epsilon-HCH.

*RCRA 40 CFR 261
EPA hazardous waste no. U129 (hazardous substance)

HEXACHLOROCYCLOPENTADIENE (HCCP). [77-47-4]

ACGIH-TLV
TWA: 0.01 ppm, 0.1 mg/m^3
STEL: 0.03 ppm, 0.3 mg/m^3

TSCA 4(e), ITC
Fourth Report of the Interagency Testing Committee to the administrator, Environmental Protection Agency, April 1979; 44 FR 31866, June 1, 1979, responded to by EPA administrator 47 FR 58023, December 29, 1982

*TSCA 8(d) Unpublished Health and Safety Studies Reporting, 40 CFR 716, 47 FR 387, September 2, 1982

TSCA-CHIP available

CWA 304(a)(1), 45 FR 79318, November 28, 1980, Water Quality Criteria

Freshwater Aquatic Life

The available data for hexachlorocyclopentadiene indicate that acute and chronic toxicity to freshwater aquatic life occurs at concentrations as low as 7.0 and 5.2 µg/liter, respectively, and would occur at lower concentrations among species that are more sensitive than those tested.

Saltwater Aquatic Life

The available data for hexachlorocyclopentadiene indicate that acute toxicity to saltwater aquatic life occurs at concentrations as low as 7.0 µg/liter and would occur at lower concentrations among species that are more sensitive than those tested. No data are available concerning the chronic toxicity of hexachlorocyclopentadiene to sensitive saltwater aquatic life.

Human Health

For comparison purposes, two approaches were used to derive criterion levels for hexachlorocyclopentadiene. Based on available toxicity data, for the protection of public health, the derived level is 206 µg/liter. Using available organoleptic data, for controlling undesirable taste and odor quality of ambient water, the estimated level is 1.0 µg/liter. It should be recognized that organoleptic data as a basis for establishing a water quality criterion have limitations and have no demonstrated relationship to potential adverse human health effects.

CWA 307(a)(1), Priority Pollutant

*CWA 311(b)(2)(A), 40 CFR 116, 117
Discharge RQ = 1 pound (0.454 kilogram)

CAA 112
Chemical proposed to be assessed as a Hazardous Air Pollutant by a House of Representatives bill to amend the CAA

*RCRA 40 CFR 261
EPA hazardous waste no. U130 (hazardous substance)

1,2,3,6,7,8(7,8,9)-HEXACHLORODIBENZO-p-DIOXIN (HCDD MIX).
[34465-46-8, 19408-74-3]

NCI—carcinogenic in animals

1,2,3,4,10,10-HEXACHLORO-6,7-EPOXY-1,4,4a,5,6,7,8,8a-OCTAHYDRO—endo,endo-1,4:5,8-DIMETHANONAPHTHALENE.
[72-20-8]

*RCRA 40 CFR 261
EPA hazardous waste no. P051 (hazardous substance)

1,2,3,4,10,10-HEXACHLORO-6,7-EPOXY-1,4,4a,5,6,7,8,8a-OCTAHYDRO-endo,exo-1,4:5,8-DIMETHANONAPHTHALENE.
[60-57-1]

*RCRA 40 CFR 261
EPA hazardous waste no. P037 (hazarodus substance)

HEXACHLOROETHANE. [67-72-1]

*OSHA 29 CFR 1910.1000 Table Z-1

Skin

PEL(TWA): 1 ppm, 10 mg/m^3

ACGIH-TLV (1982 Addition)
TWA: 10 ppm, 100 mg/m^3

OSHA Candidate List of Potential Occupational Carcinogens (EPA Carcinogen Assessment Group List), 45 FR 53672, August 12, 1980

NCI-carcinogenic in animals

TSCA 4(e), ITC
Eighth Report of the TSCA Interagency Testing Committee to the administrator, Environmental Protection Agency, April 1981; 46 FR 28138, May 22, 1981, EPA responded to the committee's recommendations for testing, 47 FR 18175, April 28, 1982

*TSCA 8(d) Unpublished Health and Safety Studies Reporting, 40 CFR 716, 47 FR 38800, September 2, 1982; 48 FR 13178, March 30, 1983

TSCA-CHIP available

CWA 307(a)(1), 40 CFR 125, Priority Pollutant

*RCRA 40 CFR 261
EPA hazardous waste no. U131

1,2,3,4,10,10-HEXACHLORO-1,4,4a,5,8,8a-HEXAHYDRO-1,4:5,8-endo,endo-DIMETHANONAPHTHALENE.
[465-73-6]

*RCRA 40 CFR 261
EPA hazardous waste no. P060

1,2,3,4,10,10-HEXACHLORO-1,4,4a,5,8,8a-HEXAHYDRO-1,4:5,8-endo,exo-DIMETHANONAPHTHALENE.
[309-00-2]

*RCRA 40 CFR 261
EPA hazardous waste no. P004 (hazardous substance)

HEXACHLOROHEXAHYDRO-exo,exo-DIMETHANONAPHTHALENE.

*RCRA 40 CFR 261
EPA hazardous waste no. P060

HEXACHLORONAPHTHALENE.
[1335-87-1]

*OSHA 29 CFR 1910.1000 Table Z-1

Skin

PEL(TWA): 10 mg/m^3
ACGIH-TLV

Skin

TWA: 0.2 mg/m^3
STEL: 0.6 mg/m^3

HEXACHLOROPHENE. [70-30-4]

*RCRA 40 CFR 261
EPA hazardous waste no. U132

HEXACHLOROPROPENE. [1888-71-7]

*RCRA 40 CFR 261
EPA hazardous waste no. U243

1-HEXADECANETHIOL.

See Thiols, NIOSH Criteria Document.

HEXAETHYL TETRAPHOSPHATE. [7-57-58-4]

*RCRA 40 CFR 261
EPA hazardous waste no. P062

HEXAFLUOROACETONE. [684-16-2]

ACGIH-TLV

TWA: 0.1 ppm, 0.7 mg/m^3
STEL: 0.3 ppm, 2 mg/m^3

TSCA-CHIP available

HEXAFLUOROPROPENE. See Fluoroalkenes. [116-15-4]

TSCA 4(e), ITC
Seventh Report of the TSCA Interagency Testing Committee to the administrator, Environmental Protection Agency, October 1980; 45 FR 78432, November 25, 1980

*TSCA 8(d) Unpublished Health and Safety Studies Reporting, 40 CFR 716, 47 FR 38800, September 2, 1982; 48 FR 13178, March 30, 1983

HEXAMETHYLENE DIISOCYANATE.
[822-06-0]

See Diisocyanates, NIOSH Criteria Document.

HEXAMETHYL PHOSPHORAMIDE.
See Phosphoric triamide, hexamethyl-.
[680-31-9]

HEXAMETHYLPHOSPHORIC TRIAMIDE (HMPA).
See Phosphoric triamide, hexamethyl-.
[680-31-9]

HEXANE. [110-54-3]

*OSHA 29 CFR 1910.1000 Table Z-1
PEL(TWA): 500 ppm, 1800 mg/m^3
ACGIH-TLV (1982 Addition)
TWA: 50 ppm, 180 mg/m^3

See Alkanes (C_5–C_8), NIOSH Criteria Document.

NFPA—hazardous chemical (flammable chemical)
Health—1, flammability—3, reactivity—0
TSCA-CHIP available

HEXANE (other isomers).
ACGIH-TLV (1982 Addition)
TWA: 500 ppm, 1800 mg/m^3
STEL: 1000 ppm, 3600 mg/m^3

1,6-HEXANEDIAMINE, N,N'-DIBUTYL-. [4835-11-4]
*TSCA 8(a), 40 CFR 712, 47 FR 26992, June 22, 1982, Chemical Information Rule

HEXANEDINITRILE. See Adiponitrile. [111-69-3]

1-HEXANETHIOL. [111-31-9]

See Thiols, NIOSH Criteria Document.

1-HEXANOL. [111-27-3]
NFPA—hazardous chemical (flammable chemical)
health—1, flammability—2, reactivity—0

2-HEXANONE. [591-78-6]
*OSHA 29 CFR 1910.1000 Table Z-1
PEL(TWA): 100 ppm, 410 mg/m^3
ACGIH-TLV
TWA: 5 ppm, 20 mg/m^3
See Ketones, NIOSH Criteria Document.
TSCA-CHIP available

HEXAZINONE (VELPAR®).
FIFRA—Registration Standard Development—Data Evaluation and Development of Regulatory Position (herbicide)

HEXONE (METHYL ISOBUTYL KETONE). [108-10-1]
*OSHA 29 CFR 1910.1000 Table Z-1
PEL(TWA): 100 ppm, 410 mg/m^3

sec-HEXYL ACETATE. [142-92-7]
*OSHA 29 CFR 1910.1000 Table Z-1
PEL(TWA): 50 ppm, 300 mg/m^3
ACGIH-TLV
TWA: 50 ppm, 300 mg/m^3

HEXYLENE GLYCOL. [107-41-5]
ACGIH-TLV
TWA: ceiling limit, 25 ppm, 125 mg/m^3

HMPA. See Phosphoric triamide, hexamethyl-. [680-31-9]

HYCANTHONE (MESYLATE). [23255-93-8]
IARC—carcinogenic in animals

HYDRACRYLONITRILE. See 3-Hydroxypropanenitrile. [109-78-4]

HYDRAZINE. [302-01-2]
*OSHA 29 CFR 1910.1000 Table Z-1
Skin
PEL(TWA): 1 ppm, 1.3 mg/m^3
ACGIH-TLV
Skin
TWA: 0.1 ppm, 0.1 mg/m^3
Industrial substance suspect of carcinogenic potential for humans—which is suspect of inducing cancer, based on either (1) limited epidemiologic evidence, exclusive of clinical reports of single cases, or (2) demonstration of carcinogenesis in one or more animal species by appropriate methods. Worker exposure by all routes should be carefully controlled to levels consistent with the animal and human experience data.
NIOSH Criteria Document (Pub. No. 78-172), NTIS Stock No. PB 81 225690, June 26, 1978
The criteria and recommended standard apply to exposure of workers to hydrazine and its derivatives, methylhydrazine, 1,1-dimethylhydrazine, 1,2-dimethylhydrazine, and phenylhydrazine, and their salts. Common synonyms used for methylhydrazine are monomethylhydrazine and MMH; for 1,1-dimethylhydrazine, unsymmetrical or asymmetrical dimethylhydrazine and UDMH; and for 1,2-dimethylhydrazine, symmetrical dimethylhydrazine and SDMH. "Occupational exposure to hydrazines" is defined as work in any area where one or more of the hydrazines is stored, produced, processed, transported, handled, or otherwise used and present in such a manner that vapors or aerosols may be released in workroom air or

that the material may spill or splash onto the skin or into the eyes.

Occupational exposure to hydrazines shall be controlled so that employees are not exposed at concentrations greater than those specified below, expressed as milligrams of the free base per cubic meter of air (mg/m^3), determined as ceiling concentrations in any 2-hour period:

Hydrazine	0.04 mg/m^3 (0.03 ppm)^a
Methylhydrazine	0.08 mg/m^3 (0.04 ppm)
1,1-Dimethylhydrazine	0.15 mg/m^3 (0.06 ppm)
Phenylhydrazine	0.60 mg/m^3 (0.14 ppm)

These recommended limits are the lowest concentrations measured by the recommended method of analysis with an analytical precision of at least 15%. No limit is recommended for 1,2-dimethylhydrazine, since an acceptable method of sampling and analysis is presently unavailable. The standard is designed to protect the health and provide for the safety of employees for up to a 10-hour work shift, 40-hour workweek, over a working lifetime.

OSHA Candidate List of Potential Occupational Carcinogens (EPA Carcinogen Assessment Group List), 45 FR 53672, August 12, 1980

IARC—carcinogenic in animals

NFPA—hazardous chemical
health—3, flammability—3, reactivity—2

*RCRA 40 CFR 261
EPA hazardous waste no. U133 (reactive waste, toxic waste)

HYDRAZINECARBOTHIOAMIDE.

*RCRA 40 CFR 261
EPA hazardous waste no. P116

HYDRAZINE, 1,2-DIETHYL-.
[1615-80-1]

*RCRA 40 CFR 261
EPA hazardous waste no. U086

HYDRAZINE, 1,1-DIMETHYL-.
[57-14-7]

*RCRA 40 CFR 261
EPA hazardous waste no. U098

HYDRAZINE, 1,2-DIMETHYL-.
[540-73-8]

*RCRA 40 CFR 261
EPA hazardous waste no. U099

HYDRAZINE, 1,2-DIMETHYL-, DIHYDROCHLORIDE. Syn: *sym*-dimethylhydrazine dihydrochloride. [306-37-6]

OSHA Candidate List of Potential Occupational Carcinogens, 45 FR 53672, August 12, 1980

HYDRAZINE, 1,2-DIPHENYL-.
[122-66-7]

*RCRA 40 CFR 261
EPA hazardous waste no. U109

HYDRAZINE, METHYL-. Syn: Hydrazomethane. [60-34-4]

OSHA Candidate List of Potential Occupational Carcinogens, 45 FR 53672, August 12, 1980

*RCRA 40 CFR 261
EPA hazardous waste no. P068

HYDRAZINE, METHYL-, SULFATE (1:1). Syn: Methylhydrazine monosulfate. [302-15-8]

OSHA Candidate List of Potential Occupational Carcinogens, 45 FR 53672, August 12, 1980

HYDRAZINE, PHENYL-, MONOHYDROCHLORIDE. Syn: Phenylhydrazine hydrochloride. [59-88-1]

OSHA Candidate List of Potential Occupational Carcinogens, 45 FR 53672, August 12, 1980

HYDRAZINE HYDRATE. See Hydrazine monohydrate. [7803-57-8]

HYDRAZINE HYDROGEN SULFATE.
See Hydrazine sulfate (1:1). [10034-93-2]

HYDRAZINE MONOHYDRATE. Syn: Hydrazine hydrate. [7803-57-8]

OSHA Candidate List of Potential Occupational Carcinogens, 45 FR 53672, August 12, 1980

HYDRAZINE SULFATE (1:1). Syn: Hydrazine Hydrogen sulfate. [10034-93-2]

OSHA Candidate List of Potential Occupational Carcinogens, 45 FR 53672, August 12, 1980

HYDRAZOBENZENE. [122-66-7]

TSCA-CHIP available

HYDRAZOMETHANE. See Hydrazine, methyl-. [60-34-4]

HYDRIODIC ACID. [10034-85-2]

NFPA—hazardous chemical
health—3, flammability—0, reactivity—0

HYDROBROMIC ACID. See Hydrogen bromide. [10035-10-6]

HYDROCARBONS (NONMETHANE).
See Ozone.

*CAA § 109(b); part C§ 160-178; 40 CFR 50, National Ambient Air Quality Standard (NAAQS) (not to be exceeded more than once per year)

	$\mu g/m^3$
Primary	Secondary
3 hours (6–9 A.M.)	
160 (0.24 ppm)	Same as primary

HYDROCHLORIC ACID. See Hydrogen chloride. [7647-01-0]

HYDROCYANIC ACID. See Hydrogen cyanide. [74-90-8]

See Cyanide, hydrogen, and cyanide salts, NIOSH Criteria Document.

HYDROFLUORIC ACID. [7664-39-3]

See Hydrogen fluoride, NIOSH Criteria Document.

*CWA 311(b)(2)(A), 40 CFR 116, 117
Discharge RQ = 5000 pounds (2270 kilograms)

*RCRA 40 CFR 261
EPA hazardous waste no. U134 (hazardous substance, corrosive waste, toxic waste)

HYDROGEN. [1333-74-0]

ACGIH-TLV
TWA: asphyxiants—inert gas or vapors, when present in high concentrations in air, act primarily as simple asphyxiants without other significant physiologic effects
NFPA—hazardous chemical (flammable chemical)
health—0, flammability—4, reactivity—0

HYDROGEN (LIQUEFIED). [1333-74-0]

NFPA—hazardous chemical
health—3, flammability—4, reactivity—0

HYDROGENATED TERPHENYLS.
[92-94-4]

ACGIH-TLV
TWA: 0.5 ppm, 5 mg/m^3

HYDROGEN BROMIDE (HYDROBROMIC ACID).
[10035-10-6]

*OSHA 29 CFR 1910.1000 Table Z-1
PEL(TWA): 3 ppm, 10 mg/m^3
ACGIH-TLV
TWA: 3 ppm, 10 mg/m^3
NFPA—hazardous chemical
health—3, flammability—0, reactivity—0

HYDROGEN CHLORIDE. [7647-01-1]

*OSHA 29 CFR 1910.1000 Table Z-1
PEL(Ceiling): 5 ppm, 7 mg/m^3
ACGIH-TLV
TWA: ceiling limit, 5 ppm, 7 mg/m^3
NFPA—hazardous chemical
health—3, flammability—0, reactivity—0
*CWA 311(b)(2)(A), 40 CFR 116, 117
Discharge RQ = 5000 pounds (2270 kilograms)

HYDROGEN CYANIDE (HYDROCYANIC ACID). [74-90-8]

*OSHA 29 CFR 1910.1000 Table Z-1

PEL(TWA): 10 ppm, 11 mg/m^3
ACGIH-TLV

Skin

TWA: ceiling limit, 10 ppm, 11 mg/m^3
NFPA—hazardous chemical
health—4, flammability—4, reactivity—2
*CWA 311(b)(2)(A), 40 CFR 116, 117
Discharge RQ = 10 pounds (454 kilograms)
*RCRA 40 CFR 261
EPA hazardous waste no. P063 (hazardous substance)

HYDROGEN FLUORIDE (HYDROFLUORIC ACID). [7664-39-3]

*OSHA 1910.1000 Table Z-2
3 ppm, 8-hour TWA

ACGIH-TLV
TWA: 3 ppm, 2.5 mg/m^3
STEL: 6 ppm, 5 mg/m^3

NIOSH Criteria Document (PUB. No. 76-143), NTIS Stock No. PB81 226516, March 1, 1976
"Hydrogen fluoride" is defined as: (a) Gaseous or liquefied anhydrous hydrogen fluoride and aqueous solutions thereof (hydrofluoric acid). The abbreviations HF and HF acid, as used in this document, denote the anhydrous and aqueous forms, respectively. (b) Any gaseous fluoride that is a by-product of processes using or producing hydrogen fluoride as defined above and is emitted into the air concomitantly with HF or HF acid.
"Occupational exposure to hydrogen fluoride" is defined as exposure to HF at airborne concentrations at or above one-half the recommended time-weighted average (TWA) environmental limit. (a) Occupational exposure shall be controlled so that no worker is exposed to hydrogen fluoride at a TWA concentration greater than 2.5 mg of fluoride ion (atomic weight 19) per cubic meter of air (2.5 mg F/m^3) for up to a 10-hour workday, 40-hour workweek, or greater than a ceiling of 5.0 mg of fluoride ion per cubic meter of air (5.0 mg F/m^3) as determined by a sampling time of 15 minutes. If both particulate and gaseous fluorides are present, total fluoride exposure from all occupational sources shall not exceed the recommended TWA concentration.

Control of occupational exposure to any particulate inorganic fluorides produced as a by-product or an end product of the use or production of HF shall be governed by the criteria document, Occupational Exposure to Inorganic Fluorides. The standard is designed to protect the health and provide for the safety of workers for up to a 10-hour workday, 40-hour workweek, over a working lifetime.

NFPA—hazardous chemical
health—4, flammability—0, reactivity—0
*RCRA 40 CFR 261
EPA hazardous waste no. U134 (hazardous substance, corrosive waste, toxic waste)

HYDROGEN IODIDE. See Hydriodic acid. [10034-85-2]

HYDROGEN PEROXIDE (90%). [7722-84-1]

*OSHA 29 CFR 1910.1000 Table Z-1
PEL(TWA): 1 ppm, 1.4 mg/m^3

ACGIH-TLV
TWA: 1 ppm, 1.5 mg/m^3
STEL: 2 ppm, 3 mg/m^3

NFPA—hazardous chemical (35–52% by weight)
health—2, flammability—0, reactivity—1, oxy—oxidizing chemical

NFPA—hazardous chemical (52% by weight or greater)
health—2, flammability—0, reactivity—3, oxy—oxidizing chemical

HYDROGEN PHOSPHIDE. [7803-51-2]

*RCRA 40 CFR 261
EPA hazardous waste no. P096

HYDROGEN SELENIDE. [7783-07-5]

*OSHA 29 CFR 1910.1000 Table Z-1
PEL(TWA): 0.05 ppm, 0.2 mg/m^3
ACGIH-TLV
TWA: 0.05 ppm, 0.2 mg/m^3

HYDROGEN SULFIDE. [7783-06-4]

*OSHA 1910.1000 Table Z-2

20 ppm, acceptable ceiling concentration
50 ppm/10 minutes once only if no other measurable exposure occurs
Acceptable maximum peak

ACGIH-TLV
TWA: 10 ppm, 14 mg/m^3
STEL: 15 ppm, 21 mg/m^3

NIOSH Criteria Document (Pub. No. 77-158), NTIS Stock No. PB 274196, May 4, 1977

"Hydrogen sulfide" refers to either the gaseous or liquid forms of the compound. Synonyms for hydrogen sulfide include hydrosulfuric acid, sulfurated hydrogen, sulfur hydride, rotten-egg gas, and stink damp.

"Occupational exposure to hydrogen sulfide" refers to any workplace situation in which hydrogen sulfide is stored, used, produced, or may be evolved as a consequence of the process. All sections of this standard shall apply where there is occupational exposure to hydrogen sulfide.

Exposure to hydrogen sulfide shall be controlled so that no employee is exposed to hydrogen sulfide at a ceiling concentration greater than 15 mg of hydrogen sulfide per cubic meter of air (15 mg/m^3 or approximately 10 ppm), as determined with a sampling period of 10 minutes, for up to a 10-hour work shift in a 40-hour workweek.

Evacuation of the area shall be required if the concentration of hydrogen sulfide equals or exceeds 70 mg/m^3. The standard is designed to protect the health and provide for the safety of employees for up to a 10-hour work shift, 40-hour workweek, over a working lifetime.

NFPA—hazardous chemical
health—3, flammability—4, reactivity—0

*CWA 311(b)(2)(A), 40 CFR 116, 117
Discharge RQ = 100 pounds (45.4 kilograms)

*RCRA 40 CFR 261
EPA hazardous waste no. U135

HYDROPEROXIDE, 1-METHYL-1-PHENYLETHYL-.

*RCRA 40 CFR 261
EPA hazardous waste no. U096 (reactive waste)

HYDROQUINONE. [123-31-9]

*OSHA 29 CFR 1910.1000 Table Z-1
PEL(TWA): 2 mg/m^3

ACGIH-TLV
TWA: 2 mg/m^3
STEL: 4 mg/m^3

NIOSH Criteria Document (Pub. No. 78-155) NTIS Stock No. PB 274196, April 11, 1978

Synonyms for hydroquinone include hydroquinol, *para*-hydroquinone, dihydroxybenzene, *para*-dihydroxybenzene, 1,4-dihydroxybenzene, *para*-dioxybenzene, *para* benzenediol, 1,4-bensenediol, benzoquinol, benzohydroquinone, *para*-diphenol, and *para*-hydroxyphenol. A complete list of synonyms and trade names for hydroquinone may be found in Table XI-1 of the Criteria Document.

"Occupational exposure to hydroquinone" is defined as work in an area where hydroquinone is stored, produced, processed, or otherwise used, except as a component of other materials at a concentration of 5% or less by weight. This exception is made so that the recommended standard will not interfere with those work situations in which hydroquinone occurs as a component of other materials at a concentration of 5% or less by weight, for example, in most private or commercial photographic developing facilities. In these work situations, the employer will not be required to collect and analyze air samples, provide repiratory protective equipment and protective clothing, supply the special labels and posters specified, and maintain the required records. However, both the employer and employee must recognize the potential for development of adverse effects inherent in prolonged exposure to hydroquinone. Therefore, sufficient protection of the workers' health should be ensured by avoiding excessive contact of the chemical with the skin, eyes, and respiratory and gastrointestinal systems and by following effective procedures for maintaining cleanliness. If an employee is exposed to concentrations of airborne hydroquinone in the workplace at more than the recommended ceiling value, all sections of the recommended standard shall be complied with; if the employee is exposed at or below the recommended ceiling value, all sections of the recommended standard shall be complied with except 4(b), 8(a)(5), and 8(b). See the Criteria Document for details.

The employer shall control workplace concentrations of hydroquinone so that no employee is exposed at a concentration greater than 2.0 mg/m^3 (about 0.44 ppm) of air determined as a ceiling concentration during a 15-minute collection period.

The standard is designed to protect the health and provide for the safety of employees for up to a 10-hour work shift, 40-hour workweek, over a working lifetime.

TSCA 4(e), ITC
Fifth Report of the TSCA Interagency Testing Committee to the administrator, Environmental Protection Agency, November 1979; 44 FR 70664, December 7, 1979
*TSCA 8(d) Unpublished Health and Safety Studies Reporting, 40 CFR 716, 47 FR 387, September 2, 1982

HYDROXYDIMETHYLARSINE OXIDE. [75-60-5]
*RCRA 49 CFR 172.101
EPA hazardous waste no. U136

2-HYDROXYETHYLAMINE. *See* 2-Aminoethanol. [141-43-5]

beta-HYDROXYETHYLHYDRAZINE. *See* Ethanol, 2-hydrazino-. [109-84-2]

HYDROXYLAMINE. [7803-49-8]
NFPA—hazardous chemical
health—1, flammability—3, reactivity—3

HYDROXYLAMINE, *N*-NITROSO-*N*-PHENYL-, AMMONIUM SALT. Syn: Cupferron. [135-20-6]
OSHA Candidate List of Potential Occupational Carcinogens, 45 FR 53672, August 12, 1980
NCI—carcinogenic in animals

4-HYDROXY-4-METHYL-2-PENTANONE. *See* Diacetone alcohol. [123-42-2]

2-HYDROXY-2-METHYLPROPANENITRILE. [75-86-5]
NFPA—hazardous chemical
health—4, flammability—1, reactivity—2

3-HYDROXYPROPANENITRILE. [109-78-4]
NFPA—hazardous chemical
health—2, flammability—1, reactivity—1

2-HYDROXYPROPYL ACRYLATE. [999-61-1]
ACGIH-TLV

Skin
TWA: 0.5 ppm, 3 mg/m^3

HYPOCHLORITES.
FIFRA—Registration Standard Development—Data Evaluation and Development of Regulatory Position (disinfectants)

---
^{a}Approximate equivalents in parts of free base per million parts of air (ppm).

I

ICRF-159 [21416-87-5]
NCI—carcinogenic in animals

IMIDAZOLE-4-CARBOXAMIDE, 5-(3,3'-DIMETHYL-1-TRIAZENO)-.
Syn: dacarbazine. [4342-03-4]
OSHA—Candidate List of Potential Occupational Carcinogens, 45 FR 53672, August 12, 1980

2-IMIDAZOLIDINETHIONE. [96-45-7]
*RCRA 40 CFR 261
EPA hazardous waste no. U116

INDENE. [95-13-6]
ACGIH-TLV
TWA: 10 ppm, 45 mg/m^3
STEL: 15 ppm, 70 mg/m^3

INDENO(1,2,3-cd)PYRENE. [193-39-5]
OSHA Candidate List of Potential Occupational Carcinogens (EPA Carcinogen Assessment Group List), 45 FR 53672, August 12, 1980
IARC—carcinogenic in animals
CWA 307(a)(1), 40 CFR 125, Priority Pollutant
*RCRA 40 CFR 261
EPA hazardous waste no. U137

INDIUM AND COMPOUNDS. (as in) [7440-74-6]
ACGIH-TLV
TWA: 0.1 mg/m^3
STEL: 0.3 mg/m^3

INDOMETHACIN. [53-86-1]
*RCRA 40 CFR 261
EPA hazardous waste no. U245

IODINE. [7553-56-2]
*OSHA 29 CFR 1910.1000 Table Z-1
PEL(Ceiling): 1 ppm, 1 mg/m^3
ACGIH-TLV
TWA: Ceiling limit, 0.1 ppm, 1 mg/m^3

IODOFORM. [75-47-8]
ACGIH-TLV
TWA: 0.6 ppm, 10 mg/m^3
STEL: 1 ppm, 20 mg/m^3

IODOMETHANE. [74-88-4]
OSHA Candidate List of Potential Occupational Carcinogens (EPA Carcinogen Assessment Group List), 45 FR 53672, August 12, 1980

IRON.
*National Secondary Drinking Water Regulations, 40 CFR 13; 44 FR 42198, July 19, 1979, effective January 19, 1981
Maximum contaminant level—0.3 mg/liter

IRON DEXTRAN. [9004-66-4]
IARC—carcinogenic in animals
*RCRA 40 CFR 261
EPA hazardous waste no. U139

IRON DEXTRIN. [9004-51-7]
IARC—carcinogenic in animals

IRON OXIDE FUME. [1309-37-1]
*OSHA 29 CFR 1910.1000 Table Z-1
PEL(TWA): 10 mg/m^3
ACGIH-TLV

IRON PENTACARBONYL (as Fe). [13463-40-6]
ACGIH-TLV (1982 addition)
TWA: 0.1 ppm, 0.8 mg/m^3
STEL: 0.2 ppm, 0.16 mg/m^3

IRON SALTS, SOLUBLE, as Fe.
ACGIH-TLV
TWA: 1 mg/m^3
STEL: 2 mg/m^3

ISATIDINE. [15503-86-3]
IARC—carcinogenic in animals

ISOAMYL ACETATE. [123-92-2]
*OSHA 29 CFR 1910.1000 Table Z-1
PEL(TWA): 100 ppm, 525 mg/m^3
ACGIH-TLV
TWA: 100 ppm, 525 mg/m^3
STEL: 125 ppm, 655 mg/m^3

ISOAMYL ALCOHOL. [123-51-3]
*OSHA 29 CFR 1910.1000 Table Z-1
PEL(TWA): 100 ppm, 360 mg/m^3
ACGIH-TLV
TWA: 100 ppm, 360 mg/m^3
STEL: 125 ppm, 450 mg/m^3

1,3-ISOBENZOFURANDIONE, 4,5,6,7-TETRABROMO-. [632-79-1]
*TSCA 8(a), 40 CFR 712, 47 FR 26992, June 22, 1982, Chemical Information Rule

1-ISOBUTENYL METHYL KETONE. [141-79-7]
NFPA—hazardous chemical
health—3, flammability—3, reactivity—0

ISOBUTYL ACETATE. [110-19-0]
*OSHA 29 CFR 1910.1000 Table Z-1
PEL(TWA): 150 ppm, 700 mg/m^3
ACGIH-TLV
TWA: 150 ppm, 700 mg/m^3
STEL: 187 ppm, 875 mg/m^3

ISOBUTYL ALCOHOL. [78-83-1]
*OSHA 29 CFR 1910.1000 Table Z-1
PEL(TWA): 100 ppm, 300 mg/m^3
ACGIH-TLV
TWA: 50 ppm, 150 mg/m^3
STEL: 75 ppm, 225 mg/m^3
*RCRA 40 CFR 261
EPA hazardous waste no. U140 (ignitable waste, toxic waste)

ISOMETHYL METHYL KETONE. [108-10-1]
NFPA—hazardous chemical
health—2, flammability—3, reactivity—0

ISOCYANIC ACID, METHYL ESTER. [624-83-9]
*RCRA 40 CFR 261
EPA hazardous waste no. P064
*RCRA 40 CFR 261
EPA hazardous waste no. U161 (ignitable waste)

ISOCYANURATES (some products).
40 CFR 162.11, FIFRA-RPAR, Final Action, Voluntary Cancellation
Current Status: Voluntary cancellations being processed. Registrants have either undertaken or scheduled numerous toxicological studies.
Criteria Possibly Met or Exceeded: Kidney Effects

1H-ISOINDOLE-1,3(2H)-DIONE, 2-OXIRANYLMETHYL-. [5455-98-1]
*TSCA 8(a), 40 CFR 712, 47 FR 26992, June 22, 1982, Chemical Information Rule

ISONICOTINIC ACID HYDRAZIDE. [54-85-3]
IARC—carcinogenic in animals

ISONITROPROPANE. See Propane, 2-nitro-. [79-46-9]

ISOOCTYL ALCOHOL. [26952-21-6]
ACGIH-TLV
TWA: 50 ppm, 270 mg/m^3

ISOPENTYL ALCOHOL. [123-51-3]
NFPA—hazardous chemical (flammable chemical)
health—1, flammability—2, reactivity—0

ISOPHORONE. [78-59-1]
*OSHA 29 CFR 1910.1000 Table Z-1
PEL(TWA): 25 ppm, 140 mg/m^3
ACGIH-TLV
TWA: Ceiling limit, 5 ppm, 25 mg/m^3
See Ketones, NIOSH Criteria Document.

NFPA—hazardous chemical
health—2, flammability—2, reactivity—0

CWA 304(a)(1), 45 FR 79318, November 28, 1980, Water Quality Criteria

Freshwater Aquatic Life

The available data for isophorone indicate that acute toxicity to freshwater aquatic life occurs at concentrations as low as 117,000 µg/liter and would occur at lower concentrations among species that are more sensitive than those tested. No data are available concerning the chronic toxicity of isophorone to sensitive freshwater aquatic life.

Saltwater Aquatic Life

The available data for isophorone indicate that acute toxicity to saltwater aquatic life occurs at concentrations as low as 12,900 µg/liter and would occur at lower concentrations among species that are more sensitive than those tested. Not data are available concerning the chronic toxicity of isophorone to sensitive saltwater aquatic life.

Human Health

For the protection of human health from the toxic properties of isophorone ingested through water and contaminated aquatic organisms, the ambient water criterion is determined to be 5.2 mg/liter. For the protection of human health from the toxic properties of isophorone ingested through contaminated organisms alone, the ambient water criterion is determined to be 520 mg/liter.

CWA 307(a)(1), Priority Pollutant

TSCA 4(e), ITC
Fourth Report of the Interagency Testing Committee to the administrator, Environmental Protection Agency, April 1979, 44 FR 31866, June 1, 1979; 48 FR 727, January 6, 1983

*TSCA 8(d) Unpublished Health and Safety Studies Reporting, 40 CFR 716, 47 FR 387, September 2, 1982

ISOPHORONE DIISOCYANATE.
[4098-71-9]

ACGIH-TLV

Skin

TWA: 0.01 ppm, 0.09 mg/m^3

See Diisocyanates, NIOSH Criteria Document

ISOPHOSPHAMIDE. [3778-73-2]

NCI—carcinogenic in animals

ISOPHTHALONITRILE, TETRACHLORO-.

Syn: m-Tetrachlorophthalonitrile.
[1897-45-6]

OSHA Candidate List of Potential Occupational Carcinogens, 45 FR 53672, August 12, 1980

ISOPRENE. [77-79-5]

NFPA—hazardous chemical
health—2, flammability—4, reactivity—2

*CWA 311(b)(2)(A), 40 CFR 116, 117
Discharge RQ = 1000 pounds (454 kilograms)

ISOPROPALIN. [33820-53-0]

FIFRA—Registration Standard issued August 1981, NTIS# PB82 131293 (herbicide)

ISOPROPANOLAMINE DODECYLBENZENESULFONATE.
[42504-46-1]

*CWA 311(b)(2)(A), 40 CFR 116, 117
Discharge RQ = 1000 pounds (454 kilograms)

ISOPROPOXYETHANOL. [109-59-1]

ACGIH-TLV
TWA: 25 ppm, 105 mg/m^3
STEL: 75 ppm, 320 mg/m^3

ISOPROPYL ACETATE. [108-21-4]

*OSHA 29 CFR 1910.1000 Table Z-1
PEL(TWA): 250 ppm, 950 mg/m^3

ACGIH-TLV
TWA: 250 ppm, 950 mg/m^3
STEL: 310 ppm, 1185 mg/m^3

ISOPROPYL ALCOHOL. [67-63-0]

*OSHA 29 CFR 1910.1000 Table Z-1
PEL(TWA): 400 ppm, 980 mg/m^3

ACGIH-TLV
TWA: 400 ppm, 980 mg/m^3
STEL: 500 ppm, 1225 mg/m^3

NIOSH Criteria Document (Pub. No. 76-142), NTIS Stock No. PB 273873, March 2, 1976
Because it appears that exposure to carcinogenic agent(s) may occur in the manufacture of isopropyl alcohol, it is recommended that employers engaged in the manufacture of isopropyl alcohol provide special medical surveillance procedures for employees and ensure that employees follow special work practices. Regulated areas shall be established and maintained where isopropyl alcohol is manufactured. Access to these regulated areas shall be limited to authorized persons. A daily roster shall be made of persons authorized to enter; these rosters shall be maintained for at least 30 years. Employers shall ensure that before employees leave a regulated area they remove and leave protective clothing at the point of exit. In addition, it is recommended that employers engaged in the manufacture of isopropyl alcohol install special engineering controls to prevent all exposure of employees to carcinogenic agents.

These criteria and the recommended standard apply to workplace occupational exposures of employees to isopropyl alcohol. Synonyms for isopropyl alcohol include isopropanol, avantine, 2-propanol, sec-propyl alcohol, dimethylcarbinol, lutosol, petrohol, and propan-2-ol.

"Manufacture of isopropyl alcohol" means a process involved in the production of isopropyl alcohol using sulfuric acid.

"Isopropyl alcohol-manufacturing area" is a controlled area consisting of all process equipment beginning with the reactor in which propylene feed enters and ending with the column where the refined isopropyl alcohol and other refined products emerge.

"Action level" means one-half of the time-weighted average limit (TWA) for isopropyl alcohol.

"Occupational exposure to isopropyl alcohol" means exposure at or above the action level. Exposure to isopropyl alcohol at concentrations less than one-half of the workplace environmental limit will require adherence to Sections 4(a), 5(a,b), 6(a–f), and 7. See the Criteria Document for details.

Employee exposure to airborne isopropyl alcohol shall not exceed 400 parts per million (400 ppm) parts of air by volume (approximately 984 mg/m^3 of air) determined as a TWA exposure for up to a 10-hour workday, 40-hour workweek, with a ceiling of 800 ppm (approximately 1968 mg/m^3) as determined by a sampling time of 15 minutes.

The standard is designed to protect the health and provide for the safety of employees for up to a 10-hour work shift, 40-hour workweek, over a working lifetime.

NFPA—hazardous chemical (flammable chemical)
health—1, flammability—3, reactivity—0
TSCA-CHIP available

ISOPROPYLAMINE. [75-13-0]

*OSHA 29 CFR 1910.1000
PEL(TWA): 5 ppm, 12 mg/m^3

ACGIH-TLV
TWA: 5 ppm, 12 mg/m^3
STEL: 10 ppm, 24 mg/m^3

NFPA—hazardous chemical
health—3, flammability—4, reactivity—0

n-ISOPROPYLANILINE. [643-28-7]

ACGIH-TLV

Skin

TWA: 2 ppm, 10 mg/m^3
STEL: 5 ppm, 20 mg/m^3

ISOPROPYLBENZENE. See Cumene. [98-82-8]

ISOPROPYL ETHER. See Diisopropyl ether. [108-20-3]

ISOPROPYL FORMATE. [625-55-8]

NFPA—hazardous chemical
health—2, flammability—3, reactivity—0

ISOPROPYL GLYCIDYL ETHER (IGE).
[4016-14-2]

*OSHA 29 CFR 1910.1000 Table Z-1
PEL(TWA): 50 ppm, 240 mg/m^3

ACGIH-TLV
TWA: 50 ppm, 240 mg/m^3
STEL: 75 ppm, 360 mg/m^3

See Glycidyl ethers, NIOSH Criteria Document.

ISOPROPYL OILS.

IARC—carcinogenic in humans

ISOPROPYL PERCARBONATE. See
Diisopropyl peroxydicarbonate. [105-64-6]

ISOSAFROLE. [120-58-1]

OSHA Candidate List of Potential Occupational Carcinogens (EPA Carcinogens Assessment Group List), 45 FR 53672, August 12, 1980

IARC—carcinogenic in animals

*RCRA 40 CFR 261
EPA hazardous waste no. U141

3(2H)-ISOXAZOLONE, 5-(AMINOMETHYL)-.

*RCRA 40 CFR 261
EPA hazardous waste no. P007

J

JACUTIN. *See* lindane. [58-89-9]

K

KAOLIN.

ACGIH-TLV
TWA: nuisance particulate—30 mppcf or 10 mg/m^3 of total dust < 1% quartz, or 5 mg/m^3 respirable dust
STEL: 20 mg/m^3

KELTHANE®. [115-32-2]

FIFRA—Registration Standard Development—Data Evaluation and Development of Regulatory Position (insecticide)
*CWA 311(b)(2)(A), 40 CFR 116, 117
Discharge RQ = 5000 pounds (2270 kilograms)

KEPONE. [143-50-0]

NIOSH Criteria Document—no NIOSH publication number is assigned, available only from NIOSH, January 27, 1976
"Occupational exposure to kepone" is defined as exposure to airborne kepone at concentrations greater than one-half of the workplace environmental limit for kepone. Exposure to kepone at concentrations less than one-half of the workplace environmental limit will require adherence to Sections 3, 4(a), 5, 6(b,c,e,f), and 7.

Kepone shall be controlled in the workplace so that the airborne concentration of kepone, sampled and analyzed according to the procedures in Appendices I and II, is not greater than 1 µg/m^3 of breathing zone air as a time-weighted average concentration for up to a 10-hour workday, 40-hour workweek. The standard is designed to protect the health and provide for the safety of workers for up to a 10-hour workday, 40-hour workweek, over a working lifetime.

OSHA Candidate List of Potential Occupational Carcinogens (EPA Carcinogen Assessment Group List), 45 FR 53672, August 12, 1980

NCI—carcinogenic in animals
40CFR 162.11, FIFRA-RPAR, Notice of Intent to Cancel/Suspend Issued
Current Status: PD 1 published, 41 FR 12333 (3/25/76).
PD 4 published, 42, FR 18885 (4/11/77).
Notice of Intent to Cancel, 42 FR 18885 (4/11/77).
Criteria Possibly Met or Exceeded: Oncogenicity
40 CFR 162.11, FIFRA-RPAR, Final Action, Voluntary Cancellation
Current Status: 42 FR 38205 (7/27/77), NTIS# PB80 216773
Criteria Possibly Met or Exceeded: Oncogenicity
*CWA 311(b)(2)(A), 40 CFR 116, 117
Discharge RQ = 1 pound (0.454 kilograms)
*RCRA 40 CFR 261
EPA hazardous waste no. U142 (hazardous substance)

KETENE. [463-51-4]

*OSHA 29 CFR 1910.1000 Table Z-1
PEL(TWA): 0.5 ppm, 0.9 mg/m^3
ACGIH-TLV
TWA: 0.5 ppm, 0.9 mg/m^3
STEL: 1.5 ppm, 3 mg/m^3

KETONES.

NIOSH Criteria Document (Pub. No. 78-173), NTIS Stock No. PB80 176076, June 30, 1978
These criteria and the recommended standard apply to exposure of employees in the workplace to acetone, (2-propanone), methyl ethylketone (2-butanone), methyl n-propyl ketone (2-pentanone), methyl n-butyl ketone (2-hexanone), methyl n-amyl ketone (2-heptanone), methyl isobutyl ketone (4-methyl-2-

TABLE 2 RECOMMENDED EXPOSURE LIMITS FOR THE KETONES

Ketone	TWA Concentration Limits	
	mg/m^3	Approximate ppm equivalents
Acetone	590	250
Methyl ethyl ketone	590	200
Methyl n-propyl ketone	530	150
Methyl n-butyl ketone	4	1
Methyl n-amyl ketone	465	100
Methyl iso-butyl ketone	200	50
Methyl iso-amyl ketone	230	50
Diisobutyl ketone	140	25
Cyclohexane	100	25
Mesityl oxide	40	10
Diacetone alcohol	240	50
Isophorone	23	4

pentanone), methyl isoamyl ketone (5-methyl-2-hexanone), diisobutyl ketone (2,6-dimethyl-4-heptanone), cyclohexanone, mesityl oxide (4-methyl-3-penten-2-one), diacetone alcohol (4-hydroxy-4-methyl-2-pentanone), and isophorone (3,5,5-trimethyl-2-cyclohexan-1-one), referred to as "ketones" in this document.

The action level is defined as half the time-weighted average (TWA) concentration envrionmental limit of each ketone, "Occupational exposure to ketones" is defined as exposure to ketones at a TWA concentration greater than the action level. Exposure at lower concentrations will require adherence to Sections 3(a), 4, 5, 6(b–c), 7, and 8(a).

Occupational exposure to ketones shall be controlled so that employees are not exposed at concentrations greater than the limits, in milligrams per cubic meter of air, in Table 2 as TWA concentrations for up to a 10-hour work shift, 40-hour workweek. The standard is designed to protect the health and provide for the safety of employees for up to a 10-hour work shift, 40-hour workweek, over a working lifetime.

L

LANDRIN. [12407-86-2]
FIFRA—Data Call In, 45 FR 75488, November 14, 1980
Status: agency decision reached

LANOLIN. [2686-99-9]
NFPA—hazardous chemical (flammable chemical)
health—0, flammability—1, reactivity—0

LARD OIL (commercial or animal).
NFPA—hazardous chemical (flammable chemical)
health—0, flammability—1, reactivity—0

LASIOCARPINE. [303-34-4]
OSHA Candidate List of Potential Occupational Carcinogens (EPA Carcinogen Assessment Group List), 45 FR 53672, August 12, 1980
IARC—carcinogenic in animals
NCI—carcinogenic in animals
*RCRA 40 CFR 261
EPA hazardous waste no. U143

LAUROYL PEROXIDE. *See* Dilauroyl peroxide. [105-74-8]

LEAD. [7439-92-1]
*OSHA Standard 29 CFR 1910.1025
1. *Permissible exposure limit (PEL).* The employer shall assure that no employee is exposed to lead at concentrations greater than 50 $\mu g/m^3$ of air averaged over an 8-hour period.
2. If an employee is exposed to lead for more than 8 hours in any workday, the permissible exposure limit, as a time-weighted average (TWA) for that day, shall be reduced according to the following formula:
maximum permissible limit ($\mu g/m^3$)
= 400 ÷ hours worked in the day

3. When respirators are used to supplement engineering and work practice controls to comply with the PEL and all the requirements or respiratory protection have been met, employee exposure, for the purpose of determining whether the employer has complied with the PEL, may be considered to be at the level provided by the protection factor of the respirator for those periods the respirator is worn. Those periods may be averaged with exposure levels during periods when respirators are not worn to determine the employee's daily TWA exposure.
4. "Action level" means employee exposure, without regard to the use of respirators, to an airborne concentration of lead of 30 $\mu g/m^3$ of air averaged over an 8-hour period. Shall not apply to the construction industry or to agricultural operations.

CWA 304(a)(1), 45 FR 79318, November 28, 1980, Water Quality Criteria

Freshwater Aquatic Life

For total recoverable lead the criterion (in micrograms per liter) to protect freshwater aquatic life as derived using the guidelines is the numerical value given by $e(2.35[\ln(\text{hardness})] - 9.48)$ as a 24-hour average and the concentration (in micrograms per liter) should not exceed the numerical value given by $e(1.22[\ln(\text{hardness})] - 0.47)$ at any time. For example, at hardnesses of 50, 100, and 200 mg/liter as $CaCO_3$, the criteria are 0.75, 3.8, and 20 μg/liter, respectively, as 24-hour averages, and the concentrations should not exceed 74, 170, and 400 μg/liter, respectively, at any time.

Saltwater Aquatic Life

The available data for total recoverable lead indicate that acute and chronic toxicity to saltwater aquatic life occur at concentrations as low as 668 and 25 µg/liter, respectively, and would occur at lower concentrations among species that are more sensitive than those tested.

Human Health

The ambient water quality criterion for lead is recommended to be identical to the existing drinking water standard which is 50 µg/liter. Analysis of the toxic effects data resulted in a calculated level that is protective to human health against the ingestion of contaminated water and contaminated aquatic organisms. The calculated value is comparable to the present standard. For this reason a selective criterion based on exposure solely from consumption of 6.5 g of aquatic organisms was not derived.

CWA 307(a)(1), Priority Pollutant

*National Interim Primary Drinking Water Regulations, 40 CFR 141; 40 FR 59565, December 24, 1975; amended by 41 FR 28402, July 9, 1976; 44 FR 68641, November 29, 1979; corrected by 45 FR 15542, March 11, 1980; 45 FR 57342, August 27, 1980; 47 FR 18998, March 3, 1982; corrected by 47 FR 10998, March 12, 1982

Maximum contaminant level—0.05 mg/liter

*CAA § 109 I (6); Part C 160-178; 40 CFR 50, National Ambient Air Quality Standard (NAAQS) (not to be exceeded more than once per year)

	µg/m^3	
	Primary	Secondary
Quarterly	1.5	Same as primary

LEAD, inorganic dusts and fumes as Pb. [7439-92-1]

ACGIH-TLV
TWA: 0.15 mg/m^3
STEL: 0.45 mg/m^3

NIOSH Criteria Document, revised (Pub. No. 78-158), NTIS Stock No. PB81 225278, May 1978

"Inorganic lead" means lead oxides, metallic lead, and lead salts (including organic salts such as lead soaps but excluding lead arsenate). "Exposure to inorganic lead" is defined as exposure above half the recommended workroom environmental standard. Exposures at lower environmental concentrations will require adherence to Section 7(a).

Occupational exposure to inorganic lead shall be controlled so that workers shall not be exposed to inorganic lead at a concentration greater than 0.10 mg Pb/m^3 determined as a time-weighted average (TWA) exposure for a 10-hour workday, 40-hour workweek.

Biologic monitoring shall be made available to all workers subject to exposure to inorganic lead. This consists of sampling and analysis of whole blood for lead content. Such monitoring shall be performed to ensure that no worker absorbs an unacceptable amount of lead. Unacceptable absorption of lead posing a risk of lead poisoning is demonstrated at levels of 0.060 mg Pb/100 g of whole blood or greater.

All workers subject to exposure to inorganic lead shall be offered biologic monitoring at least every 6 months. The schedule of biologic monitoring may be made more frequent if indicated by a professional industrial hygiene survey. If environmental sampling and analysis show that environmental levels are at or greater than the recommended environmental levels, the interval of biologic monitoring shall be halved, that is, blood analysis shall be conducted quarterly. This increased frequency shall be continued for at least 6 months after the high environmental level has been shown.

If a blood lead level of 0.060 mg Pb/100 g or greater is found, and confirmed by a second sample to be taken within 2 weeks, steps to reduce absorption of lead shall be taken as soon as the high levels are confirmed. Steps to be considered should include improvement of environmental controls, of personal protection of personal hygiene, and use of administrative controls. A medical examination for possible lead poisoning shall be made available to workers with unacceptable blood lead levels.

Each employee who absorbs unacceptable amounts of lead as indicated by biologic monitoring shall be examined as soon as practicable after such absorption is demonstrated and confirmed, and at least every 3 months thereafter until blood lead levels have returned to below the acceptable limit, that is, below 0.060 mg/100 g of blood. If clinical evidence of plumbism is developed from these medical examinations, the worker shall be kept under a physician's care until the worker has completely recovered or maximal improvement has occurred. The standard is designed to protect the health and provide for the safety of workers for a 10-hour workday, 40-hour workweek over a working lifetime.

LEAD ACETATE. [301-04-2]

IARC—carcinogenic in animals
*CWA 311(b)(2)(A), 40 CFR 116, 117
Discharge RQ = 5000 pounds (2270 kilograms)
*RCRA 40 CFR 261
EPA hazardous waste no. U144 (hazardous substance)

LEAD ARSENATE, as Pb.
[10102-48-4, 7784-40-9, 7645-25-2]

See Arsenic, inorganic, NIOSH Criteria Document.
ACGIH-TLV
TWA: 0.15 mg/m^3
STEL: 0.45 mg/m^3
Intended changes for 1982:
as $Pb_3(AsO_4)_2$
TWA: 0.15 mg/m^3
STEL: 0.45 mg/m^3
NFPA—hazardous chemical
health—2, flammability—0, reactivity—0
*CWA 311(b)(2)(A), 40 CFR 116, 117
Discharge RQ = 5000 pounds (2270 kilograms)

LEAD CHLORIDE. [7758-95-4]

*CWA 311(b)(2)(A), 40 CFR 116, 117
Discharge RQ = 5000 pounds (2270 kilograms)

LEAD CHROMATE, as Cr.
[18454-12-1]

ACGIH-TLV
TWA: 0.05 ppm

Industrial substance suspect of carcinogenic potential for humans—which is suspect of inducing cancer based either on (1) limited epidemiologic evidence, exclusive of clinical reports of single cases, or (2) demonstration of carcinogenesis in one or more animal species by appropriate methods. Worker exposure by all routes should be carefully controlled to levels consistent with the animal and human experience data.

LEAD DIOXIDE. [1309-60-0]

See Lead, inorganic dusts and fumes, NIOSH Criteria Document.

LEAD FLUOBORATE. [13814-96-5]

*CWA 311(b)(2)(A), 40 CFR 116, 117
Discharge RQ = 5000 pounds (2270 kilograms)

LEAD FLUORIDE. [7783-46-2]

*CWA 311(b)(2)(A), 40 CFR 116, 117
Discharge RQ = 1000 pounds (454 kilograms)

LEAD IODIDE. [1010-63-0]

*CWA 311(b)(2)(A), 40 CFR 116, 117
Discharge RQ = 5000 pounds (2270 kilograms)

LEAD MONOXIDE. [1317-36-8]

See Lead, inorganic dusts and fumes, NIOSH Criteria Document.

LEAD NAPTHENATE. [61790-14-5]

TSCA 4(e), ITC
Twelfth Report of the Interagency Testing Committee to the administrator, Environmental Protection Agency, May 1983; 48 FR 24443, June 1, 1983
*TSCA 8(a), 40 CFR 712, 48 FR 28443, June 22, 1983, Chemical Information Rule

LEAD NITRATE. [10099-74-8]

NFPA—hazardous chemical
when nonfire: health—0, flammability—0, reactivity—0, oxy—oxidizing chemical
when fire: health—1, flammability—0, reactivity—0, oxy-oxidizing chemical
*CWA 311(b)(2)(A), 40 CFR 116, 117
Discharge RQ = 5000 pounds (2270 kilograms)

LEAD PHOSPHATE. [7446-27-7]

IARC—carcinogenic in animals
*RCRA 40 CFR 261
EPA hazardous waste no. U145

LEAD STEARATE.
[7428-48-0, 1072-35-1, 52652-59-2]
*CWA 311(b)(2)(A), 40 CFR 116, 117
Discharge RQ = 5000 pounds (2270 kilograms)

LEAD SUBACETATE. [1335-32-6]

IARC—carcinogenic in animals
*RCRA 40 CFR 261
EPA hazardous waste no. U146

LEAD SULFATE. [7446-14-2]

*CWA 311(b)(2)(A), 40 CFR 116, 117
Discharge RQ = 5000 pounds (2270 kilograms)

LEAD SULFIDE. [1314-87-0]

*CWA 311(b)(2)(A), 40 CFR 116, 117
Discharge RQ = 5000 pounds (2270 kilograms)

LEAD SULFOCYANATE. See Lead thiocyanate. [592-87-0]

LEAD THIOCYANATE. [592-87-0]

NFPA—hazardous chemical
health—1, flammability—1, reactivity—1
*CWA 311(b)(2)(A), 40 CFR 116, 117
Discharge RQ = 5000 pounds (2270 kilograms)

LIGHT GREEN SF. [5141-20-8]

IARC—carcinogenic in animals

LIME (UNSLAKED). See Calcium oxide. [1305-78-8]

LIMESTONE [1317-65-3]

ACGIH-TLV
TWA: nuisance particulate—30 mppcf or 10 mg/m^3 of total dust <1% quartz, or 5 mg/m^3 respirable dust
STEL: 20 mg/m^3

LINDANE. (1,2,3,4,5,6-hexachlorocyclohexane, gamma isomer). [58-89-9]

*OSHA 29 CFR 1910.1000 Table Z-1

Skin

PEL(TWA): 0.5 mg/m^3
ACGIH-TLV

Skin

TWA: 0.5 mg/m^3
STEL: 1.5 mg/m^3
IARC—carcinogenic in animals
40 CFR 162.11, FIFRA-RPAR issued
Current Status: PD 1 published, 42 FR 9816 (2/18/77); comment period closed 6/20/77: NTIS No. PB80 213911.
PD 2/3 completed and Notice of Determination published, 45 FR 45362 (7/30/80); comment period closed 9/15/80.
PD 4 being developed.
Criteria Possibly Met or Exceeded: Oncogenicity, Teratogenicity, Reproductive Effects, and Actue Toxicity
NFPA—hazardous chemical
health—2, flammability—0, reactivity—0
when solution: health—2, flammability—1, reactivity—0
CWA 307(a)(1), 40 CFR 125, Priority Pollutant
CWA 311(b)(2)(A), 40 CFR 116, 117
Discharge RQ = 1 pound (0.454 kilogram)
*RCRA 40 CFR 261
EPA hazardous waste no. U129 (hazardous substance)
*National Interim Primary Drinking Water Regulations, 40 CFR 141; 40 FR 59565, December 24, 1975; amended by 41 FR 28402, July 9, 1976; 44 FR 68641, November 29, 1979; corrected by 45 FR 15542, March 11, 1980; 45 FR 57342, August 27, 1980; 47 FR 18998, March 3, 1982; corrected by 47 FR 10998, March 12, 1982
Maximum contaminant level—0.004 mg/liter

LINSEED OIL (BOILED).
[8001-26-1]

NFPA—hazardous chemical (flammable chemical)
health—0, flammability—1, reactivity—0

LINURON. [330-55-2]

FIFRA—Registration Standard Development—Data Evaluation and Development of Regulatory Position (herbicide)

LIQUEFIED NATURAL GAS.

NFPA—hazardous chemical
health—3, flammability—4, reactivity—1

LITHIUM. [7439-93-2]

NFPA—hazardous chemical
health—1, flammability—1, reactivity—2,
W—water may be hazardous in fire fighting

LITHIUM ALUMINIUM HYDRIDE. See Lithium tetrahydroaluminate. [16853-85-3]

LITHIUM AND LITHIUM COMPOUNDS.

[546-89-4, 554-13-2, 556-63-8, 1310-65-2, 7439-93-2, 7447-41-8, 7550-35-8, 7580-67-8, 7782-89-0, 7789-24-4, 7790-69-4, 10102-24-6, 10377-51-2, 12007-60-2, 12057-24-8, 13453-69-5, 16853-85-3]

TSCA-CHIP available

LITHIUM CHROMATE. [14307-35-8]

See Chromium (VI), NIOSH Criteria Document.

*CWA 311(b)(2)(A), 40 CFR 116, 117
Discharge RQ = 1000 pounds (454 kilograms)

LITHIUM DICHROMATE. See Dichromates.

See Chromium (VI), NIOSH Criteria Document.

LITHIUM HYDRIDE. [7580-67-8]

*OSHA 29 CFR 1910.1000 Table Z-1
PEL(TWA): 0.025 mg/m^3
ACGIH-TLV
TWA: 0.025 mg/m^3
NFPA—hazardous chemical
health—3, flammability—4, reactivity—2,
W—water may be hazardous in fire fighting

LITHIUM TETRAHYDROALUMINATE. [16853-85-3]

NFPA—hazardous chemical
health—3, flammability—1, reactivity—2,
W—water may be hazardous in fire fighting

LNG. See Liquefied natural gas, l.p.g. (liquefied petroleum gas).

*OSHA 29 CFR 1910.1000 Table Z-1
PEL(TWA): 1000 ppm, 1800 mg/m^3
ACGIH-TLV
TWA: 1000 ppm, 1800 mg/m^3
STEL: 1250 ppm, 2250 mg/m^3

LOREXANE. See Lindane. [58-89-9]

LOX. See Oxygen (liquid). [7782-44-7]

LPG. See Liquified Natural Gas.

LUBRICATING OIL (MINERAL).

NFPA—hazardous chemical (flammable chemical)
health—0, flammability—1, reactivity—0

LUTEOSKYRIN. [21884-44-6]

IARC—carcinogenic in animals

LYE. See Sodium hydroxide, Potassium hydroxide. [1310-58-3, 1310-73-2]

M

MAGENTA. [632-99-5]
IARC—carcinogenic in animals

MAGNESITE. [546-90-0]
ACGIH-TLV
TWA: nuisance particulate—30 mppcf or 10 mg/m^3 of total dust <1% quartz, or 5 mg/m^3 respirable dust
STEL: 20 mg/m^3

MAGNESIUM (including alloys). [7439-95-4]
NFPA—hazardous chemical
health—0, flammability—1, reactivity—2, W —water may be hazardous in fire fighting

MAGNESIUM DICHROMATE. See Dichromates.

MAGNESIUM NITRATE. [10377-60-3]
NFPA—hazardous chemical
when nonfire: health—0, flammability—0, reactivity—0, oxy—oxidizing chemical
when fire: health—1, flammability—0, reactivity—0, oxy—oxidizing chemical

MAGNESIUM OXIDE FUME. [1309-48-4]
*OSHA 29 CFR 1910.1000 Table Z-1
PEL(TWA): 15 mg/m^3
ACGIH-TLV
TWA: 10 mg/m^3

MAGNESIUM PERCHLORATE. [10034-81-8]
NFPA—hazardous chemical
health—1, flammability—0, reactivity—0, oxy—oxidizing chemical

MAGNESIUM PHOSPHIDE. [12057-74-8]

FIFRA—Registration Standard Development—Data Evaluation and Development of Regulatory Position (insecticide)

MALATHION. [121-75-5]
*OSHA 29 CFR 1910.1000 Table Z-1
Skin
PEL(TWA): 15 mg/m^3
ACGIH TLV
Skin
TWA: 10 mg/m^3
NIOSH Criteria Document (Pub. No. 76-205), NTIS Stock No. PB 267070, June 30, 1976

"Malathion" is defined as O,O-dimethyl S-(1,2-dicarboethoxyethyl) dithiophosphate, regardless of production process, alone or in combination with other compounds. "Action level" is defined as one-half the recommended time-weighted average (TWA) environmental exposure limit for malathion. "Occupational exposure to malathion" is defined as exposure to airborne malathion at concentrations greater that the action level.

The criteria and recommended standard apply to any area in which malathion or materials containing malathion, alone or in combination with other substances, is produced, packaged, processed, mixed, blended, handled, or stored in large quantities, or applied.

"Overexposure" is defined as either known or suspected exposure that leads to the development of signs or symptoms of absorption of organophosphorous compounds and cholinesterase (ChE) inhibition. Exposure to malathion at concentrations less than or equal to the action level will not require adherence to the recommended standard except for Sections 2, 3, 4(a), 5, 6, 7(a–c), and 8(b).

When skin exposure is prevented, exposure to malathion in the workplace shall be controlled so that employees are not exposed to malathion at a TWA concentration greater than 15 mg/m^3 of air for up to a 10-hour work shift, 40-hour workweek. The standard is designed to protect the health of employees for up to a 10-hour work shift, 40-hour workweek, during a working lifetime.

*CWA 311(b)(2)(A), 40 CFR 116, 117
Discharge RQ = 10 pounds (4.54 kilograms)

MALEIC ACID. [110-16-7]

*CWA 311(b)(2)(A), 40 CFR 116, 117
Discharge RQ = 5000 pounds (2270 kilograms)

MALEIC ANHYDRIDE. [108-31-6]

*OSHA 29 CFR 1910.1000 Table Z-1
PEL(TWA): 0.25 ppm, 1 mg/m^3

ACGIH-TLV
TWA: 0.25 ppm, 1 mg/m^3

NFPA—hazardous chemical
health—3, flammability—1, reactivity—1
TSCA-CHIP available

*CWA 311(b)(2)(A), 40 CFR 116, 117
Discharge RQ = 5000 pounds (2270 kilograms)

CAA 112
Chemical proposed to be assessed as a Hazardous Air Pollutant by a House of Representatives bill to amend the CAA

*RCRA 40 CFR 261
EPA hazardous waste no. U147 (hazardous substance)

MALEIC HYDRAZIDE (MH). [123-33-1]

IARC—carcinogenic in animals
40 CFR 162.11, FIFRA-RPAR issued
Current Status: PD 1 published, 42 FR 56920 (10/28/77), comment period closed 12/14/77: NTIS# PB80 216470.
PD 2/3 being developed. Decision document being developed to return the potassium formulation to the registration process. Reproduction and mutagenicity studies are being generated to comply with requirements imposed under FIFRA Section 3(c)(2)(B). Criteria Possibly Met or Exceeded: Oncogenicity, Mutagenicity and Reproductive Effects

40 CFR 162.11, FIFRA-RPAR issued
Current Status: Notices of Intent to Suspend the registrations of all manufacturers of the DEA formulation of maleic hydrazide were issued on October 9, 1981 for noncompliance with requirements imposed under FIFRA Section 3(c)(2)(B). Final resolution in process.
Criteria Possibly Met or Exceeded: Oncogenicity, Mutagenicity, and Reproductive Effects

*RCRA 40 CFR 261
EPA hazardous waste no. U148

MALONIC ACID MONONITRILE. See
Cyanoacetic acid. [372-09-8]

MALONONITRILE. [109-77-3]

See Nitriles, NIOSH Criteria Document
*RCRA 40 CFR 261
EPA hazardous waste no. U149

MANGANESE. [7439-96-5]

*OSHA 29 CFR 1910.1000 Table Z-1
PEL(Ceiling): 5 mg/m^3

*National Secondary Drinking Water Regulations, 40 CFR 143; 44 FR 42198, July 19, 1979, effective January 19, 1981
Maximum contaminant level—0.05 mg/liter, odor—3 threshold odor number, pH—6.5–8.5

CAA 112
Chemical proposed to be assessed as a Hazardous Air Pollutant by a House of Representatives bill to amend the CAA

MANGANESE, as Mn, dust and compounds. [7439-96-5]

ACGIH-TLV
TWA: ceiling limit, 5 mg/m^3

MANGANESE, as Mn, fume. [7439-96-5]

ACGIH-TLV
TWA: 1 mg/m^3
STEL: 3 mg/m^3

MANGANESE CYCLOPENTADIENYL TRICARBONYL, as Mn. [12108-13-3]

Skin

TWA: 0.1 mg/m^3
STEL: 0.3 mg/m^3

MANGANESE TETROXIDE. [1317-35-7]

ACGIH-TLV
TWA: 1 mg/m^3

MANNOMUSTINE DIHYDROCHLORIDE. [551-74-6]

IARC—carcinogenic in animals

MARBLE/CALCIUM CARBONATE. [1317-65-3]

ACGIH-TLV
TWA: nuisance particulate—30 mggcf or 10 mg/m^3 of total dust < 1% quartz, or 5 mg/m^3 respirable dust.

MCPA. [94-74-6]

FIFRA—Registration Standard Development—Data Evaluation and Development of Regulatory Position (herbicide)

MECHLORETHAMINE HYDROCHLORIDE. See Diethylamine, 2,2'-dichloro-N-methyl-, hydrochloride. [55-86-7]

MEDROXYPROGESTERONE ACETATE. [71-58-9]
IARC—carcinogenic in animals

MELAMINE. [108-78-1]

NCI—carcinogenic in animals
TSCA-CHIP available

MELPHALAN. [148-82-3]

OSHA Candidate List of Potential Occupational Carcinogens (EPA Carcinogen Assessment Group List), 45 CFR 53672, August 12, 1980

IARC—carcinogenic in humans
*RCRA 40 CFR 261
EPA hazardous waste no. U150

MERCAPTOBENZOTHIAZOLE DISULFIDE. [120-78-5]
TSCA-CHIP available

MERCAPTODIMETHUR. [2032-65-7]
*CWA 311(b)(2)(A), 40 CFR 116, 117
Discharge RQ = 100 pounds (45.4 kilograms)

2-MERCAPTO-4-HYDROXYPYRIMIDINE. See Uracil, 2-thio-. [141-90-2]

MERCURIC CYANIDE. [592-04-1]
NFPA—hazardous chemical
health—3, flammability—0, reactivity—0
*CWA 311(b)(2)(A), 40 CFR 116, 117
Discharge RQ = 1 pound (0.454 kilogram)

MERCURIC NITRATE.
[10045-94-0, 7782-86-7, 10415-75-5]
*CWA 311(b)(2)(A), 40 CFR 116, 117
Discharge RQ = 10 pounds (4.54 kilograms)

MERCURIC SULFATE. [7783-35-9]
*CWA 311(b)(2)(A), 40 CFR 116, 117
Discharge RQ = 10 pounds (4.54 kilograms)

MERCURIC THIOCYANATE. [592-85-8]
*CWA 311(b)(2)(A), 40 CFR 116, 117
Discharge RQ = 10 pounds (4.54 kilograms)

MERCURY. [7439-97-6]
*OSHA 1910.1000 Table Z-2
1 mg/10 m^3 Acceptable ceiling concentration
CWA 304(a)(1), 45 FR 79318, November 28, 1980, Water Quality Criteria

Freshwater Aquatic Life

For total recoverable mercury the criterion to protect freshwater aquatic life as derived using the guidelines is 0.00057 μg/liter as a 24-hour average and the concentration should not exceed 0.0017 μg/liter at any time.

Saltwater Aquatic Life

For total recoverable mercury the criterion to protect saltwater aquatic life as derived using the guidelines is 0.025 µg/liter as a 24-hour average and the concentration should not exceed 3.7 µg/liter at any time.

Human Health

For the protection of human health from the toxic properties of mercury ingested through water and contaminated aquatic organisms, the ambient water criterion is determined to be 144 ng/liter.

For the protection of human health from the toxic properties of mercury ingested through contaminated aquatic organisms alone, the ambient water criterion is determined to be 146 ng/liter.

Note: These values include the consumption of freshwater, estuarine, and marine species.

CWA 307(a)(1), Priority Pollutant

*National Interim Primary Drinking Water Regulations, 40 CFR 141; 40 FR 59565, December 24, 1975; amended by 41 FR 28402, July 9, 1976; 44 FR 68641, November 29, 1979; corrected by 45 FR 15542, March 11, 1980; 45 FR 57342, August 27, 1980; 47 FR 18998, March 3, 1982; corrected by 47 FR 10998, March 12, 1982
Maximum contaminant level—0.002 mg/liter

*CAA 112, 40 CFR 61, Hazardous Air Pollutant

a. Emissions to the atmosphere from mercury ore processing facilities and mercury cell chlor-alkali plants shall not exceed 2300 g of mercury per 24-hour period.

b. Emissions to the atmosphere from sludge incineration plants, sludge drying plants, or a combination of these that process wastewater treatment plant sludges shall not exceed 3200 g of mercury per 24-hour period.

*RCRA 40 CFR 261
EPA hazardous waste no. U151

MERCURY, (ACETATO-O) PHENYL-. [62-38-4]

*RCRA 40 CFR 261
EPA hazardous waste no. P092

MERCURY, as Hg, alkyl compounds. [7439-97-6]

ACGIH-TLV

Skin

TWA: 0.01 mg/m^3
STEL: 0.03 mg/m^3

MERCURY, as Hg, all forms except alkyl vapor. [7439-97-6]

ACGIH-TLV (1982 addition)

Skin

TWA: 0.05 mg/m^3

MERCURY, as Hg, aryl and inorganic compounds. [7439-97-6]

ACGIH-TLV (1982 addition)
TWA: 0.1 mg/m^3

MERCURY, INORGANIC. [7439-97-6]

NIOSH Criteria Document (Pub. No. 73-11024), NTIS Stock No. PB 222223, August 13, 1973

"Inorganic mercury" in this document includes elemental mercury, and all inorganic mercury compounds and organic mercury compounds other than ethyl and methyl mercury compounds.

"Exposure to inorganic mercury" is defined as exposure to a concentration of inorganic mercury greater than 40% of the recommended level in the workplace. Exposure at lower environmental concentrations will require adherence to Section 7(a).

Occupational exposure to mercury shall be controlled so that workers are not exposed to inorganic mercury at a concentration greater than 0.05 mg Hg/m^3 determined as a time-weighted average (TWA) exposure for an 8-hour workday. The standard is designed to protect the health and provide for the safety of workers for an 8-hour workday, 40-hour workweek, over a working lifetime.

MERCURY FULMINATE. [628-86-4]

*RCRA 40 CFR 261

EPA hazardous waste no. P065 (reactive waste, toxic waste)

MERPHALAN. [531-76-0]
IARC—carcinogenic in animals

MESITYL OXIDE. *See* 1-Isobutenyl methyl ketone. [141-79-7]
*OSHA 29 CFR 1910.1000 Table Z-1
PEL(TWA): 25 ppm, 100 mg/m^3
ACGIH-TLV
TWA: 15 ppm, 60 mg/m^3
STEL: 25 ppm, 100 mg/m^3
See ketones, NIOSH Criteria Document.
TSCA 4(e), ITC
Fourth Report of the Interagency Testing Committee to the administrator, Environmental Protection Agency, April 1979, 44 FR 31866, June 1, 1979
*TSCA 8(d), Unpublished Health and Safety Studies Reporting, 40 CFR 716, 47 FR 387, September 2, 1982

MESTRANOL. [72-33-3]
IARC—carcinogenic in animals

METALANYL (RIDOMIL^R).
FIFRA—Registration Standard—Issued August 1981 (fungicide)

METHACRYLIC ACID. [79-41-4]
ACGIH-TLV
TWA: 20 ppm, 70 mg/m^3
NFPA—hazardous chemical
health—3, flammability—2, reactivity—2

METHACRYLONITRILE. [126-98-7]
*RCRA 40 CFR 261
EPA hazardous waste no. U152 (ignitable waste, toxic waste)

METHALLYL 2-NITROPHENYL ETHER. *See* Carbofuran intermediates. [13414-54-5]
TSCA 4(e), ITC
Eleventh Report of the Interagency Testing Committee to the administrator, Environmental Protection Agency, November 1982; 47 FR 54626, December 3, 1982, recommended but not designated for response within 12 months
TSCA 8(a), 40 CFR 712, 48 FR 22697, May 19, 1983, Chemical Information Rule
TSCA 8(d), Unpublished Health and Safety Studies Reporting, 40 CFR 716, 48 FR 28483, June 22, 1983

METHAMIDOPHOS (MONITOR^R). [10265-92-6]
FIFRA—Registration Standard Development—Data Evaluation and Development of Regulatory Position (insecticide)

METHANAL. *See* Formaldehyde. [50-00-0]

METHANAMINE, N-METHYL-. [124-40-3]
*RCRA 40 CFR 261
EPA hazardous waste no. U092 (hazardous substance, ignitable waste)

METHANE. [74-82-8]
ACGIH-TLV
TWA:
Asphyxiant—gas or vapors, when present in high concentration in air, act primarily as simple asphyxiants without other significant physiologic effects.

METHANEARSONATES. Includes amine methanearsonate, calcium acid methanearsonate, monoammonium methanearsonate (MAMA), monosodium methanearsonate (MSMA), disodium methanearsonate (DSMA).
40 CFR 162.11, FIFRA Pre-RPAR review
Current Status: Pre-RPAR evaluation under way
Criteria Possibly Met or Exceeded: Oncogenicity and Mutagenicity

METHANE, BROMO-. [74-83-9]
*RCRA 40 CFR 261
EPA hazardous waste no. U029

METHANECARBOXAMIDE. *See* Acetamide. [60-35-5]

METHANE, CHLORO-. [74-87-3]

*TSCA 8(a), 40 CFR 712, 47 FR 26992, June 22, 1982, Chemical Information Rule

*RCRA 40 CFR 261
EPA hazardous waste no. U045 (ignitable waste, toxic waste)

METHANE, CHLOROMETHOXY-.

*RCRA 40 CFR 261
EPA hazardous waste no. U046

METHANE, DIBROMO-. [74-61-6]

*RCRA 40 CFR 261
EPA hazardous waste no. U068

METHANE, DICHLORO-. [75-09-2]

*TSCA 8(a), 40 CFR 712, 47 FR 26992, June 22, 1982, Chemical Information Rule

*RCRA 40 CFR 261
EPA hazardous waste no. U080

METHANE, DICHLORODIFLUORO-. [75-71-8]

*RCRA 40 CFR 261
EPA hazardous waste no. U075

METHANE, IODO-. [74-88-4]

*RCRA 40 CFR 261
EPA hazardous waste no. U138

METHANE, OXYBIS(CHLORO)-. [542-88-1]

*RCRA 40 CFR 261
EPA hazardous waste no. P016

METHANESULFONIC ACID, ETHYL ESTER. [62-50-0]

*RCRA 40 CFR 261
EPA hazardous waste no. U119

METHANE, TETRACHLORO-. [56-23-5]

*RCRA 40 CFR 261
EPA hazardous waste no. U211 (hazardous substance)

METHANE, TETRANITRO-. [509-14-8]

*RCRA 40 CFR 261
EPA hazardous waste no. P112 (reactive waste)

METHANETHIOL. See Methyl mercaptan. [74-93-1]

See Thiols, NIOSH Criteria Document.

METHANETHIOL, TRICHLORO-.

*RCRA 40 CFR 261
EPA hazardous waste no. P118

METHANE, TRIBROMO-. [75-25-2]

*RCRA 40 CFR 261
EPA hazardous waste no. U225

METHANE, TRICHLORO-. [67-66-3]

*RCRA 40 CFR 261
EPA hazardous waste no. U044 (hazardous substance)

METHANE, TRICHLOROFLUORO-. [75-69-4]

*RCRA 40 CFR 261
EPA hazardous waste no. U121

METHANOIC ACID.

*RCRA 40 CFR 261
EPA hazardous waste no. U123 (hazardous substance, corrosive waste, toxic waste)

4,7-METHANOINDAN, 1,2,3,4,5,6,8,8-OCTACHLORO-3a,4,-7,7a-TETRAHYDRO-. [5103-71-9]

*RCRA 40 CFR 261
EPA hazardous waste no. U036 (hazardous substance)

4,7-METHANO-1H-INDENE, 1,4,5,6,7,8,8-HEPTACHLORO-3a,4,7,7a-TETRAHYDRO-. [76-44-8]

*RCRA 40 CFR 261
EPA hazardous waste no. P059 (hazardous substance)

2,4-METHANO-2H-INDENO [1,2-b:5,6,-b'] BISOXIRENE, OCTAHYDRO. [81-21-0]

*TSCA 8(a), 40 CFR 712, 47 FR 26992, June 22, 1982, Chemical Information Rule

4,7-METHANOISOBENZOFURAN-1, 3-DIONE, 4,5,6,7,8,8-HEXACHLORO-3a,4,7,7a-TETRAHYDRO-. [115-27-5]

*TSCA 8(a), 40 CFR 712, 47 FR 26992, June 22, 1982, Chemical Information Rule

4,7-METHANOISOBENZOFURAN-1,3-DIONE, 3a,4,7,7a-TETRAHYDROMETHYL-. [25134-21-8]

*TSCA 8(a), 40 CFR 712, 47 FR 26992, June 22, 1982, Chemical Information Rule

METHANOL. See Methyl alcohol. [67-56-1]

METHAPYRILENE. [91-80-5]

OSHA Candidate List of Potential Occupational Carcinogens (EPA Carcinogen Assessment Group List), 45 FR 53672, August 12, 1980

*RCRA 40 CFR 261
EPA hazardous waste no. U155

1,3,4-METHENO-1H-CYCLOBUTA(cd) PENTALENE, 1,1a,2,2,3,3a,4,5,5,5a,5b,6-DODECACHLOROOCTAHYDRO-. Syn: Mirex. [2385-85-5]

OSHA Candidate List of Potential Occupational Carcinogens, 45 FR 53672, August 12, 1980

IARC—carcinogenic in animals

40 CFR 162.11, FIFRA-RPAR, Notice of Intent to Cancel/Suspend Issued
Current Status: 41 FR 56703 (12/29/76) (most uses cancelled)

METHIDATION. [950-37-8]

FIFRA—Registration Standard Development—Data Evaluation and Development of Regulatory Position (insecticide)

METHOMYL. [16752-77-5]

ACGIH-TLV

Skin

TWA: 2.5 mg/m^3
FIFRA—Registration Standard—Issued September 1981 (insecticide)

*RCRA 40 CFR 261
EPA hazardous waste no. P066

METHOPRENE.

FIFRA—Registration Standard Development—Data Evaluation and Development of Regulatory Position (insecticide)

METHOXY CHLOR (1,1,1-TRICHLORO-1,1-BIS[p-METHOXY PHENYL] ETHANE. [72-43-5]

*OSHA 29 CFR 1910.1000 Table Z-1
PEL(TWA): 15 mg/m^3
ACGIH-TLV
TWA: 2.5 mg/m^3

*CWA 311(b)(2)(A), 40 CFR 116, 117
Discharge RQ = 1 pound (0.454 kilogram)

*National Interim Primary Drinking Water Regulations, 40 CFR 141; 40 FR 59565, December 24, 1975; amended by 41 FR 28402, July 9, 1976; 44 FR 68641, November 29, 1979; corrected by 45 FR 15542, March 11, 1980; 45 FR 57342, August 27, 1980; 47 FR 18998, March 3, 1982; corrected by 47 FR 10998, March 12, 1982
Maximum contaminant level—0.1 mg/liter

2-METHOXYETHANOL. See Methyl cellosolve. [109-86-4]

2-METHOXYETHANOL ACETATE. See Methyl cellosolve acetate. [110-49-6]

METHOXYFLURANE. [76-38-0]

See Waste anesthetic gases and vapors, NIOSH Criteria Document.

METHOXYMETHANE. See Dimethyl ether. [115-10-6]

4-METHOXYPHENOL. [150-76-5]

ACGIH-TLV (1982 Addition)
TWA: 5 mg/m^3

METHYL ACETATE. [79-20-9]

*OSHA 29 CFR 1910.1000 Table Z-1
PEL(TWA): 200 ppm, 610 mg/m^3
ACGIH-TLV
TWA: 200 ppm, 610 mg/m^3
STEL: 250 ppm, 760 mg/m^3
NFPA—hazardous chemical (flammable chemical)
health—1, flammability—3, reactivity—0

METHYL ACETYLENE (PROPYNE). [74-99-7]

*OSHA 29 CFR 1910.1000 Table Z-1
PEL(TWA): 1000 ppm, 1650 mg/m³
ACGIH-TLV
TWA: 1000 ppm, 1650 mg/m³
STEL: 1250 ppm, 2040 mg/m³

METHYL ACETYLENE-PROPADIENE MIXTURE (MAPP).

*OSHA 29 CFR 1910.1000 Table Z-1
PEL(TWA): 1000 ppm, 1800 mg/m³
ACGIH-TLV
TWA: 1000 ppm, 1800 mg/m³
STEL: 1250 ppm, 2250 mg/m³

METHYL ACRYLATE. [96-33-3]

*OSHA 29 CFR 1910.1000 Table Z-1

Skin

PEL(TWA): 10 ppm, 35 mg/m³
ACGIH-TLV

Skin

TWA: 10 ppm, 35 mg/m³
NFPA—hazardous chemical
health—2, flammability—3, reactivity—2

METHYLACRYLONITRILE. [126-98-7]
ACGIH-TLV

Skin

TWA: 1 ppm, 3 mg/m³
STEL: 2 ppm, 6 mg/m³

METHYLAL (DIMETHOXYMETHANE). [109-87-5]

*OSHA 29 CFR 1910.1000 Table Z-1
PEL(TWA): 1000 ppm, 3100 mg/m³
ACGIH-TLV
TWA: 1000 ppm, 3100 mg/m³
STEL: 1250 ppm, 3875 mg/m³

METHYL ALCOHOL. [67-56-1]

*OSHA 29 CFR 1910.1000 Table Z-1
PEL(TWA): 200 ppm, 260 mg/m³
ACGIH-TLV

Skin

TWA: 200 ppm, 260 mg/m³
STEL: 250 ppm, 310 mg/m³

NIOSH Criteria Document (Pub. No. 76-148), NTIS Stock No. PB 273806, March 12, 1976

Synonyms for methyl alcohol include wood spirit, carbinol, wood alcohol, wood naphtha, Colombian spirit, colonial spirit, methylol, pyroxylic spirit, monohydroxymethane, methyl hydroxide, and methanol. "Action level" means half of the time-weighted average (TWA) environmental exposure limit for methyl alcohol. "Occupational exposure to methyl alcohol" means exposure at or above the action level.

Occupational exposure to methyl alcohol shall be controlled so as not to exceed 200 parts per million parts of air by volume (262 mg/m³ of air) determined as a time-weighted average (TWA) exposure for up to a 10-hour workday, 40-hour workweek, with a ceiling of 800 ppm (1048 mg/m³) as determined by a sampling time of 15 minutes. The standard is designed to protect the health and provide for the safety of workers for up to a 10-hour workday, 40-hour workweek, over a working lifetime.

NFPA—hazardous chemical (flammable chemical)
health—1, flammability—3, reactivity—0
TSCA-CHIP available
*RCRA 40 CFR 261
EPA hazardous waste no. U154 (ignitable waste)

METHYLAMINES (MONO-, DI-, AND TRIMETHYLAMINE). [74-89-5, 75-50-3, 124-40-3]

*OSHA 29 CFR 1910.1000 Table Z-1
PEL(TWA): 10 ppm, 12 mg/m³
ACGIH-TLV
TWA: 10 ppm, 12 mg/m³
NFPA—hazardous chemical
health—3, flammability—4, reactivity—0
TSCA-CHIP available

METHYL AMYL ALCOHOL. See Methyl isobutyl carbinol. [108-11-2]

METHYL n-AMYL KETONE (2-HEPTANONE). [591-78-6, 110-43-0]

*OSHA 29 CFR 1910.1000 Table Z-1
PEL(TWA): 100 ppm, 465 mg/m³

ACGIH-TLV
TWA: 50 ppm, 235 mg/m³
STEL: 100 ppm, 465 mg/m³
See Ketones, NIOSH Criteria Document.
TSCA-CHIP available

N-METHYL ANILINE. [100-61-8]
*OSHA 29 CFR 1910.1000 Table Z-1

Skin

PEL(TWA): 2 ppm, 9 mg/m³
ACGIH-TLV (1982 Addition)

Skin

TWA: 0.5 ppm, 2 mg/m³
STEL: 1 ppm, 5 mg/m³

2-METHYLAZIRIDINE. [75-55-8]

IARC—carcinogenic in animals
*RCRA 40 CFR 261
EPA hazardous waste no. P067

METHYLAZOXYMETHANOL ACETATE. [592-62-1]

IARC—carcinogenic in animals

METHYLBENZENE. See Toluene. [108-88-3]

METHYL BROMIDE. [74-83-9]

*OSHA 29 CFR 1910.1000 Table Z-1

Skin

PEL(Ceiling): 20 ppm, 80 mg/m³
ACGIH-TLV

Skin

TWA: 5 ppm, 20 mg/m³
STEL: 15 ppm, 60 mg/m³
FIFRA—Data Call In, 45 FR 75488, November 14, 1980
Status: letter issued July 2, 1981
NFPA—hazardous chemical
health—3, flammability—1, reactivity—1

CWA 307(a)(1), 40 CFR 125, Priority Pollutant
*RCRA 40 CFR 261
EPA hazardous waste no. U029

1-METHYLBUTADIENE. See 1,3-Pentadiene. [504-60-9]

2-METHYL-1,3-BUTADIENE. See Isoprene. [78-79-5]

2-METHYLBUTANAL. See 2-Methylbutyraldehyde. [96-17-3]

2-METHYL-2-BUTANOL. [75-85-4]

NFPA—hazardous chemical (flammable chemical)
health—1, flammability—2, reactivity—0

METHYL BUTYL KETONE. See 2-Hexanone. [591-78-6]

2-METHYLBUTYRALDEHYDE. [96-17-3]

NFPA—hazardous chemical
health—2, flammability—3, reactivity—0

METHYL CELLOSOLVE. [109-86-4]
*OSHA 29 CFR 1910.1000 Table Z-1

Skin

PEL(TWA): 25 ppm, 80 mg/m³
ACGIH-TLV

Skin

TWA: 25 ppm, 80 mg/m³
STEL: 35 ppm, 120 mg/m³
Intended changes for 1982 (1982 Revision or Addition):

Skin

TWA: 5 ppm, 16 mg/m³
NIOSH Current Intelligence Bulletin No. 39—cancer-related bulletin
TSCA-CHIP available

METHYL CELLOSOLVE ACETATE. [110-49-6]

*OSHA 29 CFR 1910.1000 Table Z-1

184 METHYL CHLORIDE

Skin
PEL(TWA): 25 ppm, 120 mg/m^3
ACGIH-TLV

Skin
TWA: 25 ppm, 120 mg/m^3
STEL: 35 ppm, 170 mg/m^3
Intended changes for 1982 (1982 Revision or Addition):

Skin
TWA: 5 ppm, 24 mg/m^3
TSCA-CHIP available

METHYL CHLORIDE. [74-87-3]

*OSHA 1910.1000 Table Z-2
100 ppm, 8-hour TWA
200 ppm, acceptable ceiling concentration
300 ppm/5 minutes in any 3 hours, acceptable maximum peak
ACGIH-TLV
TWA: 50 ppm, 105 mg/m^3
STEL: 100 ppm, 205 mg/m^3
NFPA—hazardous chemical
health—2, flammability—4, reactivity—0
TSCA 4(e), ITC
Initial Report to the administrator, Environmental Protection Agency, TSCA Interagency Testing Committee, October 1, 1977; 42 FR 55026, October 12, 1977, responded to by the EPA administrator 45 FR 48524, July 18, 1980, removed from the Priority List in the Seventh Report of the TSCA Interagency Testing Committee to the administrator, Environmental Protection Agency, October 1980, 45 FR 78432, November 25, 1980
*TSCA 8(d), Unpublished Health and Safety Studies Reporting, 40 CFR 716, 47 FR 387, September 2, 1982
TSCA-CHIP available
CWA 307(a)(1), 40 CFR 125, Priority Pollutant
*RCRA 40 CFR 261
EPA hazardous waste no. U045 (ignitable waste, toxic waste)

METHYL CHLOROCARBONATE. [79-22-1]

*RCRA 40 CFR 261
EPA hazardous waste no. U156 (ignitable waste, toxic waste)

METHYL CHLOROFORM. See 1,1,1-Trichloroethane. [71-55-6]

METHYL CHLOROMETHYL ETHER. [107-30-2]

*OSHA Standard 29 CFR 1910.1006
Cancer-Suspect Agent
Shall not apply to solid and liquid mixtures containing less than 0.1% by weight or volume of methyl chloromethyl ether.

3-METHYLCHOLANTHRENE. [56-49-5]

OSHA Candidate List of Potential Occupational Carcinogens (EPA Carcinogen Assessment Group List), 45 FR 53672, August 12, 1980
*RCRA 40 CFR 261
EPA hazardous waste no. U157

METHYL CYANIDE. See Acetonitrile. [75-05-8]

METHYL 2-CYANOACRYLATE. [137-05-3]

ACGIH-TLV
TWA: 2 ppm, 8 mg/m^3
STEL: 4 ppm, 16 mg/m^3

METHYLCYCLOHEXANE. [108-87-2]

*OSHA 29 CFR 1910.1000 Table Z-1
PEL(TWA): 500 ppm, 2000 mg/m^3
ACGIH-TLV
TWA: 400 ppm, 1600 mg/m^3
STEL: 500 ppm, 2000 mg/m^3
TSCA-CHIP available

METHYLCYCLOHEXANOL. [25639-42-3]

*OSHA 29 CFR 1910.1000 Table Z-1
PEL(TWA): 100 ppm, 470 mg/m^3
ACGIH-TLV
TWA: 50 ppm, 235 mg/m^3
STEL: 75 ppm, 350 mg/m^3

2-METHYLCYCLOHEXANONE. [583-60-8]

*OSHA 29 CFR 1910.1000 Table Z-1

Skin

PEL(TWA): 100 ppm, 460 mg/m^3

ACGIH-TLV

Skin

TWA: 50 ppm, 230 mg/m^3
STEL: 75 ppm, 345 mg/m^3

METHYLCYCLOPENTADIENYL MANGANESE TRICARBONYL, as Mn. [12108-13-3]

ACGIH-TLV

Skin

TWA: 0.2 mg/m^3
STEL: 0.6 mg/m^3

METHYLCYCLOPENTANE. [96-37-7]

NFPA—hazardous chemical
health—2, flammability—3, reactivity—0

METHYL DEMETON. [8022-00-2]

ACGIH-TLV

Skin

TWA: 0.5 mg/m^3
STEL: 1.5 mg/m^3

METHYLDICHLOROSILANE. [563-41-7]

NFPA—hazardous chemical
health—3, flammability—3, reactivity—2, W
—water may be hazardous in fire fighting

4,4'-METHYLENE BIS (2-CHLOROANILINE). [101-14-4]

ACGIH-TLV

Skin

TWA: 0.02 ppm, 0.22 mg/m^3
Industrial substance suspect of carcinogenic potential for humans—which is suspect of inducing cancer, based on either (1) limited epidemiologic evidence, exclusive of clinical reports of single cases, or (2) demonstration of carcinogenesis in one or more animal species by appropriate methods. Worker exposure by all routes should be carefully controlled to levels consistent with the animal and human experience data.

OSHA Candidate List of Potential Occupational Carcinogens (EPA Carcinogen Assessment Group List), 45 FR 53672, August 12, 1980

IARC—carcinogenic in animals

TSCA-CHIP available

*RCRA 40 CFR 261
EPA hazardous waste no. U158

METHYLENE BIS(4-CYCLOHEXYLISOCYANATE). [101-68-8]

ACGIH-TLV
TWA: ceiling limit, 0.01 ppm, 0.11 mg/m^3

4,4'-METHYLENEBIS(N,N-DIMETHYLANILINE). [101-61-1]

NCI—carcinogenic in animals

TSCA-CHIP available

METHYLENE BISPHENYL ISOCYANATE (MDI). [101-68-8]

*OSHA 29 CFR 1910.1000 Table Z-1
PEL(Ceiling): 0.02 ppm, 0.2 mg/m^3
ACGIH-TLV
TWA: ceiling limit, 0.02 ppm, 0.2 mg/m^3

2,2'-METHYLENEBIS(3,4,6-TRICHLOROPHENOL). [70-30-4]

*RCRA 49 CFR 172.101
EPA hazardous waste no. U132

METHYLENE BROMIDE. [74-95-3]

*RCRA 40 CFR 261
EPA hazardous waste no. U068

METHYLENE CHLORIDE. Syn: 1,1-Dichloromethane. [75-09-2]

*OSHA 1910.1000 Table Z-2
500 ppm, 8-hour TWA
1000 ppm, acceptable ceiling concentration
2000 ppm/5 minutes in any 2 hours, acceptable maximum peak
ACGIH-TLV
TWA: 100 ppm, 360 mg/m^3
STEL: 500 ppm, 1700 mg/m^3

METHYLENE CHLORIDE

NIOSH Criteria Document (Pub. No. 76-138), NTIS Stock No. PB81 227027, March 4, 1976

Methylene chloride is the common name for dichloromethane.

"Occupational exposure to methylene chloride" is defined as exposure above one-half the daily time-weighted average (TWA) exposure limit, except when there is also exposure to carbon monoxide (CO) at more than 9 ppm.

Because the toxicities of CO and methylene chloride are additive, the appropriate environmental limit and action level of methylene chloride must be reduced in the presence of CO. When CO levels are more than 9 ppm "occupational exposure to methylene chloride" shall be determined from the following formula:

$$\frac{C(CO)}{L(CO)} + \frac{C(CH_2Cl_2)}{L(CH_2Cl_2)} \leq 0.5$$

where:

$C(CO)$ = TWA exposure concentration of CO, ppm
$L(CO)$ = the recommended TWA exposure limit of CO = 35 ppm
$C(CH_2Cl_2)$ = TWA exposure concentration of methylene chloride, ppm
$L(CH_2Cl_2)$ = the recommended TWA exposure limit of methylene chloride = 75 ppm

Exposure to methylene chloride below the appropriate occupational exposure level as defined above will require adherence to Section 4(a)(4), Section 6(d), and Sections 7(a) and (b).

1. In the absence of occupational exposure to CO above a time-weighted average (TWA) of 9 ppm for up to a 10-hour workday, occupational exposure to methylene chloride shall be controlled so that workers are not exposed to methylene chloride in excess of 75 ppm (261 mg/m^3) determined as a TWA for up to a 10-hour workday, 40-hour workweek.

2. In the presence of exposure to CO in the work environment at more than 9 ppm determined as a TWA exposure for up to a 10-hour workday, exposure limits of CO, or methylene chloride, or both shall be reduced to satisfy the relationship:

$$\frac{C(CO)}{L(CO)} + \frac{C(CH_2Cl_2)}{L(CH_2Cl_2)} \leq 0.5$$

where:

$C(CO)$ = TWA exposure concentration of CO, ppm
$L(CO)$ = the recommended TWA exposure limit of CO = 35 ppm
$C(CH_2Cl_2)$ = TWA exposure concentration of methylene chloride, ppm
$L(CH_2Cl_2)$ = the recommended TWA exposure limit of methylene chloride = 75 ppm

3. Occupational exposure shall be controlled so that workers are not exposed to methylene chloride above a peak concentration of 500 ppm (1740 mg/m^3) as determined by any 15-minute sampling period. The standard is designed to protect the health and safety of workers for up to a 10-hour workday, 40-hour workweek, over a working lifetime.

NFPA—hazardous chemical
health—2, flammability—1, reactivity—0

TSCA 4(e), ITC
Second Report of the TSCA Interagency Testing Committee to the administrator, Environmental Protection Agency, April 1978; 43 FR 16684, April 19, 1978, responded to by the EPA administrator, 46 FR 30300, June 5, 1981, removed from the Priority List in the Ninth Report of the TSCA Interagency Testing Committee to the administrator, Environmental Protection Agency, October 30, 1981, 47 FR 5456, February 5, 1982

*TSCA 8(d), Unpublished Health and Safety Studies Reporting, 40 CFR 716, 47 FR 387, September 2, 1982

TSCA-CHIP available

CWA 307(a)(1), 40 CFR 125, Priority Pollutant

CAA 112
Chemical proposed to be assessed as a Haz-

ardous Air Pollutant by a House of Representatives bill to amend the CAA.

*RCRA 40 CFR 261
EPA hazardous waste no. U080

4,4'-METHYLENE DIANILINE (MDA). [101-77-9]

ACGIH-TLV

Skin

TWA: 0.1 ppm, 0.8 mg/m^3
STEL: 0.5 ppm, 4 mg/m^3
TSCA 4(f) (proposed)—significant risk, 48 FR 19078, April 27, 1983

TSCA 4(e), ITC
Fourth Report of the Interagency Testing Committee to the administrator, Environmental Protection Agency, April 1979; 44 FR 31866, June 1, 1979

*TSCA 8(d) Unpublished Health and Safety Studies Reporting, 40 CFR 716, 47 FR 387, September 2, 1982

4,4'-METHYLENE DIANILINE DIHYDROCHLORIDE. [13552-44-8]

NCI—carcinogenic in animals

METHYLENE OXIDE.

*RCRA 40 CFR 261
EPA hazardous waste no. U122 (hazardous substance)

METHYL ETHER. See Dimethyl ether. [115-10-6]

METHYL ETHYL KETONE (MEK). [78-93-3]

*OSHA 29 CFR 1910.1000 Table Z-1
PEL(TWA): 200 ppm, 590 mg/m^3

ACGIH-TLV
TWA: 200 ppm, 590 mg/m^3
STEL: 300 ppm, 885 mg/m^3

See Ketones, NIOSH Criteria Document.

TSCA 4(e), ITC
Fourth Report of the Interagency Testing Committee to the administrator, Environmental Protection Agency, April 1979; 44 FR 31866, June 1, 1979, responded to by administrator EPA 47 FR 58025, December 29, 1982

*TSCA 8(a), 40 CFR 712, 47 FR 26992, June 22, 1982, Chemical Information Rule

*TSCA 8(d) Unpublished Health and Safety Studies Reporting, 40 CFR 716, 47 FR 387, September 2, 1982

*RCRA 40 CFR 261
EPA hazardous waste no. U159 (ignitable waste, toxic waste)

METHYL ETHYL KETONE PEROXIDE. [1338-23-4]

ACGIH-TLV
TWA: ceiling limit, 0.2 ppm, 1.5 mg/m^3

TSCA-CHIP available

*RCRA 40 CFR 261
EPA hazardous waste no. U160 (reactive waste, toxic waste)

METHYL FORMATE. [107-31-3]

*OSHA 29 CFR 1910.1000 Table Z-1
PEL(TWA): 100 ppm, 250 mg/m^3

ACGIH-TLV
TWA: 100 ppm, 250 mg/m^3
STEL: 150 ppm, 375 mg/m^3

NFPA—hazardous chemical
health—2, flammability—4, reactivity—0

5-METHYL-3-HEPTANONE. See Ethyl amyl ketone. [541-85-5]

METHYL HYDRAZINE. [60-34-4]

*OSHA 29 CFR 1910.1000 Table Z-1

Skin

PEL(Ceiling): 0.2 ppm, 0.35 mg/m^3

ACGIH-TLV

Skin

TWA: Ceiling limit, 0.2 ppm, 0.35 mg/m^3
Industrial substance suspect of carcinogenic potential for humans—which is suspect of inducing cancer, based on either (1) limited epidemiologic evidence, exclusive of clinical reports of single cases, or (2) demonstration of carcinogenesis in one or more animal species by appropriate methods. Worker ex-

posure by all routes should be carefully controlled to levels consistent with the animal and human experience data.
See Hydrazine, NIOSH Criteria Document.
*RCRA 40 CFR 261
EPA hazardous waste no. P068
NFPA—hazardous chemical
health—3, flammability—3, reactivity—2

METHYLHYDRAZINE MONO SULFATE. *See* Hydrazine methyl-, sulfate (1:1)

METHYL IODIDE. [74-88-4]

*OSHA 29 CFR 1910.1000 Table Z-1

Skin

PEL(TWA): 5 ppm, 28 mg/m^3
ACGIH-TLV

Skin

TWA: 2 ppm, 10 mg/m^3
STEL: 5 ppm, 30 mg/m^3
Industrial substance suspect of carcinogenic potential for humans—which is suspect of inducing cancer, based on either (1) limited epidemiologic evidence, exclusive of clinical reports of single cases, or (2) demonstration of carcinogenesis in one or more animal species by appropriate methods. Worker exposure by all routes should be carefully controlled to levels consistent with the animal and human experience data.
IARC—carcinogenic in animals
*RCRA 40 CFR 261
EPA hazardous waste no. U138

METHYL ISOAMYL KETONE.
[110-12-3]
ACGIH-TLV (1982 Addition)
TWA: 50 ppm, 240 mg/m^3
See Ketones, NIOSH Criteria Document.

METHYL ISOBUTYL CARBINOL.
[105-30-6]
*OSHA 29 CFR 1910.1000 Table Z-1

Skin

PEL(TWA): 25 ppm, 100 mg/m^3
ACGIH-TLV

Skin

TWA: 25ppm, 100 mg/m^3
STEL: 40 ppm, 165 mg/m^3

METHYL ISOBUTYL KETONE.
See Hexone. [108-10-1]
ACGIH-TLV
TWA: 50 ppm, 205 mg/m^3
STEL: 75 ppm, 300 mg/m^3
See Ketones, NIOSH Criteria Document.
TSCA 4(3), ITC
Fourth Report of the Interagency Testing Committee to the administrator, Environmental Protection Agency, April 1979; 44 FR 31866, June 1, 1979; responded to by administrator EPA 47 FR 58025, December 29, 1982
*TSCA 8(d) Unpublished Health and Safety Studies Reporting, 40 CFR 716, 47 FR 387, September 2, 1982
*RCRA 40 CFR 261
EPA hazardous waste no. U161 (ignitable waste)

METHYL ISOCYANATE. [624-83-9]
*OSHA 29 CFR 1910.1000 Table Z-1

Skin

PEL(TWA): 0.02 ppm, 0.05 mg/m^3
ACGIH-TLV

Skin

TWA: 0.02 ppm, 0.05 mg/m^3
*RCRA 40 CFR 261
EPA hazardous waste no. P064

METHYL ISOPROPYL KETONE.
[563-80-4]
ACGIH-TLV
TWA: 200 ppm, 705 mg/m^3

2-METHYLLACTONITRILE. [75-86-5]
*RCRA 40 CFR 261
EPA hazardous waste no. P069 (hazardous substance)

METHYL MERCAPTAN. [74-93-1]
*OSHA 29 CFR 1910.1000 Table Z-1
PEL(Ceiling): 10 ppm, 20 mg/m^3
ACGIH-TLV
TWA: 0.5 ppm, 1 mg/m^3
*CWA 311(b)(2)(A), 40 CFR 116, 117
Discharge RQ = 100 pounds (45.4 kilograms)
*RCRA 40 CFR 261
EPA hazardous waste no. U153 (ignitable waste, toxic waste)

METHYL METHACRYLATE.
[80-62-6]
*OSHA 29 CFR 1910.1000 Table Z-1
PEL(TWA): 100 ppm, 410 mg/m^3
ACGIH-TLV
TWA: 100 ppm, 410 mg/m^3
STEL: 125 ppm, 510 mg/m^3
NFPA—hazardous chemical
health—2, flammability—3, reactivity—2
*CWA 311(b)(2)(A), 40 CFR 116, 117
Discharge RQ = 5000 pounds (2270 kilograms)
*RCRA 40 CFR 261
EPA hazardous waste no. U162 (hazardous substance, ignitable waste, toxic waste)

METHYL METHANESULFONATE.
[66-27-3]
OSHA Candidate List of Potential Occupational Carcinogens (EPA Carcinogen Assessment Group List), 45 FR 53672, August 12, 1980
IARC—carcinogenic in animals

2-METHYL-1-NITROANTHRAQUINONE. [129-15-7]
NCI—carcinogenic in animals

N-METHYL-N-NITROSOBENZENAMINE. See Aniline, N-methyl-N-nitroso-. [614-0-6]

N-METHYL-N'-NITRO-N-NITROSOGUANIDINE. [70-25-7]
OSHA Candidate List of Potential Occupational Carcinogens (EPA Carcinogen Assessment Group List), 45 FR 53672, August 12, 1980
IARC—carcinogenic in animals
*RCRA 40 CFR 261
EPA hazardous waste no. U163

METHYLOLUREA. [1000-82-4]
TSCA 4(e), ITC
Twelfth Report of the Interagency Testing Committee to the administrator, Environmental Protection Agency, May 1983; 48 FR 24443, June 1, 1983
*TSCA 8(a), 40 CFR 712, 48 FR 28443, June 22, 1983, Chemical Information Rule

METHYLOXIDE. See
Dimethyl ether. [115-10-6]

METHYL OXIRANE. See Alkyl epoxides. [75-56-9]
TSCA 8(d) Unpublished Health and Safety Studies Reporting, 40 CFR 716, 47 FR 387, September 2, 1982

METHYL PARATHION. [298-00-0]
ACGIH-TLV

Skin
TWA: 0.2 mg/m^3
STEL: 0.6 mg/m^3
NIOSH Criteria Document (Pub. No. 77-106), NTIS Stock No. PB 274191, September 30, 1976
"Methyl parathion" is defined as O,O-dimethyl O-p-nitrophenyl phosphorothioate, regardless of production process, alone or in combination with other compounds. "Occupational exposure to methyl parathion" is defined as employment in any area in which methyl parathion or materials containing methyl parathion, alone or in combination with other substances, is produced, packaged, processed, mixed, blended, handled, stored in large quantities, or applied. Adherence to all provisions of the standard is required in workplaces using methyl parathion regardless of the airborne methyl parathion concentration because of serious effects produced by contact with the skin, mucous membranes, or eyes.

Occupational exposure to methyl parathion shall be controlled so that no employee is exposed to methyl parathion at a concentration greater than 0.2 mg/m³ of air determined as a time-weighted average (TWA) exposure for up to a 10-hour workday in a 40-hour workweek.

The standard is designed to protect the health and safety of employees for up to a 10-hour workday, 40-hour workweek, over a working lifetime.

FIFRA—Data Call In, 45 FR 75488, November 14, 1982
Status: letter issued September 30, 1981
NFPA—hazardous chemical
when solid: health—4, flammability—1, reactivity—2
when xylene solution: health—4, flammability—3, reactivity—2
*CWA 311(b)(2)(A), 40 CFR 116, 117
Discharge RQ = 100 pounds (45.4 kilograms)
*RCRA 40 CFR 261
EPA hazardous waste no. P071 (hazardous substance)

2-METHYL-4-PENTANONE.
See Isobutyl methyl ketone. [108-10-1]

4-METHYL-2-PENTANONE.
See Isobutyl methyl ketone. [108-10-1]

METHYL PHOSPHATE. See Phosphoric acid, trimethyl ester. [512-56-1]

2-METHYL-2-PROPANOL.
See tert-Butyl alcohol. [75-65-0]

METHYL PROPYL KETONE.
See 2-Pentanone. [107-87-9]

2-METHYLPYRIDINE. [109-06-8]
TSCA-CHIP available

3-METHYLPYRIDINE. [108-99-6]
TSCA-CHIP available

4-METHYLPYRIDINE. [108-89-4]
TSCA-CHIP available

METHYL SALICYLATE. [119-36-8]
NFPA—hazardous chemical (flammable chemical)
health—1, flammability—1, reactivity—0

METHYL SILICATE. [681-84-5]
ACGIH-TLV
TWA: 1 ppm, 6 mg/m³
STEL: 5 ppm, 30 mg/m³

α-METHYL STYRENE. [98-83-9]
*OSHA 29 CFR 1910.1000 Table Z-1
PEL(Ceiling): 100 ppm, 480 mg/m³
ACGIH-TLV
TWA: 50 ppm, 240 mg/m³
STEL: 100 ppm, 485 mg/m³

METHYL SULFATE. See Dimethyl sulfate. [77-78-1]

METHYL SULFIDE. See Dimethyl sulfide. [75-18-3]

METHYLTHIOURACIL. [56-04-2]
OSHA Candidate List of Potential Occupational Carcinogens (EPA Carcinogen Assessment Group List), 45 FR 53672, August 12, 1980
IARC—carcinogenic in animals
*RCRA 40 CFR 261
EPA hazardous waste no. U164

METHYLTRICHLOROSILANE.
[75-79-6]
NFPA—hazardous chemical
health—3, flammability—3, reactivity—2,
W—water may be hazardous in fire fighting

METHYLTRINITROBENZENE. See Trinitrotoluene. [118-96-7]

METHYL VINYL KETONE. [78-94-4]
NFPA—hazardous chemical
health—3, flammability—3, reactivity—2

METOLACHLOR. [51218-45-2]
FIFRA—Registration Standard—Issued September 1980, NTIS# PB81 123820 (herbicide)

METRIBUZIN. [21087-64-9]
FIFRA—Data Call In, 45 FR 75488, November 14, 1980
Status: letter being drafted

METRONIDAZOLE. [443-48-1]
IARC—carcinogenic in animals

MEVINPHOS. [7786-34-7]
ACGIH-TLV

Skin

TWA: 0.01 ppm, 0.1 mg/m^3
STEL: 0.03 ppm, 0.3 mg/m^3
*CWA 311(b)(2)(A), 40 CFR 116, 117
Discharge RQ = 1 pound (0.454 kilogram)

MEXACARBATE. [315-18-4]
*CWA 311(b)(2)(A), 40 CFR 116, 117
Discharge RQ = 1000 pounds (454 kilograms)

MICA. [12001-26-2]
ACGIH-TLV
20 mppcf

MICHLER'S BASE. See Aniline, 4,4'-methylenebis (N,N-dimethyl-. [101-61-1]

MICHLER'S KETONE. See Benzophenone, 4,4'-bis(dimethylamino)-. [90-94-8]

MINERAL DUSTS. See Silica, Silicates, or Coal dust.

*OSHA 1910.1000 Table Z-3

Substances	mppcf[a]	mg/m^3
Silica:		
Crystalline:		
Quartz (respirable)	$\frac{250^b}{\% SiO_2 + 5}$	$\frac{10 \text{ mg/m}^{3c}}{\% SiO_2 + 2}$
Quartz (total dust)		$\frac{30 \text{ mg/m}^3}{\% SiO_2 + 2}$
Cristobalite: Use half the value calculated from the count or mass formulae for quartz.		
Tridymite: Use half the value calculated from the formulae for quartz.		
Amorphous, including natural diatomaceous earth	20	$\frac{80 \text{ mg/m}^3}{\% SiO_2}$
Silicates (less than 1% crystalline silica):		
Mica	20	
Soapstone	20	
Talc (nonasbestos-form)	20[d]	
Talc (fibrous). Use asbestos limit		
Tremolite (see talc, fibrous)		
Portland cement	50	
Graphite (natural)	15	
Coal dust (respirable fraction less than 5% SiO$_2$)		2.4 mg/m^3 or
For more than 5% SiO$_2$		$\frac{10 \text{ mg/m}^3}{\% SiO_2 + 2}$
Inert or nuisance dust:		
Respirable fraction	15	5 mg/m^3
Total dust	50	15 mg/m^3

NOTE: Conversion factors—mppcf × 35.3 = million particles per cubic meter = particles per cubic centimeter.

[a]Millions of particles per cubic foot of air, based on impinger samples counted by light-field technics.
[b]The percentage of crystalline silica in the formula is the amount determined from airborne samples, except in those instances in which other methods have been shown to be applicable.
[c]Both concentration and percent quartz for the application of this limit are to be determined from the fraction passing a size selector with the following characteristics:
[d]Containing < 1% quartz; if > 1% quartz, use quartz limit.

Computation formulae:
The cumulative exposure for an 8-hour work shift shall be computed as follows:

$$E = \frac{C_a T_a + C_b T_b + \ldots C_n T_n}{8}$$

where:

E is the equivalent exposure for the working shift

C is the concentration during any period of time T where the concentration remains constant

T is the duration in hours of the exposure at the concentration C.

The value of E shall not exceed the 8-hour time-weighted average limit in Table Z-1, Z-2, or Z-3 for the material involved.

In case of a mixture of air contaminants an employer shall compute the equivalent exposure as follows:

$$E_m = \frac{C_1}{L_1} + \frac{C_2}{L_2} + \cdots \frac{C_n}{L_n}$$

where:

E_m is the equivalent exposure for the mixture

C is the concentration of a particular contaminant

L is the exposure limit for that contaminant, from Table Z-1, Z-2, or Z-3.

The value of E_m shall not exceed unity (1).

MINERAL OIL. [8012-95-1]
NFPA—hazardous chemical (flammable chemical)
health—0, flammability—1, reactivity—0

MINERAL SPIRITS. [8030-30-6]
See Petroleum solvents, refined, NIOSH Criteria Document.

MINERAL WOOL FIBER.
ACGIH-TLV
10 mg/m³

MIREX. See 1,3,4-Metheno-1H-cyclobuta(cd)pentalene, 1,1a,2,2,3,3a, 4,5,5a,5b,6-dodecachlorooctahydro-. [2385-85-5]

MITOMYCIN C. See 6-Amino-1,1a,2,8, 8a,8b-hexahydro-8-(hydroxymethyl)-8a-methoxy-5-methylcarbamate azirino (2',3':3,4)pyrrolo(1,2-a)indole-4,7-iodine (ester). [50-07-7]

MMH. See Monomethyl hydrazine. [60-34-4]

MOLYBDENUM. [7439-98-7]
*OSHA 29 CFR 1910.1000 TableZ-1
PEL(TWA): soluble compounds 5 mg/m³, insoluble compounds 15 mg/m³

As Mo, soluble compounds:
ACGIH-TLV
TWA: 5 mg/m³
STEL: 10 mg/m³

As Mo, insoluble compounds:
ACGIH-TLV
TWA: 10 mg/m³
STEL: 20 mg/m³

MONOBUTYLTIN TRIS(ISOOCTYL MERCAPTOACETATE). [25852-70-4, 54849-38-6]
TSCA 4(e), ITC
Eleventh Report of the Interagency Testing Committee to the administrator, Environmental Protection Agency, November 1982; 47 FR 54626, December 3, 1982
*TSCA 8(d) Unpublished Health and Safety Studies Reporting, 40 CFR 716, 47 FR 54624, December 3, 1982

MONOCHLOROBENZENE. [108-90-7]
TSCA 4(e), ITC
Initial Report to the administrator, Environmental Protection Agency, TSCA Interagency Testing Committee, October 1, 1977; 42 FR 55026, October 12, 1977, responded to by the EPA administrator Vol. 43 FR 50134, October 26, 1978 and 45 FR 48524, July 18, 1980

MONOCHLORO-S-TRIAZINETRIONE ACID.
NFPA—hazardous chemical
health—3, flammability—0, reactivity—2, oxy—oxidizing chemical

MONOCROTALINE. [315-22-0]
IARC—carcinogenic in animals

MONOCROTOPHOS. [6923-22-4]
ACGIH-TLV
TWA: 0.25 mg/m^3
FIFRA—Data Call In, 45 FR 75488, November 14, 1982
Status: letter issued 9/30/81

MONO(2,2-DIMETHYLHYDRAZIDE) SUCCINIC ACID. [1596-84-5]
NCI—carcinogenic in animals

MONOETHYLAMINE. See Ethylamines. [75-04-7]
*CWA 311(b)(2)(A), 40 CFR 116, 117
Discharge RQ = 1000 pounds (454 kilograms)

MONOMETHYLAMINE. See Ethylamines. [74-89-5]
CWA 311(b)(2)(A), 40 CFR 116, 117
Discharge RQ = 1000 pounds (454 kilograms)

MONOMETHYL ANILINE. See Methylaniline. [100-61-8]

MONOMETHYL HYDRAZINE. See Methylhydrazine. [60-34-4]

MONOMETHYLTIN TRIS(ISOOCTYL MERCAPTOACETATE). [54849-38-6]
TSCA 4(e), ITC
Eleventh Report of the Interagency Testing Committee to the administrator, Environmental Protection Agency, November 1982; 47 FR 54626, December 3, 1982

MONO-(TRICHLORO)TETRA(MONO-POTASSIUMDICHLORO)-PENTA-S-TRIAZINETRIONE.
NFPA—hazardous chemical
health—3, flammability—0, reactivity—2, oxy—oxidizing chemical

MONURON. See Urea, 3-(p-chlorophenyl)-1,1-dimethyl-. [150-68-5]
OSHA Candidate List of Potential Occupational Carcinogens, 45 FR 53672, August 12, 1980
NCI—carcinogenic in animals
40 CFR 162.11, FIFRA-RPAR, Final Action, Voluntary Cancellation
Current Status: 42 FR 41320, August 16, 1977
Criteria Possibly Met or Exceeded: Oncogenicity
FIFRA—Registration Standard Development—Data Evaluation and Development of Regulatory Position (herbicide)

MONURON TRICHLOROACETATE. [140-41-0]
FIFRA—Registration Standard Development—Data Evaluation and Development of Regulatory Position (herbicide)

MORPHINAN-3,6-DIOL, 7,8-DIDEHYDRO-4,5-EPOXY-17-METHYL-(5.alpha.,6.alpha.)-,(Z)-9-OCTADECENOATE (SALT). [6033-05-2]
*TSCA 8(a), 40 CFR 712, 47 FR 26992, June 22, 1982, Chemical Information Rule

MORPHOLINE. [110-91-8]
*OSHA 29 CFR 1910.1000 Table Z-1

Skin
PEL(TWA): 20 ppm, 70 mg/m^3
STEL: 30 ppm, 105 mg/m^3
NFPA—hazardous chemical
health—2, flammability—3, reactivity—0
TSCA-CHIP available

5-(MORPHOLINOMETHYL)-3-[(5-NITROFURFURYLIDENE)-AMINO]-2-OXAZOLIDINONE.
IARC—carcinogenic in animals

MOTOR FUEL ANTIKNOCK COMPOUNDS (containing LEAD).
NFPA—hazardous chemical
health—3, flammability—3, reactivity—3

MURIATIC ACID. See Hydrogen chloride. [7647-01-0]

MUSTARD GAS. [505-60-2]
OSHA Candidate List of Potential Occupational Carcinogens (EPA Carcinogen Assessment Group List), 45 FR 53672, August 12, 1980
IARC—carcinogenic in humans

N

NaK. *See* Sodium-potassium alloys. [11135-81-2]

NALED. [300-76-5]

*OSHA 29 CFR 1910.1000 Table Z-1
PEL(TWA): 3 mg/m³
ACGIH-TLV
TWA: 3 mg/m³
STEL: 6 mg/m³
FIFRA—Data Call In, 45 FR 75488, November 14, 1980
Status: letter issued April 7, 1981
FIFRA—Registration Standard Development—Data Collection (insecticide)
*CWA 311(b)(2)(A), 40 CFR 116, 117
Discharge RQ = 10 pounds (4.54 kilograms)

NAPHTHA (COAL TAR). [8030-30-6]

*OSHA 29 CFR 1910.1000 Table Z-1
PEL(TWA): 100 ppm, 400 mg/m³

4',4'''-bi-2NAPHTH-*O*-ANISIDINE, 3,3''-DIHYDROXY-. *See* Bisazobiphenyl dyes. [91-92-9]

*TSCA 8(d) Unpublished Health and Safety Studies Reporting, 40 CFR 716, 47 FR 387, September 2, 1982

5,12-NAPHTHACENEDIONE, (8*S-cis*)-8-ACETYL-10-[(3-AMINO-2,3,6-TRIDEOXY—alpha-L-LYXO-HEXOPYRANOSYL) OXY]-7,8,9,10-TETRAHYDRO-6,8,11-TRIHYDROXY-1-METHOXY-.

*RCRA 40 CFR 261
EPA hazardous waste no. U059

NAPHTHALENE. [91-20-3]

*OSHA 29 CFR 1910.1000 Table Z-1
PEL(TWA): 10 ppm, 50 mg/m³
ACGIH-TLV
TWA: 10 ppm, 50 mg/m³
STEL: 15 ppm, 75 mg/m³
FIFRA—Registration Standard—Issued September 1981 (insecticide)
NFPA—hazardous chemical
health—2, flammability—2, reactivity—0
CWA 304(a)(1), 45 FR 79318, November 28, 1980, Water Quality Criteria

Freshwater Aquatic Life

The available data for naphthalene indicate that acute and chronic toxicity to freshwater aquatic life occur at concentrations as low as 2300 and 620 µg/liter, respectively, and would occur at lower concentrations among species that are more sensitive than those tested.

Saltwater Aquatic Life

The available data for naphthalene indicate that acute toxicity to saltwater aquatic life occurs at concentrations as low as 2350 µg/liter and would occur at lower concentrations among species that are more sensitive than those tested. No data are available concerning the chronic toxicity of naphthalene to sensitive saltwater aquatic life.

Human Health

Using the present guidelines, a satisfactory criterion cannot be derived at this time due to the insufficiency in the available data for naphthalene.

CWA 307(a)(1), Priority Pollutant
*CWA 311(b)(2)(A), 40 CFR 116, 117
Discharge RQ = 5000 pounds (2270 kilograms)
*RCRA 40 CFR 261
EPA hazardous waste no. U165 (hazardous substance)

NAPHTHALENEACETIC ACID.
[86-87-3]
FIFRA—Registration Standard—Issued August 1981, NTIS#PB82 131145 (growth regulator)

NAPHTHALENE, 1-CHLORO-.
[90-13-1]
*TSCA 8(a), 40 CFR 712, 47 FR 26992, June 22, 1982, Chemical Information Rule

NAPHTHALENE, 2-CHLORO-.
[91-58-7]
*RCRA 40 CFR 261
EPA hazardous waste no. U047

1,5-NAPHTHALENEDIAMINE.
[2243-62-1]
NCI—carcinogenic in animals

NAPHTHALENE DIISOCYANATE.
[3173-72-6]
See Diisocyanates, NIOSH Criteria Document.

1,4-NAPHTHALENEDIONE. [130-15-4]
*RCRA 40 CFR 261
EPA hazardous waste no. U166

2,7-NAPHTHALENEDISULFONIC ACID, 5-AMINO-3-[[4'-[(7-AMINO-1-HYDROXY-3-SULFO-2-NAPHTHALENYL)AZO][1,1'-BIPHENYL]-4-YL]AZO]-4-HYDROXY-, TRISODIUM SALT. See Bisazobiphenyl dyes. [2429-73-4]
*TSCA 8(d) Unpublished Health and Safety Studies Reporting, 40 CFR 716, 47 FR 287, September 2, 1982

2,7-NAPHTHALENEDISULFONIC ACID, 4-AMINO-3-[[4'-[(2,4-DIAMINO-5-METHYLPHENYL)AZO][1,1'-BIPHENYL]-4-YL]AZO]-5-HYDROXY-6(PHENYLAZO)-, DISODIUM SALT.
See Bisazobiphenyl dyes. [2429-83-6]
*TSCA 8(d) Unpublished Health and Safety Studies Reporting, 40 CFR 716, 47 FR 387, September 2, 1982

2,7-NAPHTAHALENEDISULFONIC ACID, 4-AMINO-3-[(2,4-DIAMINOPHENYL)AZO][1,1'-BIPHENYL]-4-YL]AZO]-5-HYDROXY-6-[PHENYLAZO]-, DISODIUM SALT.
See C. I. direct black 38, disodium salt.
[1937-37-7]

1,3-NAPHTHALENEDISULFONIC ACID, 4-AMINO-5-HYDROXY-6-[[4'-[2-HYDROXY-1-NAPHTHALENYL)AZO]-3,3'-DIMETHOXY]1,1'BIPHENYL[-4-YL]AZO]-, DISODIUM SALT. See Bisazobiphenyl dyes. [2586-57-4]
*TSCA 8(d) Unpublished Health and Safety Studies Reporting, 40 CFR 716, 47 FR 387 September 2, 1982

2,7-NAPHTHALENEDISULFONIC ACID, 4-AMINO-5-HYDROXY-6-[[4'-[(4-HYDROXYPHENYL)AZO][1,1'-BIPHENYL]-4-YL]AZO]-3-[(4-NITROPHENYL)AZO]-, DISODIUM SALT. See Bisazobiphenyl dyes.
[4335-09-5]
*TSCA 8(d) Unpublished Health and Safety Studies Reporting, 40 CFR 716, 47 FR 387, September 2, 1982

2,7-NAPHTHALENEDISULFONIC ACID, 4-AMINO-5-HYDROXY-3-[[4'-[(4-HYDROXYPHENYL)AZO][1,1'-BIPHENYL]-4-YL]AZO]-6-(PHENYLAZO)-, DISODIUM SALT. See Bisazobiphenyl dyes. [3626-28-6]
*TSCA 8(d) Unpublished Health and Safety Studies Reporting, 40 CFR 716, 47 FR 387, September 2, 1982

2,7-NAPHTHALENEDISULFONIC ACID, 3,3'-((4,4'-BIPHENYLENE)BIS(AZO))BIS(5-AMINO-4-HYDROXY-, TETRASODIUM SALT. See C.I. direct blue 6, tetrasodium salt. [2602-46-2]

1,3-NAPTHALENEDISULFONIC ACID, 6,6'[(3,3'-DIMETHOXY[1,1'-BIPHENYL]-4,4-DIYL)BIS(AZO)]BIS[4-AMINO-5-HYDROXY-, TETRASODIUM SALT. See Bisazobiphenyl dyes.
[2610-05-1]

*TSCA 8(a), 40 CFR 712, 47 FR 26992, June 22, 1982, Chemical Information Rule

*TSCA 8(d) Unpublished Health and Safety Studies Reporting, 40 CFR 716, 47 FR 387, September 2, 1982

2,7-NAPHTHALENEDISULFONIC ACID, 3,3'-[(3,3-DIMETHOXY[1,1'-BIPHENYL]-4,4'-DIYL)BIS(AZO)]BIS[5-AMINO-4-HYDROXY-], TETRASODIUM SALT.
See Bisazobiphenyl dyes. [2429-74-5]

*TSCA 8(a) Unpublished Health and Safety Studies Reporting, 40 CFR 716, 47 FR 387, September 2, 1982

2,7-NAPHTHALENEDISULFONIC ACID, 3,3'-[(3,3'-DIMETHYL[1,1'-BIPHENYL]-4,4'-DIYL)BIS(AZO)]BIS[5-AMINO-4-HYDROXY-, TETRASODIUM SALT.
See Bisazobiphenyl dyes. [72-57-1]

*TSCA 8(a), 40 CFR 712, 47 FR 26992, June 22, 1982, Chemical Information Rule

*TSCA 8(d) Unpublished Health and Safety Studies Reporting, 40 CFR 716, 47 FR 387, September 2, 1982

*RCRA 40 CFR 261
EPA hazardous waste no. U236

2,7-NAPHTHALENEDISULFONIC ACID, 3,3'-[(3,3-DIMETHYL[1,1'-BIPHENYL]-4,4'-DIYL)BIS(AZO)]BIS[4,5-DIHYDROXY-, TETRASODIUM SALT.
See Bisazobiphenyl dyes. [2150-54-1]

*TSCA 8(d) Unpublished Health and Safety Studies Reporting, 40 CFR 716, 47 FR 387, September 2, 1982

1,3-NAPHTHALENEDISULFONIC ACID, 8-[(4-ETHOXYPHENYL)AZO][1,1'-BIPHENYL]-4-YL]AZO]-7-HYDROXY-, DISODIUM SALT. See Bisazobiphenyl dyes. [3530-19-6]

*TSCA 8(d) Unpublished Health and Safety Studies Reporting, 40 CFR 716, 47 FR 387, September 2, 1982

1,3-NAPHTHALENEDISULFONIC ACID, 8-[[4'[(4-ETHOXYPHENYL)AZO]-3,3'-DIMETHYL[1,1'-BIPHENYL]-4-YL]AZO]-7-HYDROXY-, DISODIUM SALT. See Bisazobiphenyl dyes. [6358-29-8]

TSCA 8(d) Unpublished Health and Safety Studies Reporting, 40 CFR 716, 47 FR 387, September 2, 1982

1,3-NAPHTHALENEDISULFONIC ACID, 7-HYDROXY-8-[[4'-[[4-[[(4-METHYLPHENYL)SULFONYL]OXY]-PHENYL]AZO][1,1'-BIPHENYL]-4-YL]AZO]-, DISODIUM SALT.
See Bisazobiphenyl dyes. [3567-65-5]

*TSCA 8(d) Unpublished Health and Safety Studies Reporting, 40 CFR 716, 47 FR 387, September 2, 1982

NAPHTHALENE, HEXACHLORO-.
[1335-87-1]

*TSCA 8(a), 40 CFR 712, 47 FR 26992, June 22, 1982, Chemical Information Rule

NAPHTHALENE, OCTACHLORO-.
[2234-13-1]

*TSCA 8(a), 40 CFR 712, 47 FR 26992, June 22, 1982, Chemical Information Rule

NAPHTHALENE, PENTACHLORO-.
[1321-64-8]

*TSCA 8(a), 40 CFR 712, 47 FR 26992, June 22, 1982, Chemical Information Rule

NAPHTHALENE, TETRACHLORO-.
[1335-88-2]

*TSCA 8(a), 40 CFR 712, 47 FR 26992, June 22, 1982, Chemical Information Rule

NAPHTHALENE, TRICHLORO-.
[1321-65-9]

*TSCA 8(a), 40 CFR 712, 47 FR 26992, June 22, 1982, Chemical Information Rule

1-NAPHTHALENESULFONIC ACID, 3-[[4'-[(6-AMINO-1-HYDROXY-3-SULFO-2-NAPHTHALENYL)AZO]-3,3'-DIMETHOXY[1,1'-BIPHENYL-4-YLAZO-4-HYDROXY-], DISODIUM SALT.
See Bisazobiphenyl dyes. [6449-35-0]

*TSCA 8(d) Unpublished Health and Safety Studies Reporting, 40 CFR 716, 47 FR 387, September 2, 1982

1-NAPTHALENESULFONIC ACID, 3,3'-[[1,1'-BIPHENYL]-4,4'DIYLBIS(AZO)BIS[4-AMINO-], DISODIUM SALT. See Bisazobiphenyl dyes. [575-58-0]

*TSCA 8(d) Unpublished Health and Safety Studies Reporting, 40 CFR 716, 47 FR 387, September 2, 1982

1-NAPTHALENESULFONIC ACID, 3,3'-DIMETHOXY[1,1'-BIPHENYL]-4,4'-DIYL)BIS(AZO)BIS[4-HYDROXY-, DISODIUM SALT. See Bisazobiphenyl dyes. [2429-71-2]

*TSCA 8(d) Unpublished Health and Safety Studies Reporting, 40 CFR 716, 47 FR 387, September 2, 1982

1-NAPHTHALENESULFONIC ACID, 3,3'-[3,3'-DIMETHYL[1,1'-BIPHENYL]-4,4'-DIYL)BIS(AZO)BIS[4-AMINO-, DISODIUM SALT. See Bisazobiphenyl dyes. [992-59-6]

*TSCA 8(d) Unpublished Health and Safety Studies Reporting, 40 CFR 716, 47 FR 387, September 2, 1982

1,4-NAPHTHAQUINONE.
*RCRA 40 CFR 261
EPA hazardous waste no. U166

NAPHTHENIC ACID. [1338-24-5]
*CWA 311(b)(2)(A), 40 CFR 116, 117
Discharge RQ = 100 pounds (45.4 kilograms)

alpha-NAPHTHYLAMINE. [134-32-7]
*OSHA Standard 29 CFR 1910.1004
Cancer-Suspect Agent
Shall not apply to solid or liquid mixtures containing less than 1.0% by weight or volume of alpha-naphthylamine. Shall not apply to operations involving the destructive distillation of carbonaceous materials, such as occurs in coke ovens.

OSHA Candidate List of Potential Occupational Carcinogens (EPA Carcinogen Assessment Group List), 45 FR 53672, August 12, 1980

IARC—carcinogenic in humans
*RCRA 40 CFR 261
EPA hazardous waste no. U167

beta-NAPHTHYLAMINE. [91-59-8]
*OSHA Standard 29 CFR 1910.1009
Cancer-Suspect Agent
Shall not apply to solid and liquid mixtures containing less than 0.1% by weight or volume of beta-naphthylamine. Shall not apply to operations involving the destructive distillation of carbonaceous materials, such as occurs in coke ovens.

ACGIH-TLV
TWA
Human carcinogen—no exposure or contact by any route—respiratory, skin, or oral, as detected by the most sensitive methods—shall be permitted. The worker should be properly equipped to insure virtually no contact with the carcinogen.

OSHA Candidate List of Potential Occupational Carcinogens (EPA Carcinogen Assessment Group List), 45 FR 53672, August 12, 1980

IARC—carcinogenic in animals
*RCRA 40 CFR 261
EPA hazardous waste no. U168

beta-NAPHTHYLAMINE, METABOLIC PRECURSORS OF.
NIOSH Current Intelligence Bulletin No. 16—cancer-related bulletin

1-NAPHTHYLAMINE. See alpha-Naphthylamine. [134-32-7]

2-NAPHTHYLAMINE. See beta-Naphthylamine. [91-59-8]

2-NAPHTHYLAMINE, N,N-BIS(2-CHLOROMETHYL)-. Syn: Naphthylamine mustard. [494-03-1]
OSHA Candidate List of Potential Occupational Carcinogens, 45 FR 53672, August 12, 1980

*RCRA 40 CFR 261
EPA hazardous waste no. U026

NAPHTHYLAMINE MUSTARD. See 2-Naphthylamine, N,N-bis(2-chloromethyl)-. [494-03-1]

alpha-NAPHTHYLTHIOUREA. See ANTU. [86-88-4]

NAPTALAM. [132-66-1]

FIFRA—Data Call In, 45 FR75488, November 14, 1980
Status: letter being drafted

NEODECANOIC ACID, OXIRANYL-METHYL ESTER. [26761-45-5]

*TSCA 8(a), 40 CFR 712, 47 FR 26992, June 22, 1982, Chemical Information Rule

NEON. [7440-01-9]

ACGIH-TLV
TWA
Asphyxiant—gas or vapors, when present in high concentrations in air, act primarily as simple asphyxiant without other significant physiologic effects

NEOPENTYL GLYCOL DIACRYLATE. [2223-82-7]

TSCA—CHIP available

NEOPENTYL GLYCOL DIGLYCIDYL ETHER. [17557-23-2]

TSCA-CHIP available

NIAX CATALYST ESN.

NIOSH Current Intelligence Bulletin No. 26

NICKEL. [7440-02-0]

CWA 304(a)(1), 45 FR 78318, November 28, 1980, Water Quality Criteria

Freshwater Aquatic Life

For total recoverable nickel the criterion (in micrograms per liter) to protect freshwater aquatic life as derived using the guidelines is the numerical value given by $e(0.76[\ln(\text{hardness})] + 1.06)$ as a 24-hour average and the concentration (in micrograms per liter) should not exceed the numerical value given by $e(0.76[\ln(\text{hardness})] + 4.02)$ at any time. For example, at hardness of 50, 100, and 200 mg/liter as $CaCO_3$, the criteria are 56, 96, and 160 µg/liter, respectively, as 24-hour averages, and the concentrations should not exceed 1100, 1800, and 3100 µg/liter, respectively, at any time.

Saltwater Aquatic Life

For total recoverable nickel the criterion to protect saltwater aquatic life as derived using the guidelines is 7.1 µg/liter as a 24-hour average and the concentration should not exceed 140 µg/liter at any time.

Human Health

For the protection of human health from the toxic properties of nickel ingested through water and contaminated aquatic organisms, the ambient water criterion is determined to be 13.4 µg/liter.

For the protection of human health from the toxic properties of nickel ingested through contaminated aquatic organisms alone, the ambient water criterion is determined to be 100 µg/liter.

NIOSH Criteria Document (Pub. No. 77-164), May 13, 1977

"Nickel" is defined in this standard as elemental nickel and all nickel compounds except organonickel compounds with a covalent carbon-nickel bond, such as nickel carbonyl. "Occupational exposure to nickel" is defined as working with compounds, solutions, or metals containing nickel that can become airborne or can spill or splash on the skin or in the eyes. Occupational exposure to nickel does not include the handling of solid products, such as stainless-steel tools, provided no particle-generating operations, such as grinding or cutting, occur. The recommended method of sampling and analysis does not differentiate between individual nickel particles; thus, the standard applies to all particulate nickel. Where no occupational exposure to nickel occurs, but nickel is present in the workplace, adherence is required only to Section 8(a).

Occupational exposure to nickel shall be controlled so that no employee is exposed to nickel at a concentration greater that 15 µg, measured as nickel, per cubic meter of air (15 µg Ni/m^3) determined as a time-weighted average (TWA) concentration for up to a 10-hour work shift, 40-hour workweek.

The standard is designed to protect the health and provide for the safety of employees for up to a 10-hour work shift, 40-hour workweek, over a working lifetime.

CWA 307(a)(1), Priority Pollutant

CAA 112
Chemical proposed to be assessed as a Hazardous Air Pollutant by a House of Representatives bill to amend the CAA

NICKEL, METAL. [7440-02-0]
ACGIH-TLV
TWA: 1 mg/m^3

NICKEL, METAL and SOLUBLE COMPOUNDS. [7440-02-0]
*OSHA 39 CFR 1910.1000 Table Z-1
PEL(TWA): 1 mg/m^3 (as Ni)
ACGIH-TLV
TWA: 0.1 mg/m^3
STEL: 0.3 mg/m^3

NICKEL (NICKEL REFINING).
[7440-02-0]
IARC—carcinogenic in humans

NICKEL and COMPOUNDS, N.O.S.
OSHA Candidate List of Potential Occupational Carcinogens (EPA Carcinogen Assessment Group List), 45 FR 53672, August 12, 1980

NICKEL AMMONIUM SULFATE.
[15699-18-0]
*CWA 311(b)(2)(A), 40 CFR 116, 117
Discharge RQ = 5000 pounds (2270 kilograms)

NICKEL CARBONYL. [13463-39-3]
*OSHA 29 CFR 1910.1000 Table Z-1
PEL(TWA): 0.001 ppm, 0.007 mg/m^3
ACGIH-TLV
TWA: 0.05 ppm, 0.35 mg/m^3
OSHA Candidate List of Potential Occupational Carcinogens (EPA Carcinogen Assessment Group List), 45 FR 53672, August 12, 1980
IARC—carcinogenic in animals

NFPA—hazardous chemical
health—4, flammability—3, reactivity—3
*RCRA 40 CFR 261
EPA hazardous waste no. P073

NICKEL CATALYST (finely divided, activated, or spent). [7440-02-0]
NFPA—hazardous chemical
health—2, flammability—4, reactivity—0

NICKEL CHLORIDE.
[37211-05-5, 7718-54-9]
*CWA 311(b)(2)(A), 40 CFR 116, 117
Discharge RQ = 5000 pounds (2270 kilograms)

NICKEL(II) CYANIDE. [557-19-7]
OSHA Candidate List of Potential Occupational Carcinogens (EPA Carcinogen Assessment Group List), 45 FR 53672, August 12, 1980
*RCRA 40 CFR 261
EPA hazardous waste no. P074

NICKEL HYDROXIDE. [12054-48-7]
*CWA 311(b)(2)(A), 40 CFR 116, 117
Discharge RQ = 1000 pounds (454 kilograms)

NICKEL NITRATE. [14216-75-2]
NFPA—hazardous chemical
when fire: health—1, flammability—0, reactivity—0, oxy-oxidizing chemical
when nonfire: health—0, flammability—0, reactivity—0, oxy—oxidizing chemical
*CWA 311(b)(2)(A), 40 CFR 116, 117
Discharge RQ = 5000 pounds (2270 kilograms)

NICKELOCENE. [1271-28-9]
IARC—carcinogenic in animals

NICKEL OXIDE. [1314-06-3, 1313-99-1]
IARC—carcinogenic in animals

NICKEL POWDER. [7440-02-0]
IARC—carcinogenic in animals

NICKEL SUBSULFIDE. [12035-72-2]
IARC—carcinogenic in animals

NICKEL SULFATE. [7786-81-4]
*CWA 311(b)(2)(A), 40 CFR 116, 117
Discharge RQ = 5000 pounds (2270 kilograms)

NICKEL SULFIDE ROASTING, FUME and DUST, as Ni.
ACGIH-TLV
TWA: 1 mg/m^3
Human carcinogen—recognized to have carcinogenic or cocarcinogenic potential

NICKEL TETRACARBONYL.
[13463-39-3]
*RCRA 40 CFR 261
EPA hazardous waste no. P073

NICOTINE. [54-11-5]
*OSHA 29 CFR 1910.1000 Table Z-1

Skin

PEL(TWA): 0.5 mg/m^3
ACGIH-TLV
TWA: 0.5 mg/m^3
STEL: 1.5 mg/m^3

NICOTINE and SALTS.
*RCRA 40 CFR 261
EPA hazardous waste no. P075

NIRIDAZOLE. [61-57-4]
IARC—carcinogenic in animals

NITER. See Potassium nitrate. [7757-79-1]

NITHIAZIDE. [139-94-6]
NCI—carcinogenic in animals

NITRALIN. [4726-14-1]
FIFRA—Data Call In, 45 FR 75488, November 14, 1980
Status: letter being drafted

NITRATE (as N).
*National Interim Primary Drinking Water Regulations, 40 CFR 141; 40 FR 59565, December 24, 1975; amended by 41 FR 28402, July 9, 1976; 44 FR 68641, November 29, 1979; corrected by 45 FR 15542, March 11, 1980; 45 FR 57342, August 27, 1980; 47 FR 18998, March 3, 1982; corrected by 47 FR 10998, March 12, 1982
Maximum contaminant level—10 mg/liter. At the discretion of the state, nitrate levels not to exceed 20 mg/liter may be allowed in a noncommunity water system if the supplier of the water demonstrates to the satisfaction of the state that:

1. Such water will not be available to children under 6 months of age; and
2. There will be continuous posting of the fact that nitrate levels exceed 10 mg/liter and of the potential health effects of exposure; and
3. Local and state public health authorities will be notified annually of nitrate levels that exceed 10 mg/liter; and
4. No adverse health effects shall result.

NITRATE OF POTASH. See Potassium nitrate. [7757-79-1]

NITRIC ACID. [7697-37-2]
*OSHA 29 CFR 1910.1000 Table Z-1
PEL(TWA): 2 ppm, 5 mg/m^3
ACGIH-TLV
TWA: 2 ppm, 5 mg/m^3
STEL: 4 ppm, 10 mg/m^3
NIOSH Criteria Document (Pub. No. 76-141), NTIS Stock No. PB81 227217, March 9, 1976
"Nitric acid" is defined as the aqueous solutions of nitric acid and the vapor and mist thereof. "Occupational exposure to nitric acid" is defined as exposure to airborne concentrations of nitric acid equal to or exceeding one-half the recommended workroom environmental limit. Adherence only to Sections 3, 4(a), 4(b), 4(c)(1)(C), 5, and 6 is required when workplace environmental concentrations of nitric acid are not greater than one-half of the recommended workplace environmental limit. It must be recognized that a potential exposure to oxides of nitrogen exists whenever airborne nitric acid is present in the workplace environment. Therefore, sampling for oxides of nitrogen, as specified in *Criteria for a Recommended Standard . . . Occupational Exposure to Ox-*

ides of Nitrogen, must accompany measurements for airborne nitric acid.

"Occupational exposure to nitric acid shall be controlled so that no worker is exposed to a concentration of nitric acid vapor in excess of 2 ppm of air (5 mg/m^3 air) determined as a time-weighted average (TWA) exposure for up to a 10-hour workday, 40-hour workweek.

The standard is designed to protect the health and safety of workers for up to a 10-hour workday, 40-hour workweek, over a working lifetime.

NFPA—hazardous chemical
health—3, flammability—0, reactivity—0, oxy—oxidizing chemical

*CWA 311(b)(2)(A), 40 CFR 116, 117
Discharge RQ = 1000 pounds (454 kilograms)

NITRIC OXIDE. [10102-43-9]

*OSHA 29 CFR 1910.1000 Table Z-1
PEL(TWA): 25 ppm, 30 mg/m^3
ACGIH-TLV
TWA: 25 ppm, 30 mg/m^3
STEL: 35 ppm, 45 mg/m^3
See Nitrogen, oxides of, NIOSH Criteria Document.

*RCRA 40 CFR 261
EPA hazardous waste no. P076

NITRILES.

NIOSH Criteria Document (Pub. No. 78-212), NTIS Stock No. PB81 225534, September 27, 1978

Nitriles are defined as organic compounds that contain a cyano group, C $\equiv$ N, as the characteristic functional group. They may react to release cyanide. Ten nitriles are included in the recommended standard, namely the mononitriles (acetonitrile, propionitrile, *n*-butyronitrile, and isobutyronitrile); the cyanohydrins (glycolonitrile and acetone cyanohydrin); and the dinitriles (malononitrile, succinonitrile, adiponitrile, and tetramethylsuccinonitrile).

Acrylonitrile was omitted because NIOSH recently recommended an Emergency Temporary Standard for the chemical. The term "selected nitriles" will be used to refer to these compounds. "Occupational exposure to selected nitriles" is defined as exposure to airborne concentrations at or above the action level. "Action level" is defined as one-half the time-weighted average (TWA) or ceiling workplace environmental limit, whichever is appropriate. The criteria and recommended standards apply to any area in which nitriles are produced, packaged, processed, mixed, blended, handled, or stored.

Adherence to all provisions of the recommended standard is required if any employee is exposed to airborne nitriles at concentrations above the action level. If any employee is occupationally exposed at concentrations equal to or below the action level, then all sections of the recommended standard except Section 8(a) shall be complied with.

Workplace exposure to nitriles shall be controlled so that employees are not exposed at concentrations greater than the limits, in milligrams per cubic meter (mg/m^3) of air, shown in Table 3 as either TWA concentrations for up to a 10-hour work shift, 40-hour workweek, or as ceiling concentrations based on a 15-minute sampling period.

When there is simultaneous exposure to several nitriles or other sources of cyanide, the exposures shall be regarded as additive, and the environmental concentration limit for equivalent exposure to a mixture (E_m) shall be determined as follows:

$$E_m = \frac{C_1}{L_1} + \ldots + \frac{C_n}{L_n}$$

Where:

C_1 = concentration of the main component of the mixture

C_n = concentration of other constituents of the mixture, with *n* having values from 2 to *n*

L_1 = permissible exposure limit for the main component

L_n = permissible exposure limits for other constituents, with *n* having values from 2 to *n*

E_m = shall not exceed 1

Compounds with ceiling concentration limits are additive independently from those that have TWA exposure limits. When the additive formula exceeds 1, exposure to the mixture shall be reduced even if none of the individual TWA or ceiling concentration limits is exceeded.

TABLE 3. Recommended Workplace Exposure Limits

Nitrile	mg/m^3	Approximate ppm equivalents	Type of limit
Acetonitrile	34	20	TWA
Propionitrile	14	6	TWA
n-Butyronitrile	22	8	TWA
Isobutyronitrile	22	8	TWA
Acetone cyanohydrin	4	1	Ceiling
Glycolonitrile	5	2	Ceiling
Malononitrile	8	3	TWA
Adiponitrile	18	4	TWA
Succinonitrile	20	6	TWA
Tetramethyl-succinonitrile	6	1	Ceiling

The recommended standard is designed to protect the health and provide for the safety of employees for up to a 10-hour work shift, 40-hour workweek, over a working lifetime.

NITRILOTRIACETIC ACID. [139-13-9]
NCI—carcinogenic in animals

5-NITROACENAPHTHENE. [602-87-9]
IARC—carcinogenic in animals
NCI—carcinogenic in animals

3-NITRO-p-ACETOPHENETIDE. See p-Acetophenetide, 3'-nitro-. [1777-84-0]

m-NITROANILINE. See Aniline and chloro-, bromo-, and/or nitro- anilines. [99-09-2]

o-NITROANILINE. See Aniline and chloro-, bromo-, and/or nitro- anilines. [88-74-4]

p-NITROANILINE. See Aniline and chloro-, bromo-, and/or nitro- anilines. [100-01-6]

*OSHA 29 CFR 1910.1000 Table Z-1

Skin

PEL(TWA): 1 ppm, 6 mg/m^3
ACGIH-TLV (1982 Addition)

Skin

TWA: 3 mg/m^3
NFPA—hazardous chemical
health—3, flammability—1, reactivity—3
*RCRA 40 CFR 261
EPA hazardous waste no. P077

5-NITRO-o-ANISIDINE. [See o-Anisidine, 5-nitro-.] [99-59-2]

NITROBARITE. See Barium nitrate.

2-NITROBENZENAMINE. See Aniline and chloro-, bromo-, and/or nitro- anilines. [88-74-4]

3-NITROBENZENAMINE. See Aniline and chloro-, bromo-, and/or nitro- anilines. [99-09-2]

4-NITROBENZENAMINE. See p-Nitroaniline. [100-01-6]

NITROBENZENE. [98-95-3]

*OSHA 29 CFR 1910.1000 Table Z-1

Skin

PEL(TWA): 1 ppm, 5 mg/m^3
ACGIH-TLV

Skin

TWA: 1 ppm, 5 mg/m^3
STEL: 2 ppm, 10 mg/m^3
NFPA—hazardous chemical
health—3, flammability—2, reactivity—0
TSCA 4(e), ITC
Initial Report to the administrator, Environmental Protection Agency, TSCA Interagency Testing Committee, October 1, 1977; 44 FR 55026, October 12, 1977, responded to by the EPA administrator 46 FR 30300, June 5, 1981, removed from the Priority List in the Ninth Report of the TSCA Interagency Testing Committee to the administrator, En-

vironmental Protection Agency, October 30, 1981; 47 FR 5456, February 5, 1982

*TSCA 8(d) Unpublished Health and Safety Studies Reporting, 40 CFR 716, 47 FR 387, September 2, 1982

TSCA-CHIP available

CWA 304(a)(1), 45 FR 79318, November 28, 1980, Water Quality Criteria

Freshwater Aquatic Life

The available data for nitrobenzene indicate that acute toxicity to freshwater aquatic life occurs at concentrations as low as 27,000 µg/liter and would occur at lower concentrations among species that are more sensitive than those tested. No definitive data are available concerning the chronic toxicity of nitrobenzene to sensitive freshwater aquatic life.

Saltwater Aquatic Life

The available data for nitrobenzene indicate that acute toxicity to saltwater aquatic life occurs at concentrations as low as 6680 µg/liter and would occur at lower concentrations among species that are more sensitive than those tested. No data are available concerning the chronic toxicity of nitrobenzene to sensitive aquatic life.

Human Health

For comparison purposes, two approaches were used to derive criterion levels for nitrobenzene. Based on available toxicity data, for the protection of public health, the derived level is 19.8 mg/liter. Using available organoleptic data, for controlling undesirable taste and odor quality of ambient water, the estimated level is 30 µg/liter. It should be recognized that organoleptic data as a basis for establishing a water quality criterion have limitations and have no demonstrated relationship to potential adverse human health effects.

CWA 307(a)(1), Priority Pollutant

*CWA 311(b)(2)(A), 40 CFR 116, 117
Discharge RQ = 1000 pounds (454 kilograms)

CAA 112
Chemical proposed to be assessed as a Hazardous Air Pollutant by a House of Representatives bill to amend the CAA

*RCRA 40 CFR 261
EPA hazardous waste no. U169 (hazardous substance, ignitable waste, toxic waste)

6-NITROBENZIMIDAZOLE. See Benzimidizole, 5-nitro-. [94-52-0]

4-NITROBIPHENYL. [92-93-3]
*OSHA Standard 29 CFR 1910.1003
Cancer-Suspect Agent
Shall not apply to solid and liquid mixtures containing less than 0.1% by weight or volume of 4-nitrobiphenyl.
IARC—carcinogenic in animals

NITROCELLULOSE. See Cellulose nitrate. [9004-70-0]

p-NITROCHLOROBENZENE. [100-00-5]

NITROCHLOROFORM. See Chloropicrin. [76-06-2]

*OSHA 29 CFR 1910.1000 Table Z-1

Skin

PEL(TWA): 1 mg/m^3

ACGIH-TLV

Skin

TWA: 1 mg/m^3
STEL: 2 mg/m^3
Intended changes for 1982:

Skin

TWA: 0.5 ppm, 3 mg/m^3

NITROCOTTON. See Cellulose nitrate. [9004-70-0]

7-NITRO-2,2-DIMETHYL-2,3-DIHYDROBENZOFURAN. See Carbofuran intermediates. [13414-55-6]

TSCA 4(e), ITC

Eleventh Report of the Interagency Testing Committee to the administrator, Environmental Protection Agency, November 1982; 47 FR 54626, December 3, 1982, recommended but not designated for response within 12 months.

TSCA 8(a), 40 CFR 712, 48 FR 22697, May 19, 1983, Chemical Information Rule

TSCA 8(d) Unpublished Health and Safety Studies Reporting, 40 CFR 716, 48 FR 28483, June 22, 1983

4-NITRODIPHENYL. [92-93-3]

ACGIH-TLV
TWA
Human carcinogen—no exposure or contact by any route—respiratory, skin, or oral, as detected by the most sensitive methods—shall be permitted. The worker should be properly equipped to insure virtually no contact with the carcinogen.

NITROETHANE. [79-24-3]

*OSHA 29 CFR 1910.1000 Table Z-1
PEL(TWA): 100 ppm, 310 mg/m^3

ACGIH-TLV
TWA: 100 ppm, 310 mg/m^3
STEL: 150 ppm, 465 mg/m^3

NFPA—hazardous chemical
health—1, flammability—3, reactivity—3

NITROFEN. [1836-75-5]

NCI—carcinogenic in animals

2-NITRO-9h-FLUORENE. See Fluorene, 2-nitro-. [607-57-8]

NITROFURAZONE. See 2-Furaldehyde, 5-nitro-, semicarbazone. [59-87-0]

1-[(5-NITROFURFURYLIDENE) AMINO]-2-IMIDAZOLIDINONE.
[555-84-0]

IARC—carcinogenic in animals

N-[4-(5-NITRO-2-FURYL)-2-THIAZOLY]ACETAMIDE. [531-82-8]

IARC—carcinogenic in animals

NITROGEN (LIQUEFIED). [7727-37-9]

NFPA—hazardous chemical
health—3, flammability—0, reactivity—0

NITROGEN, OXIDES OF. [10102-44-0, 10024-97-2, 10102-43-9, 10544-72-6]

NIOSH Criteria Document (Pub. No. 76-149), NTIS Stock No. PB81 226995, March 19, 1976

For the purpose of this standard, "oxides of nitrogen" refers to nitric oxide and nitrogen dioxide. Since nitrogen dioxide in the working environment results, at least in part, from oxidation of nitric oxide, occupational exposures are customarily to mixtures of these gases rather than to either gas alone. "Occupational exposure to the oxides of nitrogen" is defined as exposure to concentrations of the oxides of nitrogen equal to or above one-half the recommended workroom limit. Adherence only to sections 3, 4(a)(1)(B), 4(a)(1)(C), 4(a)(2)(G), 4(a)(2)(L), 4(a)(2)(M), 4(b)(2), 5, 6(d), 6(e), 6(f), 7(a)(2), 7(a)(3), 7(b)(6), and 7(b)(7) is required when workplace environmental concentrations of oxides of nitrogen are not greater than one-half of the recommended workplace environmental limit.

a. Nitrogen Dioxide (NO$_2$)

Occupational exposure to nitrogen dioxide shall be controlled so that workers are not exposed to nitrogen dioxide at greater than a ceiling concentration of 1 ppm by volume (1.8 mg/m^3) as determined by a sampling time of 15 minutes.

b. Nitric Oxide (NO)

Occupational exposure to nitric oxide shall be controlled so that workers are not exposed to nitric oxide at a concentration greater than 25 ppm of air (30 mg/m^3) determined as a time-weighted average (TWA) exposure for up to a 10-hour workday, 40-hour workweek.

The standard is designed to protect the health and safety of workers for up to a 10-hour workday, 40-hour workweek, over a working lifetime.

NITROGEN DIOXIDE. [10102-44-0]

*OSHA 29 CFR 1910.1000 Table Z-1
PEL(Ceiling): 5 ppm, 9 mg/m^3

ACGIH-TLV
TWA: 3 ppm, 6 mg/m^3
STEL: 5 ppm, 10 mg/m^3

See Nitrogen, oxides of, NIOSH Criteria Document.

*CWA 311(b)(2)(A), 40 CFR 116, 117 Discharge RQ = 1000 pounds (454 kilograms)

*CAA § 109 (b); Part C § 160–178; 40 CFR 50, National Ambient Air Quality Standard (NAAQS) (not to be exceeded more than once per year)

	$\mu g/m^3$	
	Primary	Secondary
Annual	100 (0.05ppm)	Same as primary

*RCRA 40 CFR 261
EPA hazardous waste no. P078 (hazardous substance)

NITROGEN MUSTARD and HYDROCHLORIDE SALT. [51-75-2, 55-86-7]

*OSHA Candidate List of Potential Occupational Carcinogens (EPA Carcinogen Assessment Group List), 45 FR 53672, August 12, 1980

IARC—carcinogenic in animals

NITROGEN MUSTARD N-OXIDE and HYDROCHLORIDE SALT.[302-70-5]

OSHA Candidate List of Potential Occupational Carcinogens (EPA Carcinogen Assessment Group List), 45 FR 53672, August 12, 1980

IARC—carcinogenic in animals

NITROGEN(II) OXIDE. [10102-43-9]

*RCRA 40 CFR 261
EPA hazardous waste no. P076

NITROGEN(IV) OXIDE. [10102-44-0]

*RCRA 40 CFR 261
EPA hazardous waste no. P078 (hazardous substance)

NITROGEN PEROXIDE. [10102-44-0]

NFPA—hazardous chemical
health—3, flammability—0, reactivity—0, oxy—oxidizing chemical

NITROGEN TETROXIDE. See Nitrogen peroxide. [10544-72-6]

NITROGEN TRIFLUORIDE. [7783-54-2]

*OSHA 29 CFR 1910.1000 Table Z-1
PEL(TWA): 10 ppm, 29 mg/m^3

ACGIH-TLV
TWA: 10 ppm, 30 mg/m^3
STEL: 15 ppm, 45 mg/m^3

NITROGLYCERIN (NG). [55-63-0]

*OSHA 29 CFR 1910.1000 Table Z-1

Skin

PEL(Ceiling): 0.2 ppm, 2 mg/m^3

ACGIH-TLV

Skin

TWA: 0.02 ppm, 0.2 mg/m^3
STEL: 0.05 ppm, 0.5 mg/m^3

*RCRA 40 CFR 261
EPA hazardous waste no. P081 (reactive waste)

NITROGLYCERIN and ETHYLENE GLYCOL DINITRATE. [535-64-5]

NIOSH Criteria Document (Pub. No. 78-167), NTIS Stock No. PB81 225526, June 30, 1978

These criteria and the recommended standard apply to exposure of employees in the workplace to NG [1,2,3-trinitropropanetriol; $C_3H_5O_3(NO_2)_3$], to EGDN [1,2-dinitroethanediol, $C_2H_4O_2(NO_2)_2$], or to mixtures of these compounds. Workers employed where NG:EGDN explosives are used may also be exposed to by-products from explosions of these compounds, including oxides of carbon, hydrogen, and nitrogen. Applicable health and safety standards for these compounds should also be observed.

"Occupational exposure" is defined as exposure to airborne NG, EGDN, or a mixture of these compounds at a concentration above the action level or as any dermal contact with these compounds. The "action level" is defined as a concentration in the air of the workplace equal to one-half the recommended environmental limit for NG,

EGDN, or a mixture of these compounds (see Section 1). If there is no skin contact, and if exposures are below the action level, adherence to the recommended standard will not be required except for Sections 2, 3, 4(a), 5, 6(a,b,f,h), 7, and 8. "Overexposure" to NG or EGDN is defined as exposure above the recommended environmental limit or as any exposure, including dermal contact with these substances, resulting in throbbing headache, substantial decreases in blood pressure, or other observed effects on health.

Occupational exposure to NG and EGDN shall be controlled so that employees are not exposed to NG, EGDN, or a mixture of these two substances at concentrations greater than 0.1 mg/m^3 air measured as a ceiling concentration during any 20-minute sampling period.

The recommended standard is designed to protect the health and provide for the safety of employees for up to a 10-hour work shift, 40-hour workweek, throughout a working lifetime.

NITROMETHANE. [75-52-5]

*OSHA 29 CFR 1910.1000 Table Z-1
PEL(TWA): 100 ppm, 250 mg/m^3

ACGIH-TLV
TWA: 100 ppm, 250 mg/m^3
STEL: 150 ppm, 3750 mg/m^3

NFPA—hazardous chemical
health—1, flammability—3, reactivity—3

p-NITROPHENOL. [100-02-7]

NFPA—hazardous chemical
health—3, flammability—1, reactivity—0

*RCRA 40 CFR 261
EPA hazardous waste no. U170 (hazardous substance)

2-NITROPHENOL. [88-75-5]

CWA 307(a)(1), Priority Pollutant

4-NITROPHENOL. [100-02-7]

CWA 307(a)(1), 40 CFR 125, Priority Pollutant

NITROPHENOLS. [25154-55-6]

*CWA 311(b)(2)(A), 40 CFR 116, 117
Discharge RQ = 1000 pounds (454 kilograms)
CWA 304(a)(1), 45 FR 79318, November 28, 1980, Water Quality Criteria

Freshwater Aquatic Life

The available data for nitrophenols indicate that acute toxicity to freshwater aquatic life occurs at concentrations as low as 230 µg/liter and would occur at lower concentrations among species that are more sensitive than those tested. No data are available concerning the chronic toxicity of nitrophenols to sensitive freshwater aquatic life but toxicity to one species of algae occurs at concentrations as low as 150 µg/liter.

Saltwater Aquatic Life

The available data for nitrophenols indicate that acute toxicity to saltwater aquatic life occurs at concentrations as low as 4850 µg/liter and would occur at lower concentrations among species that are more sensitive than those tested. No data are available concerning the chronic toxicity of nitrophenols to sensitive saltwater aquatic life.

Human Health

For the protection of human health from the toxic properties of 2,4-dinitro-o-cresol ingested through water and contaminated aquatic organisms, the ambient water criterion is determined to be 13.4 µg/liter.

For the protection of human health from the toxic properties of 2,4-dinitro-o-cresol ingested through contaminated aquatic organisms alone, the ambient water criterion is determined to be 765 µg/liter.

For the protection of human health from the toxic properties of dinitrophenol ingested through water and contaminated aquatic organisms, the ambient water criterion is determined to be 70 µg/liter.

For the protection of human health from the toxic properties of dinitrophenol ingested through contaminated aquatic organisms alone, the ambient water criterion is determined to be 14.3 mg/liter.

Using the present guidelines, a satisfactory criterion cannot be derived at this time due to the insufficiency in the available data for mononitrophenol.

Using the present guidelines, a satisfactory criterion cannot be derived at this time due to the insufficiency in the available data for trinitrophenol.

2-NITRO-*p*-PHENYLENEDIAMINE.
[5307-14-2]

NCI—carcinogenic in animals

1-NITROPROPANE. [108-03-2]
*OSHA 29 CFR 1910.1000 Table Z-1
PEL(TWA): 25 ppm, 90 mg/m^3
ACGIH-TLV
TWA: 25 ppm, 90 mg/m^3
STEL: 35 ppm, 135 mg/m^3

2-NITROPROPANE. [79-46-9]
*OSHA 29 CFR 1910.1000 Table Z-1
PEL(TWA): 25 ppm, 90 mg/m^3
ACGIH-TLV
TWA: ceiling limit, 25 ppm, 90 mg/m^3
Industrial substance suspect of carcinogenic potential for humans—which is suspect of inducing cancer, based on either (1) limited epidemiologic evidence, exclusive of clinical reports of single cases, or (2) demonstration of carcinogenesis in one or more animal species by apropriate methods. Worker exposure by all routes should be carefully controlled to levels consistent with the animal and human experience data.
Intended changes for 1982:
TWA: 10 ppm, 35 mg/m^3
STEL: 20 ppm, 70 mg/m^3
NIOSH Current Intelligence Bulletin No. 17—cancer-related bulletin
IARC—carcinogenic in animals
TSCA-CHIP available
*RCRA 40 CFR 261
EPA hazardous waste no. U171 (ignitable waste)

NITROPROPANES (1-NITROPROPANE and 2-NITROPROPANE).
[108-03-2, 79-46-9]

NFPA—hazardous chemical
when nonfire: health—1, flammability—3, reactivity—1
when fire: health—2, flammability—3, reactivity—1

NITROPYRIN. [1929-82-4]
ACGIH-TLV
TWA: 10 mg/m^3
STEL: 20 mg/m^3

4-NITROQUINOLINE-1-OXIDE.
[56-57-5]

OSHA Candidate List of Potential Occupational Carcinogens (EPA Carcinogen Assessment Group List), 45 FR 53672, August 12, 1980

NITROSAMINES.

OSHA Candidate List of Potential Occupational Carcinogens (EPA Carcinogen Assessment Group List), 45 FR 53672, August 12, 1980

CWA 304(a)(1), 45 FR 79318, November 28, 1980, Water Quality Criteria

Freshwater Aquatic Life

The available data for nitrosamines indicate that acute toxicity to freshwater aquatic life occurs at concentrations as low as 5,850 µg/liter and would occur at lower concentrations among species that are more sensitive than those tested. No data are available concerning the chronic toxicity of nitrosamines to sensitive freshwater aquatic life.

Saltwater Aquatic Life

The available data for nitrosamines indicate that acute toxicity to saltwater aquatic life occurs at concentrations as low as 3,300,000 µg/liter and would occur at lower concentrations among species that are more sensitive than those tested. No data are available concerning the chronic toxicity of nitrosamines to sensitive saltwater aquatic life.

Human Health

For the maximum protection of human health from the potential carcinogenic effects due to exposure of *N*-nitrosodimethylamine through ingestion of contaminated water and contaminated aquatic organisms, the ambient water concentration should be zero based on the nonthreshold assumption for this chemical. However, zero level may not be attainable at the present time. Therefore, the levels that may result in incremental increase of cancer risk over the lifetime are estimated at 10^{-5}, 10^{-6}, and 10^{-7}. The corresponding criteria are 14 ng/liter, 1.4 ng/liter, and 0.14 ng/liter, respectively. If the above estimates are made for consumption of aquatic organisms only, excluding consumption of water, the levels are 160,000 ng/liter, 16,000 ng/liter, and 1600 ng/liter, respectively. Other concentrations representing different risk levels may be calculated by use of the guidelines. The risk estimate range is presented for information purposes and does not represent an agency judgment on an "acceptable" risk level.

For the maximum protection of human health from the potential carcinogenic effects due to exposure of N-nitrosodiethylamine through ingestion of contaminated water and contaminated aquatic organisms, the ambient water concentration should be zero based on the nonthreshold assumption for this chemical. However, zero level may not be attainable at the present time. Therefore, the levels that may result in incremental increase of cancer risk over the lifetime are estimated at 10^{-5}, 10^{-6}, and 10^{-7}. The corresponding criteria are 8 ng/liter, 0.8 ng/liter, 0.08 ng/liter, respectively. If the above estimates are made for consumption of aquatic organisms only, excluding consumption of water, the levels are 12,400 ng/liter, 1240 ng/liter, and 124 ng/liter, respectively. Other concentrations representing different risk levels may be calculated by use of the guidelines. The risk etimate range is presented for information purposes and does not represent an agency judgment on an "acceptable" risk level.

For the maximum protection of human health from the potential carcinogenic effects due to exposure to N-nitrosodi-N-butylamine through ingestion of contaminated water and contaminated aquatic organisms, the ambient water concentration should be zero based on the nonthreshold assumption for this chemical. However, zero level may not be attainable at the present time. Therefore, the levels that may resutl in incremental increase of cancer riskover the lifetime are estimated at 10^{-5}, 10^{-6}, and 10^{-7}. The corresponding criteria are 64 ng/liter, 6.4 ng/liter, and 0.64 ng/liter, respectively. If the above estimates are made for consumption of aquatic organisms only, excluding consumption of water, the levels are 5868 ng/liter, 587 ng/liter, and 58.7 ng/liter, respectively. Other concentrations representing different risk levels may be calculated by use of the guidelines. The risk estimate range is presented for information purposes and does not represent an agency judgment on an "acceptable" risk level.

For the maximum protection of human health from the potential carcinogenic effects due to exposure of N-nitrosodiphenylamine through ingestion of contaminated water and contaminated aquatic organisms, the ambient water concentration should be zero based on the nonthreshold assumption for this chemical. However, zero level may not be attainable at the present time. Therefore, the levels that may result in incremental increase of cancer risk over the lifetime are estimated at 10^{-5}, 10^{-6}, and 10^{-7}. The corresponding criteria are 49,000 ng/liter, 4900 ng/liter, and 490 ng/liter, respectively. If the above estimates are made for consumption of aquatic organisms only, excluding consumption of water, the levels are 161,000 ng/liter, 16,100 ng/liter, and 1610 ng/liter, respectively. Other concentrations representing different risk levels may be calculated by use of the guidelines. The risk estimate range is presented for information purposes and does not represent an agency judgment on an "acceptable" risk level.

For the maximum protection of human health from the potential carcinogenic effects due to exposure in N-nitrosopyrrolidine through ingestion of contaminated water and contaminated aquatic organisms, the ambient water concentration should be zero based on the nonthreshold assumption for this chemical. However, zero level may not be attainable at the present time. Therefore, the levels that may result in incremental increase of cancer risk over the lifetime are estimated at 10^{-5}, 10^{-6}, and 10^{-7}. The corresponding criteria are 160 ng/liter, 16.0 ng/liter, and 1.60 ng/liter, respectively. If the above estimates are made for consumption of aquatic organisms only, excluding consumption of water, the levels are 919,000 ng/liter, 91,900 ng/liter and 9190 ng/liter, respectively. Other concentrations representing different risk levels may be calculated by use of the guidelines. The risk estimate range is presented for information purposes and does not represent an agency judgment on an "acceptable" risk level.

NITROSAMINES in cutting fluids.
NIOSH Current Intelligence Bulletin No. 15—cancer-related bulletin

N-NITROSARCOSINE. [20661-60-3]
OSHA Candidate List of Potential Occupational Carcinogens (EPA Carcinogen Assessment Group List), 45 FR 53672, August 12, 1980

NITROSO-o-ANISIDINE. [99-59-2]
TSCA-CHIP available

NITROSO COMPOUNDS. [55-18-5, 62-75-9, 86-30-6, 100-75-4, 614-00-6, 621-64-7, 684-93-5, 759-73-9, 930-55-2, 1116-54-7, 53609-64-6]
TSCA-CHIP available

N-NITROSODI-N-BUTYLAMINE.
[924-16-3]
OSHA Candidate List of Potential Occupational Carcinogens (EPA Carcinogen Assessment Group List), 45 FR 53672, August 12, 1980
*RCRA 40 CFR 261
EPA hazardous waste no. U172

N-NITROSODIETHANOLAMINE.
[1116-54-7]
OSHA Candidate List of Potential Occupational Carcinogens (EPA Carcinogen Assessment Group List), 45 FR 53672, August 12, 1980
IARC—carcinogenic in animals
*RCRA 40 CFR 261
EPA hazardous waste no. U173

N-NITROSODIETHYLAMINE.
[55-18-5]
OSHA Candidate List of Potential Occupational Carcinogens (EPA Carcinogen Assessment Group List), 45 FR 53672, August 12, 1980
*RCRA 40 CFR 261
EPA hazardous waste no. U174

N-NITROSODIMETHYLAMINE.
[62-75-9]
*OSHA Standard 29 CFR 1910.1016
Cancer-Suspect Agent
Shall not apply to solid or liquid mixtures containing less than 1.0% by weight or volume of N-nitrosodimethylamine.
ACGIH-TLV

Skin

TWA
Industrial substance suspect of carcinogenic potential for humans—which is suspect of inducing cancer, based on either (1) limited epidemiologic evidence, exclusive of clinical reports of single cases, or (2) demonstration of carcinogenesis in one or more animal species by appropriate methods. Worker exposure by all routes should be carefully controlled to levels consistent with the animal and human experience data.

OSHA Candidate List of Potential Occupational Carcinogens (EPA Carcinogen Assessment Group List), 45 FR 53672, August 12, 1980
CWA 307(a)(1), 40 CFR 125, Priority Pollutant
*RCRA 40 CFR 261
EPA hazardous waste no. P082

4-NITROSODIMETHYLANILINE. See Aniline, N,N-dimethyl-p-nitroso-.
[138-89-6]

N-NITROSODIPHENYLAMINE.
[86-30-6]
NCI—carcinogenic in animals
CWA 307(a)(1), 40 CFR 125, Priority Pollutant
TSCA-CHIP available

p-NITROSODIPHENYLAMINE.
[156-10-5]
NCI—carcinogenic in animals

N-NITROSODI-N-PROPYLAMINE.
[621-64-7]
OSHA Candidate List of Potential Occupational Carcinogens (EPA Carcinogen Assessment Group List), 45 FR 53672, August 12, 1980
IARC—carcinogenic in animals
CWA 307(a)(1), 40 CFR 125, Priority Pollutant
*RCRA 40 CFR 261
EPA hazardous waste no. U111

N-NITROSO-N-ETHYLUREA.
[759-73-9]
OSHA Candidate List of Potential Occupational Carcinogens (EPA Carcinogen Assessment Group List), 45 FR 53672, August 12, 1980
IARC—carcinogenic in animals (used as positive carcinogenic control by NCI)
*RCRA 40 CFR 261
EPA hazardous waste no. U176

N-NITROSOMETHYLETHYLAMINE.
[10595-95-6]
IARC—carcinogenic in animals

N-NITROSO-N-METHYLUREA.
[684-93-5]
IARC—carcinogenic in animals (used as positive carcinogenic control by NCI)
*RCRA 40 CFR 261
EPA hazardous waste no. U177

N-NITROSO-N-METHYLURETHANE.
[615-53-2]
OSHA Candidate List of Potential Occupational Carcinogens (EPA Carcinogen Assessment Group List), 45 FR 53672, August 12, 1980
IARC—carcinogenic in animals
*RCRA 40 CFR 261
EPA hazardous waste no. U178

N-NITROSOMETHYLVINYLAMINE.
[4549-40-0]
OSHA Candidate List of Potential Occupational Carcinogens (EPA Carcinogen Assessment Group List), 45 FR 53672, August 12, 1980
IARC—carcinogenic in animals
*RCRA 40 CFR 261
EPA hazardous waste no. P084

N-NITROSOMORPHOLINE. [59-89-2]
OSHA Candidate List of Potential Occupational Carcinogens (EPA Carcinogen Assessment Group List), 45 FR 53672, August 12, 1980
IARC—carcinogenic in animals
CAA 112
Chemical proposed to be assessed as a Hazardous Air Pollutant by a House of Representatives bill to amend the CAA

N-NITROSONORNICOTINE.
[16543-55-8]
OSHA Candidate List of Potential Occupational Carcinogens (EPA Carcinogen Assessment Group List), 45 FR 53672, August 12, 1980
IARC—carcinogenic in animals

p-NITROSO-N-PHENYLANILINE. See
Diphenylamine, 4-nitroso-. [156-10-5]

N-NITROSOPIPERIDINE. [100-75-4]
OSHA Candidate List of Potential Occupational Carcinogens (EPA Carcinogen Assessment Group List), 45 FR 53672, August 12, 1980
IARC—carcinogenic in animals
*RCRA 40 CFR 261
EPA hazardous waste no. U179

N-NITROSOPYRROLIDINE. [930-55-2]
OSHA Candidate List of Potential Occupational Carcinogens (EPA Carcinogen Assessment Group List), 45 FR 53672, August 12, 1980
IARC—carcinogenic in animals

*RCRA 40 CFR 261
EPA hazardous waste no. U180

N-NITROSOSARCOSINE.
[13256-22-9]
IARC—carcinogenic in animals

NITROTOLUENE. [99-08-1, 1321-12-6]
*OSHA 29 CFR 1910.1000 Table Z-1

Skin

PEL(TWA): 5 ppm, 30 mg/m^3
ACGIH-TLV

Skin

TWA: 2 ppm, 11 mg/m^3
*CWA 311(b)(2)(A), 40 CFR 116, 117
Discharge RQ = 1000 pounds (454 kilograms)

p-NITROTOLUENE. [99-99-0]
NFPA—hazardous chemical
health—3, flammability—1, reactivity—0

5-NITRO-o-TOLUIDINE. [99-55-8]
*RCRA 40 CFR 261
EPA hazardous waste no. U181
OSHA Candidate List of Potential Occupational Carcinogens (EPA Carcinogen Assessment Group List), 45 FR 53672, August 12, 1980
NCI—carcinogenic in animals
TSCA-CHIP available

p-NITROTOLUOL. See p-nitrotoluene.
[99-99-0]

NITROTRICHLOROMETHANE. See Chloropicrin. [76-06-2]

NITROUS OXIDE. [10024-97-2]
See waste anesthetic gases and vapors, NIOSH Criteria Document.

NONANE. [111-84-2]
ACGIH-TLV
TWA: 200 ppm, 1050 mg/m^3
STEL: 250 ppm, 1300 mg/m^3

1-NONANETHIOL.
See Thiols, NIOSH Criteria Document.

5-NORBORENE-2,3-DIMETHANOL,1,4,5,6,7,7-HEXACHLORO-CYCLIC SULFATE.
[115-29-7]
*RCRA 40 CFR 261
EPA hazardous waste no. P050 (hazardous substance)

NORETHISTERONE. [68-22-4]
IARC—carcinogenic in animals

NORETHISTERONE ACETATE.
[52-98-9]
IARC—carcinogenic in animals

NORETHYNODREL. [68-23-5]
IARC—carcinogenic in animals

NORFLURAZON. [27314-13-2]
FIFRA—Data Call In, 45 FR 75488, November 14, 1980
Status: letter being drafted

O

OBPA (10,10'-OXYBISPHENARSAZINE).
FIFRA—Registration Standard—Issued September 1981 (fungicide)

OCTACHLORONAPHTHALENE. [2234-13-1]
*OSHA 29 CFR 1910.1000 Table Z-1

Skin

PEL(TWA): 0.1 mg/m^3

ACGIH-TLV

Skin

TWA: 0.1 mg/m^3
STEL: 0.3 mg/m^3

1-OCTADECANETHIOL.
See Thiols, NIOSH Criteria Document.

OCTALENE. See Aldrin. [309-00-2]

OCTAMETHYLPYROPHOSPHO-RAMIDE (OMPA). [152-16-9]
40 CFR 162.11, FIFRA—RPAR, Final Action, Voluntary Cancellation
Current Status: 41 FR 21859, May 28, 1976
Criteria Possibly Met or Exceeded: Oncogenicity
*RCRA 40 CFR 261
EPA hazardous waste no. P085

OCTANE. [111-65-9]
*OSHA 29 CFR 1910.1000 Table Z-1
PEL(TWA): 500 ppm, 2350 mg/m^3
ACGIH-TLV
TWA: 300 ppm, 1450 mg/m^3
STEL: 375 ppm, 1800 mg/m^3
See Alkanes (C$_5$–C$_8$), NIOSH Criteria Document.

NFPA—hazardous chemical (flammable chemical)
health—0, flammability—3, reactivity—0

1-OCTANETHIOL.
See Thiols, NIOSH Criteria Document.

OCTYL ACRYLATE. See 2-Ethylhexyl acrylate. [103-11-7]

***n*-OCTYL ALCOHOL.** [111-87-5]

NFPA—hazardous chemical (flammable chemical)
health—1, flammability—2, reactivity—0

17B-OESTRADIOL. [50-28-2]
IARC—carcinogenic in animals

OESTRONE. [53-16-7]
IARC—carcinogenic in animals

OIL MIST, MINERAL.
*OSHA 29 CFR 1910.1000 Table Z-1
PEL(TWA): 5 mg/m^3
ACGIH-TLV
TWA: 5 mg/m^3
As sampled by method that does not collect vapor:
STEL: 10 mg/m^3

OIL ORANGE SS. [2646-17-5]
IARC—carcinogenic in animals

OLEIC ACID. [112-80-1]
NFPA—hazardous chemical (flammable chemical)
health—0, flammability—1, reactivity—0

OLEO OIL.
NFPA—hazardous chemical (flammable chemical)
health—0, flammability—1, reactivity—0

ORANGE I. [523-44-4]
IARC—carcinogenic in animals

ORGANIC ARSENIC COMPOUNDS.
OSHA 29 CFR 1910.1000 Table Z-1
PEL(TWA): 0.5 mg/m^3 (as As)

ORGANO(ALKYL)MERCURY.
*OSHA 1910.1000 Table Z-2
0.01 mg/m^3, 8-hour TWA
0.04 mg/m^3, acceptable ceiling concentration

ORGANOTIN COMPOUNDS.
NIOSH Criteria Document (Pub. No. 77-115), NTIS Stock No. PB 274766, November 12, 1976
Organotin is the common name assigned to the group of compounds having at least one covalent bond between carbon and tin. The "action level" is set at half the recommended time-weighted average (TWA) workplace concentration limit. An employee is exposed or potentially exposed to organotins if that employee is involved in the occupational handling of the compounds or works in a plant containing organotins. "Occupational exposure" occurs when exposure exceeds the action level or if skin or eye contact with organotins occurs. "Overexposure" to organotins occurs if an employee is known to be exposed to the organotins at a concentration in excess of the TWA concentration limit, or is exposed at any concentration sufficient to produce irritation of eyes, skin, or upper or lower respiratory tract.

The employer shall control exposure to organotin compounds so that no employee is exposed at a concentration greater than 0.1 mg, measured as tin, per cubic meter of air, determined as a TWA concentration for up to a 10-hour work shift in a 40-hour workweek.

The standard is designed to protect the health and provide for the safety of employees for up to a 10-hour work shift in a 40-hour workweek over a normal working life.

ORTHOPHOSPHORIC ACID. *See* Phosphoric acid. [7664-38-2]

OSMIUM TETROXIDE. [20816-12-0]
*OSHA 29 CFR 1910.1000 Table Z-1
PEL(TWA): 0.002 mg/m^3
ACGIH-TLV
TWA: 0.0002 ppm, 0.002 mg/m^3
STEL: 0.0006 ppm, 0.006 mg/m^3
*RCRA 40 CFR 261
EPA hazardous waste no. P087

3-OXA-9-AZONIATRICYCLO[3.3.1.02.4]NONANE, 7-(3-HYDROXY-1-OXO-2-PHENYLPROPOXY)-9,9-DIMETHYL-, BROMIDE[7(S)-(1.ALPHA.,2BETA.,4.BETA.,5.ALPHA.,7.BETA.)]-
[155-41-9]
*TSCA 8(a), 40 CFR 712, 47 FR 26992, June 22, 1982, Chemical Information Rule

7-OXABICYCLO[4.1.0]HEPTANE.
[286-20-4]
*TSCA 8(a), 40 CFR 712, 47 FR 26992, June 22, 1982, Chemical Information Rule

7-OXABICYCLO[2.2.1]HEPTANE-2,3-DICARBOXYLIC ACID. [145-73-3]
*TSCA 8(a), 40 CFR 712, 47 FR 26992, June 22, 1982, Chemical Information Rule
*RCRA 40 CFR 261
EPA hazardous waste no. P088

7-OXABICYCLO[2.2.1]HEPTANE,1-METHYL-4-(1-METHYLETHYL)-.
[470-67-7]
*TSCA 8(a), 40 CFR 712, 47 FR 26992, June 22, 1982, Chemical Information Rule

7-OXABICYCLO[4.1.0]HEPTANE, 1-METHYL-4-(2-METHYLOXIRANYL)-.
[96-08-2]
*TSCA 8(a), 40 CFR 712, 47 FR 26992, June 22, 1982, Chemical Information Rule

7-OXABICYCLO[4.1.0]HEPTANE, 3-OXIRANYL. [106-87-6]
*TSCA 8(a), 40 CFR 712, 47 FR 26992, June 22, 1982, Chemical Information Rule

6-OXABICYCLO[3.1.0]HEXANE.
[285-67-6]

*TSCA 8(a), 40 CFR 712, 47 FR 26992, June 22, 1982, Chemical Information Rule

OXALIC ACID. [144-62-7]

*OSHA 29 CFR 1910.1000 Table Z-1
PEL(TWA): 1 mg/m^3

ACGIH-TLV
TWA: 1 mg/m^3
STEL: 2 mg/m^3

NFPA—hazardous chemical
when nonfire: health—1, flammability—1, reactivity—0
when fire: health—2, flammability—1, reactivity—0

OXALIC ACID, DIHYDRATE. See Oxalic acid. [144-62-7]

1,2-OXATHIOLANE 2,2-DIOXIDE. [1120-71-4]

*RCRA 40 CFR 261
EPA hazardous waste no. U193

5-OXATRICYCLO[8.2.0.04.6] DODECANE, 4,12,12-TRIMETHYL-9-METHYLENE-,[1R-(1R*,4R*,6R*, 10S*)]. [1139-30-6]

*TSCA 8(a), 40 CFR 712, 47 FR 26992, June 22, 1982, Chemical Information Rule

2H-1,3,2-OXAZAPHOSPHORINE, 2[BIS(2-CHLOROETHYL)AMINO] TETRAHYDRO-2-OXIDE.

*RCRA 40 CFR 261
EPA hazardous waste no. U058

OXAZEPAM. [604-75-1]
IARC—carcinogenic in animals

OXIRANE. See Ethylene oxide, Alkyl epoxides. [75-21-8]

OXIRANE, (BROMOMETHYL)-. [3132-64-7]

*TSCA 8(a), 40 CFR 712, 47 FR 26992, June 22, 1982, Chemical Information Rule

OXIRANE, (BUTOXYMETHYL)-. [2426-08-6]

*TSCA 8(a), 40 CFR 712, 47 FR 26992, June 22, 1982, Chemical Information Rule

OXIRANECARBOXYLIC ACID, 3-METHYL-3-PHENYL-, ETHYL ESTER. [77-83-8]

*TSCA 8(a), 40 CFR 712, 47 FR 26992, June 22, 1982, Chemical Information Rule

OXIRANECARBOXYLIC ACID, 3-PHENYL-, ETHYL ESTER. [121-39-1]

*TSCA 8(a), 40 CFR 712, 47 FR 26992, June 22, 1982, Chemical Information Rule

OXIRANE, (CHLOROMETHYL)-. See Epichlorohydrin. [106-89-8]

OXIRANE, DECYL-. [2855-19-8]

*TSCA 8(a), 40 CFR 712, 47 FR 26992, June 22, 1982, Chemical Information Rule

OXIRANE, [(DODECYLOXY) METHYL]-. [2461-18-9]

*TSCA 8(a), 40 CFR 712, 47 FR 26992, June 22, 1982, Chemical Information Rule

OXIRANE, (ETHOXYMETHYL)-. [4016-11-9]

*TSCA 8(a), 40 CFR 712, 47 FR 26992, June 22, 1982, Chemical Information Rule

OXIRANE, ETHYL-. [106-87-7]

*TSCA 8(a), 40 CFR 712, 47 FR 26992, June 22, 1982, Chemical Information Rule

OXIRANEMETHANOL. [556-52-5]

*TSCA 8(a), 40 CFR 712, 47 FR 26992, June 22, 1982, Chemical Information Rule

OXIRANE, (METHOXYMETHYL)-. [930-37-0]

*TSCA 8(a), 40 CFR 712, 47 FR 26992, June 22, 1982, Chemical Information Rule

OXIRANE, METHYL-. [75-56-9]

*TSCA 8(a), 40 CFR 712, 47 FR 26992, June 22, 1982, Chemical Information Rule

OXIRANE, (1-METHYLETHOXY) METHYL-. [4016-14-2]

*TSCA 8(a), 40 CFR 712, 47 FR 26992, June 22, 1982, Chemical Information Rule

OXIRANE, 2,2'-[(1-METHYLETHYLIDENE)BIS(4,1-PHENYLENEOXYMETHYLENE)]BIS-, HOMOPOLYMER. [25085-99-8]

*TSCA 8(a), 40 CFR 712, 47 FR 26992, June 22, 1982, Chemical Information Rule

OXIRANE, [(METHYLPHENOXY)METHYL]-. [26447-14-3]

*TSCA 8(a), 40 CFR 712, 47 FR 26992, June 22, 1982, Chemical Information Rule

OXIRANEOCTANOIC ACID, 3-OCTYL-, BUTYL ESTER. [106-83-2]

*TSCA 8(a), 40 CFR 712, 47 FR 26992, June 22, 1982, Chemical Information Rule

OXIRANEOCTANOIC ACID, 3-OCTYL-, 2-ETHYLHEXYL ESTER. [141-38-8]

*TSCA 8(a), 40 CFR 712, 47 FR 26992, June 22, 1982, Chemical Information Rule

OXIRANEOCTANOIC ACID, 3-OCTYL-, OCTYL ESTER. [106-84-3]

*TSCA 8(a), 40 CFR 712, 47 FR 26992, June 22, 1982, Chemical Information Rule

OXIRANE, 2,2'-([OXIRANYLMETHOXY)-1,3-PHENYLENE]BIS(METHYLENE)) BIS-. [13561-08-5]

*TSCA 8(a), 40 CFR 712, 47 FR 26992, June 22, 1982, Chemical Information Rule

OXIRANE, 2,2'-[OXYBIS(METHYLENE)]BIS-. [2238-07-5]

*TSCA 8(a), 40 CFR 712, 47 FR 26992, June 22, 1982, Chemical Information Rule

OXIRANE, (PHENOXYMETHYL)-. [122-60-1]

*TSCA 8(a), 40 CFR 712, 47 FR 26992, June 22, 1982, Chemical Information Rule

OXIRANE, PHENYL-. [96-09-3]

*TSCA 8(a), 40 CFR 712, 47 FR 26992, June 22, 1982, Chemical Information Rule

OXIRANE, 2,2'-[1,3-PHENYLENEBIS(OXYMETHYLENE)]BIS-. [101-90-6]

*TSCA 8(a), 40 CFR 712, 47 FR 26992, June 22, 1982, Chemical Information Rule

OXIRANE, 2,2',2''-[1,2,3-PROPANETRIYLTRIS(OXYMETHYLENE)]TRIS-. [13236-02-7]

*TSCA 8(a), 40 CFR 712, 47 FR 26992, June 22, 1982, Chemical Information Rule

OXIRANE,[(2-PROPENYLOXY)METHYL]-. [106-92-3]

*TSCA 8(a), 40 CFR 712, 47 FR 26992, June 22, 1982, Chemical Information Rule

OXIRANE, TETRADECYL-. [7320-37-8]

*TSCA 8(a), 40 CFR 712, 47 FR 26992, June 22, 1982, Chemical Information Rule

OXIRANE, (2,2,2-TRICHLOROETHYL)-. [3083-25-8]

*TSCA 8(a), 40 CFR 712, 47 FR 26992, June 22, 1982, Chemical Information Rule

OXIRANE, TRIFLUORO(TRIFLUOROMETHYL)-. [428-59-1]

*TSCA 8(a), 40 CFR 712, 47 FR 26992, June 22, 1982, Chemical Information Rule

1,1'-OXYBIS(2-METHOXYETHANE). [111-96-6]

TSCA-CHIP available

10,10'-OXYBISPHENOXARSINE (some uses). [58-36-6]

40 CFR 162.11, FIFRA-RPAR, Notice of Intent to Register Issued
Current Status: Returned to Registration Division on April 9, 1979 based on memo indicating no evidence of oncogenicity
Criteria Possibly Met or Exceeded: (Arsenical Compound) Suspected: Oncogenicity, Mutagenicity, and Teratogenicity

4,4'-OXYDIANILINE. [101-80-4]

NCI—carcinogenic in animals

OXYFLUORFEN. [42874-03-3]
FIFRA—Data Call In, 45 FR 75488, November 14, 1980
Status: letter being drafted

OXYGEN (LIQUID). [7782-44-7]
NFPA—hazardous chemical
health—3, flammability—0, reactivity—0, oxy—oxidizing chemical

OXYGEN DIFLUORIDE. [7783-41-7]
*OSHA 29 CFR 1910.1000 Table Z-1
PEL(TWA): 0.05 ppm, 0.1 mg/m^3

ACGIH-TLV
TWA: 0.05 ppm, 0.1 mg/m^3
STEL: 0.15 ppm, 0.3 mg/m^3

OXYMETHOLONE. [434-07-1]
IARC—carcinogenic in humans

OXYZALIN.
FIFRA—Data Call In, 45 FR 75488, November 14, 1980
Status: letter being drafted

OZONE. [10028-15-6]
*OSHA 29 CFR 1910.1000 Table Z-1
PEL(TWA): 0.1 ppm, 0.2 mg/m^3

ACGIH-TLV
TWA: 0.1 ppm, 0.2 mg/m^3
STEL: 0.3 ppm, 0.6 mg/m^3

*CAA § 109(b); Part C § 160–178; 40 CFR 50, National Ambient Air Quality Standard (NAAQS) (not to be exceeded more than once per year)

	µg/m^3	
	Primary	Secondary
1 hr	235 (0.12 ppm)	Same as primary

P

PARA-ACETALDEHYDE. See Paraldehyde [123-63-7]

PARAFFIN WAX FUME. [8002-74-2]
ACGIH-TLV
TWA: 2 mg/m^3
STEL: 6 mg/m^3

PARAFORMALDEHYDE. [30525-89-4]
NFPA—hazardous chemical
health—2, flammability—1, reactivity—0
*CWA 311(b)(2)(A), 40 CFR 116, 117
Discharge RQ = 1000 pounds (454 kilograms)

PARALDEHYDE. [123-63-7]
NFPA—hazardous chemical
health—2, flammability—3, reactivity—1
*RCRA 40 CFR 261
EPA hazardous waste no. U182

PARAQUAT. [1910-42-5]
*OSHA 29 CFR 1910.1000 Table Z-1

Skin
PEL(TWA): 0.5 mg/m^3
ACGIH-TLV
TWA: 0.1 mg/m^3
40 CFR 162.11, FIFRA Pre-RPAR review
Current Status: Pre-RPAR evaluation under way
Criteria Possibly Met or Exceeded: Emergency Treatment and Chronic Effects

PARASORBIC ACID. [10048-32-5]
IARC—carcinogenic in animals

PARATHION. [56-38-2]
OSHA 29 CFR 1910.1000 Table Z-1

Skin
PEL(TWA): 0.1 mg/m^3
ACGIH-TLV
Skin
TWA: 0.1 mg/m^3
STEL: 0.3 mg/m^3
NIOSH Criteria Document (Pub. No. 76-190), NTIS Stock No. PB 274192, July 1, 1976
"Parathion" is defined as O,O-diethyl O-p-nitrophenyl phosphorothioate, regardless of production process, alone or in combination with other compounds. "Occupational exposure to parathion" is defined as employment in any area in which parathion or materials containing parathion, alone or in combination with other substances, is produced, packaged, processed, mixed, blended, handled, stored in large quantities, or applied. Adherence to all provisions of the standard is required in workplaces using parathion, regardless of the airborne parathion concentration, because of serious effects produced by contact with the skin, mucous membranes, and eyes.

Occupational exposure to parathion shall be controlled so that no employee is exposed to parathion in a concentration greater than 0.05 mg/m^3 of air determined as a time-weighted average (TWA) exposure for up to a 10-hour work shift and a 40-hour workweek. The standard is designed to protect the health and safety of workers for up to a 10-hour work shift, 40-hour workweek, during a working lifetime.

FIFRA—Data Call In, 45 FR 75488, November 14, 1980
Status: letter issued November 14, 1981
NFPA—hazardous chemical
health—4, flammability—1, reactivity—2

CWA 311(b)(2)(A), 40 CFR 116, 117
Discharge RQ = 1 pound (0.454 kilogram)
*RCRA 40 CFR 261
EPA hazardous waste no. P089 (hazardous substance)

PARTICULATE POLYCYCLIC AROMATIC HYDROCARBONS (PPAH).
See Coal tar pitch volatiles.

PARTICULATES.
*CAA § 109(b); Part C § 160–178; 40 CFR 50, National Ambient Air Quality Standard (NAAQS) (not to be exceeded more than once per year)

$\mu g/m^3$

	Primary	Secondary
Annual	75	60
24 hr	260	160

PSD Air Quality Increments, Maximum Allowable Increases, $\mu g/m^3$

	Class I	Class II	Class III	Primary vs NAAQS
Annual	5	19	37	75
24 hr	10	37	75	260

PATULIN. [149-29-1]
IARC—carcinogenic in animals

PCBS. See Polychlorinated biphenyls. [1336-36-3]

PCNB. See Pentachloronitrobenzene. [82-68-8]

PCTS. See Terphenyls, chlorinated. [61788-33-8]

PEANUT OIL. [8002-03-7]
NFPA—hazardous chemical (flammable chemical)
health—0, flammability—1, reactivity—0

PENDIMETHALIN. [40487-42-1]
FIFRA—Data Call In, 45 FR 75488, November 14, 1980
Status: letter being drafted

PENICILLIC ACID. [90-65-3]
IARC—carcinogenic in animals

PENTABORANE. [19624-22-7]
*OSHA 29 CFR 1910.1000 Table Z-1
PEL(TWA): 0.005 ppm, 0.01 mg/m^3
ACGIH-TLV
TWA: 0.005 ppm, 0.01 mg/m^3
STEL: 0.015 ppm, 0.03 mg/m^3

PENTABROMO-6-CHLOROCYCLOHEXANE. [87-84-3]
TSCA-CHIP available

PENTACHLOROBENZENE. [608-93-5]
TSCA 4(e), ITC
Third Report of the Interagency Testing Committee to the administrator, Environmental Protection Agency, October 1978; 43 FR 50630, October 30, 1978, responded to by the EPA administrator 45 FR 48524, July 18, 1980
*RCRA 40 CFR 261
EPA hazardous waste no. U183

PENTACHLOROETHANE. [76-01-7]
NCI—carcinogenic in animals
TSCA-CHIP in preparation
*RCRA 40 CFR 261
EPA hazardous waste no. U184

PENTACHLORONAPHTHALENE. [1321-64-8]
*OSHA 29 CFR 1910.1000 Table Z-1

Skin
PEL(TWA): 0.5 mg/m^3
ACGIH-TLV
TWA: 0.5 mg/m^3
STEL: 2 mg/m^3

PENTACHLORONITROBENZENE. Syn: PCNB. [82-68-8]
OSHA Candidate List of Potential Occupational Carcinogens (EPA Carcinogen Assessment Group List), 45 FR 53672, August 12, 1980
40 CFR 162.11, FIFRA-RPAR issued
Current Status: PD 1 published, 42 FR

61894 (10/13/77); comment period closed 2/6/78: NTIS# PB80 216799.
PD 2/3 in agency review.
Criteria Possibly Met or Exceeded: Oncogenicity
*RCRA 40 CFR 261
EPA hazardous waste no. U185

PENTACHLOROPHENOL. [87-86-5]
*OSHA 29 CFR 1910.1000 Table Z-1

Skin
PEL(TWA): 0.5 mg/m^3
ACGIH-TLV

Skin
TWA: 0.5 mg/m^3
STEL: 1.5 mg/m^3

40 CFR 162.11, FIFRA-RPAR issued
Current Status: PD 1 published, 43 FR 48443 (10/18/78); comment period closed 2/12/79: NTIS# PB81 109464.
PFD 2/3 (wood used only) completed and Notice of Determination published in 46 FR 13020 (2/19/81); comment period closed 5/20/81.
PD 4 being developed.
Criteria Possibly Met or Exceeded: Oncogenicity, Fetoxicity, and Teratogenicity
NFPA—hazardous chemical
when dry: health—3, flammability—0, reactivity—0
when solution: health—3, flammability—2, reactivity—0
CWA 304(a)(1), 45 FR 79318, November 28, 1980, Water Quality Criteria

Freshwater Aquatic Life
The available data for pentachlorophenol indicate that acute and chronic toxicity to freshwater aquatic life occur at concentrations as low as 55 and 3.2 µg/liter, respectively, and would occur at lower concentrations among species that are more sensitive than those tested.

Saltwater Aquatic Life
The available data for pentachlorophenol indicate that acute and chronic toxicity to saltwater aquatic life occur at concentrations as low as 53 and 34 µg/liter respectively, and would occur at lower concentrations among species that are more sensitive than those tested.

Human Health
For comparison purposes, two approaches were used to derive criterion levels for pentachlorophenol. Based on available toxicity data, for the protection of public health, the derived level is 1.01 mg/liter. Using available organoleptic data, for controlling undesirable taste and odor quality of ambient water, the estimated level is 30 µg/liter. It should be recognized that organoleptic data as a basis for establishing a water quality criterion have limitations and have no demonstrated relationship to potential adverse human health effects.

CWA 307(a)(1), Priority Pollutant
*CWA 311(b)(2)(A), 40 CFR 116, 117
Discharge RQ = 10 pounds (4.54 kilograms)
*RCRA 40 CFR 261
EPA hazardous waste no U242 (hazardous substance)

1,3-PENTADIENE. [504-60-9]
*RCRA 40 CFR 261
EPA hazardous waste no. U186 (ignitable waste)

PENTAERYTHRITOL. [115-77-5]
ACGIH-TLV
TWA: nuisance particulate—30 mppcf or 10 mg/m^3 of total dust < 1% quartz, or 5 mg/m^3 respirable dust
STEL: 20 mg/m^3

PENTANE. [109-66-0]
*OSHA 29 CFR 1910.1000 Table Z-1
PEL(TWA): 1000 ppm, 2950 mg/m^3
ACGIH-TLV
TWA: 600 ppm, 1800 mg/m^3
STEL: 750 ppm, 2250 mg/m^3

See Alkanes (C$_5$–C$_8$), NIOSH Criteria Document.

NFPA—hazardous chemical (flammable chemical)
health—1, flammability—4, reactivity—0

1-PENTANETHIOL. [110-66-7]
See Thiols, NIOSH Criteria Document.

PENTANETHIOLS.
NFPA—hazardous chemical
health—2, flammability—3, reactivity—0

1-PENTANOL. [71-41-0]
NFPA—hazardous chemical
health—1, flammability—3, reactivity—0

2-PENTANOL
NFPA—hazardous chemical (flammable chemical)
health—1, flammability—2, reactivity—0

3-PENTANOL. [542-02-1]
NFPA—hazardous chemical (flammable chemical)
health—1, flammability—2, reactivity—0

2-PENTANONE. [107-87-9]
*OSHA 29 CFR 1910.1000 Table Z-1
PEL(TWA): 200 ppm, 700 mg/m^3

ACGIH-TLV
TWA: 200 ppm, 700 mg/m^3
STEL: 250 ppm, 875 mg/m^3

See Ketones, NIOSH Criteria Document.
TSCA-CHIP available

2-PENTANONE, 4-METHYL-. [108-10-1]
*TSCA 8(a), 40 CFR 712, 47 FR 26992, June 22, 1982, Chemical Information Rule

n-PENTYL ACETATE. [628-63-7]
NFPA—hazardous chemical (flammable chemical)
health—2, flammability—3, reactivity—0

PENTYLAMINE. Syn: Amylamine.
NFPA—hazardous chemical
health—2, flammability—3, reactivity—0

PENTYL BENZENE. [29316-05-0]
NFPA—hazardous chemical (flammable chemical)
health—1, flammability—2, reactivity—0

PENTYL LAURATE.
NFPA—hazardous chemical (flammable chemical)
health—0, flammability—1, reactivity—0

PENTYLNAPHTHALENE.
NFPA—hazardous chemical (flammable chemical)
health—0, flammability—1, reactivity—0

PERACETIC ACID. See Peroxyacetic acid. [79-21-0]

PERCHLORIC ACID. [7601-90-3]
NFPA—hazardous chemical
health—3, flammability—0, reactivity—3, oxy—oxidizing chemical

PERCHLOROETHYLENE. See Tetrachloroethylene. [127-18-4]

PERCHLOROMETHYL MERCAPTAN. [594-42-3]
*OSHA 29 CFR 1910.1000 Table Z-1
PEL(TWA): 0.1 ppm, 0.8 mg/m^3
ACGIH-TLV
TWA: 0.1 ppm, 0.8 mg/m^3

PERCHLORYL FLUORIDE. [7616-94-6]
*OSHA 29 CFR 1910.1000 Table Z-1
PEL(TWA): 3 ppm, 13.5 mg/m^3
ACGIH-TLV
TWA: 3 ppm, 14 mg/m^3
STEL: 6 ppm, 28 mg/m^3

PERFLUIDONE. [37924-13-3]
FIFRA—Data Call In, 45 FR 75488, November 14, 1980
Status: letter being drafted

PERFLUOROETHYLENE. See Tetrafluoroethylene [116-14-3]

PERLITE.
ACGIH-TLV
30 mppcf

PERMETHRIN. [63937-32-6]
FIFRA—Data Call In, 45 FR 75488, November 14, 1980
Status: letter issued 8/28/81

PEROXYACETIC ACID. Syn: Peracetic acid. [79-21-0]
OSHA Candidate List of Potential Occupational Carcinogens, 45 FR 53672, August 12, 1980

PEROXYACETIC ACID (60% acetic acid solution). [79-21-0]
NFPA—hazardous chemical
health—3, flammability—2, reactivity—4

PERTHANE (many products). [72-56-0]
40 CFR 162.11, FIFRA-RPAR, Final Action, Voluntary Cancellation Current Status: 45 FR 41694 (6/20/80)
Criteria Possibly Met or Exceeded: Oncogenicity

PESTICIDES, MANUFACTURE and FORMULATION.
NIOSH Criteria Document (Pub. No. 78-174), NTIS Stock No. PB81 227001, July 17, 1978

These criteria for the pesticide standard apply only to occupations in pesticide manufacturing and formulating. This includes manufacturing, formulating, packaging, mixing, blending, or repackaging of any pesticide active ingredient. "Occupational exposure" is defined as contact with any pesticide or other chemical during the manufacture and formulation of pesticides. "Pesticide" is used herein as the generic term meaning any substance or mixture of substances intended for those uses described by definition in the Federal Environmental Pesticide Control Act (FEPCA): preventing, destroying, repelling, or mitigating any pest and regulating, defoliating, or desiccating any plant (40 CFR 162). All pesticide active ingredients registered with the EPA are covered by this recommended standard. Other chemical to which workers are potentially exposed in pesticide manufacturing and formulating operations (such as intermediates, impurities, diluents, emulsifiers, carriers, inerts, and propellants) are also covered by the provisions contained herein, as well as by provisions in any separate standards applicable for such chemicals, for example, xylene. In the case of conflicts between this recommended standard and any existing standard, NIOSH recommends that the more stringent standard apply.

The standard was not designed for the population at large, or for users of pesticides who are already regulated by other agencies such as the U.S. Department of Transportation (DOT), EPA, and the Food and Drug Administration (FDA). Application of this recommended standard to situations other than the occupational settings specified above is not warranted.

Environmental (workplace air) limits are not included in the recommended standard. Such values have been promulgated for many pesticides by the Occupational Safety and Health Administration (OSHA), and NIOSH has previously recommended such limits individually for various pesticides (see Appendices III and IV). NIOSH recommends compliance with promulgated environmental limits and adoption of new environmental limits in those cases where the NIOSH recommended environmental limits differ from those already promulgated. These include the limits recommended for substances such as parathion, methyl parathion, creosote, ethylene dibromide, and dinitro-o-cresol. In this document, emphasis has been placed on work practices, engineering controls, and medical surveillance programs to protect workers from the adverse effects of pesticide exposure in manufacturing and formulating operations.

Because of the wide range of adverse health effects that can arise from employee exposure in manufacturing and formulating facilities, it was determined that a single set of work practices and engineering controls wold not effectively protect health and safety and at the same time avoid placing unnecessary burdens on both employees and employers in such facilities. Therefore, various classification schemes used by organizations in the United States and other countries were examined for the purpose of grouping pesticides according to toxicity. A considerable degree of similarity was found among such schemes.

Given the similarity that exists among the various schemes examined and the absence of significant differences in scientific merit, the system used by the U.S. Environmental Protection Agency (EPA) was selected as the basis for the approach recommended in this document.

Each section of the recommended standard contains three sets of requirements that pertain to the three groups of pesticides listed in Appendix I. Pesticides in Group III are the least toxic and therefore require the least stringent level of control. Group II pesticides produce acute effects at lower dose levels than Group III pesticides; controls for Group II include all of the Group III controls with some additional, more stringent requirements. Pesticides in Group I produce acute effects at extremely low-dose levels or serious irreversible effects; therefore, pesticides in Group I require those controls for Group III and II and additional requirements where noted. Classification by acute toxicity is based on oral and dermal LD50s and inhalational LC50s in mammals, and includes both systemic and local effects. Certain pesticides have been classified based on irreversible effects, including probable and potential carcinogenicity, and potential teratogenicity, mutagenicity, neurotoxicity, and reproductive effects, as demonstrated in animal test systems and human epidemiologic studies. Insecticides that inhibit cholinesterase (ChE) activity are also noted. Numeric definitions of the three pesticide groups are presented in Table VI-1. By necessity, only selected portions of the published literature on chronic adverse health effects of this extremely large and diverse group of chemicals were reviewed and evaluated in this document. As new data regarding toxicity and potential hazards are developed and evaluated, it is highly probable that pesticide classifications will change over time.

It should be emphasized that pesticides are an extremely diverse group of substances. There is a potential for a wide variety of toxic effects throughout the group. For many pesticides, the possibility of dermal exposure and subsequent absorption presents a greater problem to workers than does exposure through inhalation. In addition, many establishments by their nature manufacture or formulate a vast number of pesticides with varying toxicities, the possible synergisms of these combinations of substances are unknown. Consequently, there is a need for constant surveillance and monitoring and thorough record keeping to make sure that employees are following proper procedures so that their health is not compromised. The recommended standard is designed to protect the health and safety of employees in pesticides manufacturing and formulating facilities over their working lifetime.

PETROLEUM (CRUDE). [8002-05-9]

NFPA—hazardous chemical (flammable chemical)
health—1, flammability—3, reactivity—0

PETROLEUM DISTILLATES (naptha). [8030-30-6]

*OSHA 29 CFR 1910.1000 Table Z-1
PEL(TWA): 500 ppm, 2000 mg/m^3

PETROLEUM ETHER. See Petroleum solvents, refined, NIOSH Criteria Document. [8030-30-6]

PETROLEUM PITCH. See Asphalt. [8052-42-4]

PETROLEUM SOLVENTS, REFINED.

NIOSH Criteria Document (Pub. No. 77-192), NTIS stock number is not available, July 20, 1977
These criteria and recommended standards apply to occupational exposure of workers to the following refined petroleum solvents: petroleum ether, rubber solvent, varnish makers' and painters' naphtha, mineral spirits, Stoddard solvents, and kerosene, all in-

cluded in the term "refined petroleum solvents."

The refined petroleum solvents considered in this document have a total aromatic content of less than 20%. Other hydrocarbon solvents such as thinners, whose total aromatic content may exceed 20%, are not discussed in this document. For the purposes of these recommended standards for refined petroleum solvents, the following definitions will apply to those solvents covered in this document:

"Petroleum ether" is refined petroleum solvent that has a boiling range of 30–60°C and typically has a chemical composition of 80% pentane and 20% isohexane.

"Rubber solvent" is refined petroleum solvent with a boiling range of 45–125°C and is composed of organic compounds whose carbon chain lengths range from C5 to C7. The chemical composition of a typical rubber solvent would be: 41.4% paraffins, 53.6% monocycloparaffins, 0.1% monoolefins, 1.5% benzene, and 3.4% alkyl benzene.

"Varnish makers' and painters' naphtha" is a refined petroleum solvent with a boiling range of 95–160°C and is composed of organic compounds whose carbon chain lengths range from C5 to C11. The chemical composition of typical varnish makers' and painters' naphtha would be: 55.4% paraffins, 30.3% monocycloparaffins, 2.4% dicycloparaffins, 0.1% benzene, 11.7% alkylbenzenes, and 0.1% indans and tetralins.

"Mineral spirits" is a refined petroleum solvent with a boiling range of 150–200°C. A typical chemical composition for mineral spirits would be: 80–86% saturated hydrocarbons, 1% olefins, and 13–19% aromatics.

"Stoddard solvent" is a type of mineral spirits with a boiling range of 160–210°C and is composed of organic compounds whose carbon chain lengths range from C7 to C12. The chemical composition of a typical Stoddard solvent would be: 47.7% paraffins, 26% monocycloparaffins 11.6% dicyloparaffins 0.1% benzene, 14.1% alkylbenzenes, and 0.5% indans and tetralins.

"140 flash aliphatic solvent" is a type of Stoddard solvent with a boiling range of 185–207°C and is composed of organic compounds whose carbon chain lengths range from C5 to C12. A typical chemical composition for 140 flash aliphatic solvent would be: 60.8% paraffins, 24.5% monocycloparaffins, 11.2% dicycloparaffins, 0.07% benzene, 3.03% alkylbenzenes, and 0.3% indans and tetralins.

"Kerosene" is refined petroleum solvent with a boiling range of 175–325°C. A typical chemical composition for kerosene would be: 25% normal paraffins, 11% branched paraffins, 30% monocycloparaffins, 16% mononuclear aromatics, and 5% dinuclear aromatics.

"Deodorized kerosene" is a refined petroleum solvent that has a boiling range of 209–274°C and a typical chemical composition of 55.2% paraffins, 40.9% naphthenes, and 3.9% aromatics.

If exposure to other chemicals also occurs, the employer shall comply also with applicable standards for the other chemicals. In particular, special attention shall be given to the benzene content of the refined petroleum solvents being used.

The "action level" for petroleum ether, rubber solvent, and varnish makers' and painters' naphtha is defined as an airborne time-weighted average (TWA) concentration of 200 mg/m^3 of air for up to a 10-hour work shift in a 40-hour workweek. The "action level" for mineral spirits and Stoddard solvent is defined as a TWA concentration of 350 mg/m^3 of air for up to 10-hour work shift in a 40-hour workweek. The "action level" for kerosene is defined as a TWA concentration of 100 mg/m^3 of air for up to 10-hour work shift in a 40-hour workweek. "Occupational exposure to refined petroleum solvents" is defined as exposure above the action level. Exposure to refined petroleum solvents at lower concentrations will require adherence to Sections 2(a), 3, 4(a,b), 6(a,c–i), 7, and 8(a).

1. The employer shall control occupational exposure to airborne concentrations of petroleum ether, rubber solvent, varnish makers' and painters' naphtha, mineral spirits, and Stoddard solvents so that no em-

ployee is exposed at a concentration greater than 350 mg/m^3 of air, determined as a TWA concentration for up to a 10-hour work shift, 40-hour workweek. In addition, no employee shall be exposed to any of the above refined petroleum solvents at a ceiling concentration greater than 1800 mg/m^3 as determined by a sampling time of 15 minutes.

2. The employer shall control occupational exposure to kerosene so that no employee is exposed at a concentration greater than 100 mg/m^3 of air determined as a TWA concentration for up to a 10-hour work shift, 40-hour workweek.

The standards are designed to protect the health and provide for the safety of employees for up to a 10-hour work shift, 40-hour workweek, over a working lifetime.

PHENACETIN. [62-44-2]

OSHA Candidate List of Potential Occupational Carcinogens (EPA Carcinogen Assessment Group List), 45 FR 53672, August 12, 1980
IARC—carcinogenic in humans
*RCRA 40 CFR 261
EPA hazardous waste no. U187

PHENANTHRENE. See Coal tar pitch volatiles. [85-01-8]

CWA 307(a)(1), 40 CFR 125, Priority Pollutant

PHENARSAZINE CHLORIDE.
[578-94-9]

40 CFR 162.11, FIFRA-RPAR, Final Action, Voluntary Cancellation
Current Status: 42 FR 59776 (11/21/77)
Criteria Possibly Met or Exceeded: None

PHENAZOPYRIDINE HYDROCHLORIDE.

See Pyridine, 2-6-diamino-3-(phenylazo), monohydrochloride. [136-40-3]
NCI—carcinogenic in animals

PHENESTERIN. [3546-10-9]

NCI—carcinogenic in animals

PHENICARBAZIDE. [103-03-7]

IARC—carcinogenic in animals

PHENOBARBITAL SODIUM. [57-30-7]

IARC—carcinogenic in animals

PHENOL. [108-95-2]

*OSHA 29 CFR 1910.1000 Table Z-1

Skin

PEL(TWA): 5 ppm, 19 mg/m^3
ACGIH-TLV

Skin

TWA: 5 ppm, 19 mg/m^3
STEL: 10 ppm, 38 mg/m^3
NIOSH Criteria Document (Pub. No. 76-196), NTIS Stock No. PB 266495, July 1, 1976

"Phenol" in this recommended standard includes solids, aerosols, vapor, or solutions containing phenol. "Occupational exposure to phenol" is defined as exposure to phenol at airborne concentrations exceeding one-half the recommended TWA environmental limit. Exposure at lower concentrations shall require adherence to Sections 3, 4(a), 4(b), and 6.

Occupational exposure to phenol shall be controlled so that no employee is exposed to phenol at concentrations greater than 20 mg/m^3 in air determined as a time-weighted average (TWA) concentration for up to a 10-hour workday, 40-hour workweek, or to more than 60 mg phenol/m^3 of air as a ceiling concentration for any 15-minute period.

The standard is designed to protect the health and provide for the safety of employees for up to a 10-hour workday, 40-hour workweek, over a working lifetime.

NFPA—hazardous chemical
health—3, flammability—2, reactivity—0
CWA 304(a)(1), 45 FR 79318, November 28, 1980, Water Quality Criteria

Freshwater Aquatic Life

The available data for phenol indicate that acute and chronic toxicity to freshwater aquatic life occur at concentrations as low as 10,200 and 2560 μg/liter, respectively, and would occur at lower concentrations among species that are more sensitive than those tested.

Saltwater Aquatic Life

The available data for phenol indicate that acute toxicity to saltwater aquatic life occurs at concentrations as low as 5800 μg/liter and would occur at lower concentrations among species that are more sensitive than those tested. No data are available concerning the chronic toxicity of phenol to sensitive saltwater aquatic life.

Human Health

For comparison purposes, two approaches were used to derive criterion levels for phenol. Based on available toxicity data, for the protection of public health, the derived level is 3.5 mg/liter. Using available organoleptic data, for controlling undesirable taste and odor quality of ambient water, the estimated level is 0.3 mg/liter. It should be recognized that organoleptic data as a basis for establishing a water quality criterion have limitations and have no demonstrated relationship to potential adverse human health effects.

CWA 307(a)(1), Priority Pollutant

*CWA 311(b)(2)(A), 40 CFR 116, 117
Discharge RQ = 1000 pounds (454 kilograms)

CAA 112
Chemical proposed to be assessed as a Hazardous Air Pollutant by a House of Representatives bill to amend the CAA

*RCRA 40 CFR 261
EPA hazardous waste no. U188 (hazardous substance)

PHENOL, 4-AMINO-2-NITRO-.
Syn: C.I. 76555. [119-34-6]
OSHA Candidate List of Potential Occupational Carcinogens, 45 FR 53672, August 12, 1980

PHENOL, 2-CHLORO-. [95-57-8]
*RCRA 40 CFR 261
EPA hazardous waste no. U048

PHENOL, 4-CHLORO-3-METHYL-.
*RCRA 40 CFR 261
EPA hazardous waste no. U039

PHENOL, 2-CYCLOHEXYL-4,6-DINITRO-. [131-89-5]
*RCRA 40 CFR 261
EPA hazardous waste no. P034

PHENOL, 2,4-DICHLORO. [120-83-2]
*RCRA 40 CFR 261
EPA hazardous waste no. U081

PHENOL, 2,6-DICHLORO-. [6341-97-5]
*RCRA 40 CFR 261
EPA hazardous waste no. U082

PHENOL, 2,4-DIMETHYL-.
*RCRA 40 CFR 261
EPA hazardous waste no. U101 (hazardous substance)

PHENOL, 4-(1,1-DIMETHYLETHYL)-, PHOSPHATE (3:1). See Aryl phosphates. [78-33-1]
*TSCA 8(a), 40 CFR 712, 47 FR 26992, June 22, 1982, Chemical Information Rule

*TSCA 8(d), Unpublished Health and Safety Studies Reporting, 40 CFR 716, 47 FR 387, September 2, 1982

PHENOL, DIMETHYL-, PHOSPHATE (3:1). See Aryl phosphates. [25155-23-1]
*TSCA 8(a), 40 CFR 712, 47 FR 26992, June 22, 1982, Chemical Information Rule

*TSCA 8(d), Unpublished Health and Safety Studies Reporting, 40 CFR 716, 47 FR 387, September 2, 1982

PHENOL, 2,4-DINITRO-. [51-28-5]
*RCRA 40 CFR 261
EPA hazardous waste no. P048 (hazardous substance)

PHENOL, 2,4-DINITRO-6-METHYL-.
*RCRA 40 CFR 261
EPA hazardous waste no. P047

PHENOL, 2,4-DINITRO-6-(1-METHYLPROPYL)-.
*RCRA 40 CFR 261
EPA hazardous waste no. P020

PHENOL, METHYL-. [1319-77-3]
*TSCA 8(a), 40 CFR 712, 47 FR 26992, June 22, 1982, Chemical Information Rule

PHENOL, 2-METHYL-. [95-48-7]
*TSCA 8(a), 40 CFR 712, 47 FR 26992, June 22, 1982, Chemical Information Rule

PHENOL, 3-METHYL-. [108-39-4]
*TSCA 8(a), 40 CFR 712, 47 FR 26992, June 22, 1982, Chemical Information Rule

PHENOL, 4-METHYL-. [106-44-5]
*TSCA 8(a), 40 CFR 712, 47 FR 26992, June 22, 1982, Chemical Information Rule

PHENOL, 4-NITRO-. [100-02-1]
*RCRA 40 CFR 261
EPA hazardous waste no. U170 (hazardous substance)

PHENOL, PENTACHLORO-. [87-86-5]
*RCRA 40 CFR 261
EPA hazardous waste no. U242 (hazardous substance)

PHENOL, 2,3,4,6-TETRACHLORO-.
[58-90-2]
*RCRA 40 CFR 261
EPA hazardous waste no. U212

PHENOL, 2,4,6-TRIBROMO-.
[118-79-6]
*TSCA 8(a), 40 CFR 712, 47 FR 26992, June 22, 1982, Chemical Information Rule

PHENOL, 2,4,5-TRICHLORO-.
[95-95-4]
*RCRA 40 CFR 261
EPA hazardous waste no. U230 (hazardous substance)

PHENOL, 2,4,6-TRICHLORO-.
[88-06-2]
*RCRA 40 CFR 261
EPA hazardous waste no. U231 (hazardous substance)

PHENOL, 2,4,6-TRINITRO-, AMMONIUM SALT. [131-74-8]
*RCRA 40 CFR 261
EPA hazardous waste no. P009 (reactive salt)

PHENOTHIAZINE. [92-84-2]
ACGIH-TLV

Skin

TWA: 5 mg/m^3
STEL: 10 mg/m^3

PHENOXYBENZAMINE. [59-96-1]
IARC—carcinogenic in animals
NCI—carcinogenic in animals

2-PHENOXYETHANOL. [122-99-6]
*TSCA 4(e), ITC
Twelfth Report of the Interagency Testing Committee to the administrator, Environmental Protection Agency, May 1983; 48 FR 24443, June 1, 1983
*TSCA 8(a), 40 CFR 712, 48 FR 28443, June 22, 1983, Chemical Information Rule

PHENYL DICHLOROARSINE.
[696-28-6]
*RCRA 40 CFR 261
EPA hazardous waste no. P036

***p*-PHENYLENEDIAMINE.** [106-50-3]
*OSHA 29 CFR 1910.1000 Table Z-1

Skin

PEL (TWA): 0.1 mg/m^3
ACGIH-TLV

Skin

TWA: 0.1 mg/m^3
*TSCA 8(d) Unpublished Health and Safety Studies Reporting, 40 CFR 716, 47 FR 387, September 2, 1982

o-PHENYLENEDIAMINE, 4-BUTYL-.
See Phenylenediamines. [3663-23-8]

*TSCA 4(e), ITC
Sixth Report of the Interagency Testing Committee to the administrator, Environmental Protection Agency, April 1980; 45 FR 35897, May 28, 1980, responded to by the EPA administrator 47 FR 973, January 8, 1982

*TSCA 8(d) Unpublished Health and Safety Studies Reporting, 40 CFR 716, 47 FR 38800, September 2, 1982; 48 FR 13178, March 30, 1983

m-PHENYLENEDIAMINE, (4-CHLORO-. Syn: 4-Chlorophene-1,3-diamine. See Phenylenediamines.) [5131-60-2]

*OSHA Candidate List of Potential Occupational Carcinogens, 45 FR 53672, August 12, 1980

TSCA 4(e), ITC
Sixth Report of the Interagency Testing Committee to the administrator, Environmental Protection Agency, April 1980; 45 FR 35897, May 28, 1980, responded to by the EPA administrator, 47 FR 973, January 8, 1982

*TSCA 8(d), Unpublished Health and Safety Studies Reporting, 40 CFR 716, 47 FR 38800, September 2, 1982, 48 FR 13178, March 30, 1983

o-PHENYLENEDIAMINE, 4-CHLORO-.
Syn: 4-Chloro-1,2-benzenediamine. See Phenylenediamines. [95-83-0]

OSHA Candidate List of Potential Occupational Carcinogens, 45 FR 53672, August 12, 1980

*TSCA 8(d), Unpublished Health and Safety Studies Reporting, 40 CFR 716, 47 FR 38800, September 2, 1982; 48 FR 13178, March 30, 1983

p-PHENYLENEDIAMINE, 2-CHLORO, DIHYDROCHLORIDE. See Phenylenediamines. [615-46-3]

TSCA 4(e), ITC
Sixth Report of the Interagency Testing Committee to the administrator, Environmental Protection Agency, April 1980; 45 FR 35897, May 28, 1980, responded to by the EPA administrator 47 FR 973, January 8, 1982

*TSCA 8(d) Unpublished Health and Safety Studies Reporting, 40 CFR 716, 47 FR 38800, September 2, 1982; 48 FR 13178, March 30, 1983

m-PHENYLENEDIAMINE,4-CHLORO-, SULFATE. See Phenylenediamines. [68239-80-5]

TSCA 4(e), ITC
Sixth Report of the Interagency Testing Committee to the administrator, Environmental Protection Agency, April 1980; 45 FR 35897, May 28, 1980, responded to by the EPA administrator 47 FR 973, January 8, 1982

*TSCA 8(d) Unpublished Health and Safety Studies Reporting, 40 CFR 716, 47 FR 38800, September 2, 1982, 48 FR 13178, March 30, 1983

p-PHENYLENEDIAMINE, 2-CHLORO-, SULFATE. [6219-71-2]

TSCA 4(e), ITC
Sixth Report of the Interagency Testing Committee to the administrator, Environmental Protection Agency, April 1980; 45 FR 35897, May 28, 1980, responded to by the EPA administrator 47 FR 973, January 8, 1982

p-PHENYLENEDIAMINE, 2,5-DICHLORO-. See Phenylenediamines. [20103-09-7]

TSCA 4(e), ITC
Sixth Report of the Interagency Testing Committee to the administrator, Environmental Protection Agency, April 1980; 45 FR 35897, May 28, 1980, responded to by the EPA administrator 47 FR 973, January 8, 1982

*TSCA 8(d) Unpublished Health and Safety Reporting, 40 CFR 716, 47 FR 38800, September 2, 1982; 48 FR 13178, March 30, 1983

p-PHENYLENEDIAMINE, 2,6-DICHLORO. See Phenylenediamines. [609-20-1]

NCI—may be carcinogenic in animals

p-PHENYLENEDIAMINE DIHYDROCHLORIDE. See Phenylenediamines. [624-18-0]

*TSCA 4(e), ITC
Sixth Report of the Interagency Testing Committee to the administrator, Environmental Protection Agency, April 1980; 45 FR 35897, May 28 1980, responded to by the EPA administrator 47 FR 973, January 8, 1982

*TSCA 8(d) Unpublished Health and Safety Studies Reporting, 40 CFR 716, 47 FR 38800, September 2, 1982; 48 FR 13178, March 30, 1983

o-PHENYLENEDIAMINE DIHYDROCHLORIDE. See Phenylenediamines. [615-28-1]

OSHA Candidate List of Potential Occupational Carcinogens, 45 FR 53672, August 12, 1980

TSCA 4(e), ITC
Sixth Report of the Interagency Testing Committee to the administrator, Environmental Protection Agency, April 1980; 45 FR 35897, May 28, 1980, responded to by the EPA administrator 47 FR 973, January 8, 1982

*TSCA 8(d) Unpublished Health and Safety Studies Reporting, 40 CFR 716, 47 FR 38800, September 2, 1982; 48 FR 13178, March 30, 1983

o-PHENYLENEDIAMINE, 4-ETHOXY-. See Phenylenediamines. [1197-37-1]

TSCA 4(e), ITC
Sixth Report of the Interagency Testing Committee to the administrator, Environmental Protection Agency, April 1980; 45 FR 35897, May 28, 1980, responded to by the EPA administrator 47 FR 973, January 8, 1982

*TSCA 8(d) Unpublished Health and Safety Studies Reporting, 40 CFR 716, 47 FR 38800, September 2, 1982; 48 FR 13178, March 30, 1983

m-PHENYLENEDIAMINE, 4-METHOXY-. See Phenylenediamines. [615-05-4]

TSCA 4(e), ITC
Sixth Report of the Interagency Testing Committee to the administrator, Environmental Protection Agency, April 1980; 45 FR 35897, May 28, 1980, responded to by the EPA administrator 47 FR 973, January 8, 1982

*TSCA 8(d) Unpublished Health and Safety Studies Reporting, 40 CFR 716, 47 FR 38800, September 2, 1982; 48 FR 13178, March 30, 1983

m-PHENYLENEDIAMINE, 4-METHOXY-, DIHYDROCHLORIDE. See Phenylenediamines. [614-94-8]

TSCA 4(e), ITC
Sixth Report of the Interagency Testing Committee to the administrator, Environmental Protection Agency, April 1980; 45 FR 35897, May 28, 1980, responded to by the EPA administrator 47 FR 973, January 8, 1982

*TSCA 8(d) Unpublished Health and Safety Studies Reporting, 40 CFR 716, 47 FR 38800, September 2, 1982; 48 FR 13178, March 30, 1983

m-PHENYLENEDIAMINE, 4-METHOXY-, SULFATE. [6219-67-6]

TSCA 4(e), ITC
Sixth Report of the Interagency Testing Committee to the administrator, Environmental Protection Agency, April 1980; 45 FR 35897, May 28, 1980, responded to by the EPA administrator 47 FR 973, January 8, 1982

m-PHENYLENEDIAMINE, 2-NITRO-. See Phenylenediamines. [6219-67-6]

*TSCA 8(d) Unpublished Health and Safety Studies Reporting, 40 CFR 716, 47 FR 38800, September 2, 1982; 48 FR 13178, March 30, 1983

m-PHENYLENEDIAMINE, 4-NITRO-.
See Phenylenediamines. [5131-58-8]

TSCA 4(e), ITC
Sixth Report of the Interagency Testing Committee to the administrator, Environmental Protection Agency, April 1980; 45 FR 35897, May 28, 1980, responded to by the EPA administrator 47 FR 973, January 8, 1982

*TSCA 8(d) Unpublished Health and Safety Studies Reporting, 40 CFR 716, 47 FR 38800, September 2, 1982; 48 FR 13178, March 30, 1983

m-PHENYLENEDIAMINE, 5-NITRO-.
See Phenylenediamines. [5042-55-7]

TSCA 4(e), ITC
Sixth Report of the Interagency Testing Committee to the administrator, Environmental Protection Agency, April 1980; 45 FR 35897, May 28, 1980, responded to by the EPA administrator 47 FR 973, January 8, 1982

*TSCA 8(d) Unpublished Health and Safety Studies Reporting, 40 CFR 716, 47 FR 38800, September 2, 1982; 48 FR 13178, March 30, 1983

p-PHENYLENEDIAMINE, 2-NITRO-.
Syn: C.I. 76070. See Phenylenediamines. [5307-14-2]

OSHA Candidate List of Potential Occupational Carcinogens, 45 FR 53672, August 12, 1980

TSCA 4(e), ITC
Sixth Report of the Interagency Testing Committee to the administrator, Environmental Protection Agency, April 1980; 45 FR 35897, May 28, 1980, responded to by the EPA administrator 47 FR 973, January 8, 1982

*TSCA 8(d) Unpublished Health and Safety Studies Reporting, 40 CFR 716, 47 FR 38800, September 2, 1982; 48 FR 13178, March 30, 1983

o-PHENYLENEDIAMINE, 4-NITRO-.
See Phenylenediamines. [99-56-9]

TSCA 4(e), ITC
Sixth Report of the Interagency Testing Committee to the administrator, Environmental Protection Agency, April 1980; 45 CFR 35897, May 28, 1980, responded to by the EPA administrator 47 FR 973, January 8, 1982

*TSCA 8(d) Unpublished Health and Safety Studies Reporting, 40 CFR 716, 47 FR 38800, September 2, 1982; 48 FR 13178, March 30, 1983

o-PHENYLENEDIAMINE, 4-NITRO-, DIHYDROCHLORIDE. [6219-77-8]

TSCA 4(e), ITC
Sixth Report of the Interagency Testing Committee to the administrator, Environmental Protection Agency, April 1980; 45 CFR 35897, May 28, 1980, responded to by the EPA administrator 47 FR 973, January 8, 1982

p-PHENYLENEDIAMINE, 2-NITRO-, DIHYDROCHLORIDE. See Phenylenediamines. [18266-52-9]

TSCA 4(e), ITC
Sixth Report of the Interagency Testing Committee to the administrator, Environmental Protection Agency, April 1980; 45 CFR 35897, May 28, 1980, responded to by the EPA administrator 47 FR 973, January 8, 1982

*TSCA 8(d) Unpublished Health and Safety Studies Reporting, 40 CFR 716, 47 FR 38800, September 2 1982; 48 FR 13178, March 30, 1983

o-PHENYLENEDIAMINE, 4-NITRO, SULFATE. See Phenylenediamines. [6219-77-8]

*TSCA 8(d) Unpublished Health and Safety Studies Reporting, 40 CFR 716, 47 FR 38800, September 2, 1982; 48 FR 13178, March 30, 1983

PHENYLENEDIAMINES (PDAs).
[95-54-5, 106-50-3, 108-45-2]

TSCA 4(e), ITC
This category is defined as all nitrogen-unsubstituted phenylenediamines with zero to two substituents on the ring selected from

230 PHENYLENEDIAMINES (PDAs)

the same or different members of the group of halo, nitro, hydroxy, hydroxy-lower alkoxy, lower-alkyl, and lower-alkoxy. For this purpose the term "lower" is defined as a group containing between one and four carbons.

The ITC listed 50 PDAs as occuring on the TSCA Public Inventory. EPA's review has identified 47 of these chemicals (listed in Table 4) that fall within the stated definition. Note that, in the Sixth ITC Report, no. 18 and no. 32 are the same chemical and no. 29 and no. 38 are the same chemical. Number 22 [CAS 1477550] on the ITC list is a xylene derivative that does not fit the definition. The total number of CAS numbers listed in Table 4, which is based upon the original ITC list, is 49, while the total number of chemicals is 47.

Sixth report of the Interagency Testing Committee to the administrator, Environmental Protection Agency, April 1980; 45 FR 35897, May 28, 1980, responded to by the EPA administrator 47 FR 973, January 8, 1982

This category is defined as all nitrogen-unsubstituted phenylenediamines and their salts with zero to two substituents on the ring selected from the same or different members of the group of halo, nitro, hydroxy, hydroxy-lower alkoxy, lower-alkyl, and lower-alkoxy. For this purpose the term "lower" is defined as a group containing between one and four carbons. This category includes the following phenyenediamines but is not limited to them:

CAS No.	Name
[95-54-5]	o-Diaminobenzene
[95-70-5]	2,5-Diaminotoluene
[95-80-7]	1,3-Diamino-4-methylbenzene
[95-83-0]	o-Phenylenediamine, 4-chloro-
[99-56-9]	o-Phenylenediamine, 4-nitro-
[106-50-3]	p-Diaminobenzene
[108-45-2]	m-Diaminobenzene
[108-71-4]	3,5-Diaminotoluene
[137-09-7]	2,4-Diaminophenol dihydrochloride
[496-72-0]	1,2-Diamino-4-methylbenzene
[541-69-5]	m-Phenylenediammonium dichloride
[541-70-8]	m-Phenylenediamine, sulfate (1:1)
[614-94-8]	m-Phenylenediamine, 4-methoxy, dihydrochloride
[615-05-4]	m-Phenylenediamine, 4-methoxy
[615-28-1]	1,2-Phenylenediamine dihydrochloride
[615-45-2]	1,4-Benzenediamine, 2-methyl-, dihydrochloride
[615-46-3]	p-Phenylenediamine, 2-chloro-, dihydrochloride
[615-50-9]	2,5-Diaminotoluene sulfate (1:1)
[624-18-0]	p-Phenylenediamine, dihydrochloride
[823-40-5]	2,6-Diamino-1-methylbenzene
[1197-37-1]	o-Phenylenediamine, 4-ethoxy
[2687-25-4]	1,2-Diamino-3-methylbenzene
[3663-23-8]	o-Phenylenediamine, 4-butyl-
[5042-55-7]	m-Phenylenediamine, 5-nitro-
[5131-58-8]	m-Phenylenediamine, 4-nitro-
[5131-60-2]	m-Phenylenediamine, 4-chloro-
[5307-14-2]	p-Diaminoanisole
[5307-14-2]	p-Phenylenediamine, 2-nitro
[6219-67-6]	m-Phenylenediamine, 2-nitro-
[6219-71-2]	p-Phenylenediamine, 2-nitro-
[6219-77-8]	o-Phenylenediamine, 4-nitro-, sulfate
[6369-59-1]	1,4-Benzenediamine, 2-methyl-, dihydrochloride
[15872-73-8]	4,6-Diamino-o-cresol
[16245-77-5]	p-Phenylenediamine, sulfate
[18266-52-9]	p-Phenylenediamine, 2-nitro-, dihydrochloride
[20103-09-7]	p-Phenylenediamine, 2,5-dichloro-
[25376-45-8]	Diaminotoluene
[39158-41-7]	2,4-Diaminoanisole sulfate
[42389-30-0]	1,2-Benzenediamine, 5-chloro-3-nitro-

Table 4. Phenylenediamines (as adapted from the May 28, 1980 Federal Register List[a]

	CAS No.	Name
1	[95-54-5]	o-Diaminobenzene
2	[95-70-5]	2,5-Diaminotoluene
3	[95-80-7]	1,3-Diamino-4-methylbenzene
4	[95-83-0]	o-Phenylenediamine, 4-chloro-
5	[99-56-9]	o-Phenylenediamine, 4-nitro-
6	[106-50-3]	p-Diaminobenzene
7	[108-45-2]	m-Diaminobenzene
8	[108-71-4]	3,5-Diaminotoluene
9	[137-09-7]	2,4-Diaminophenol dihydrochloride
10	[496-72-0]	1,2-Diamino-4-methylbenzene
11	[541-69-5]	m-Phenylenediammonium dichloride
12	[541-70-8]	m-Phenylenediamine, sulfate (1:1)
13	[614-94-8]	m-Phenylenediamine, 4-methoxy-, dihydrochloride
14	[615-05-4]	m-Phenylenediamine, 4-methoxy-
15	[615-28-1]	1,2-Phenylenediamine dihydrochloride
16	[615-45-2]	1,4-Benzenediamine, 2-methyl-, dihydrochloride
17	[615-46-3]	p-Phenylenediamine, 2-chloro-, dihydrochloride
18	[615-50-9]	2,5-Diaminotoluene sulfate (1:1)
19	[624-18-0]	p-Phenylenediamine dihydrochloride
20	[823-40-5]	2,6-Diamino-1-methylbenzene
21	[1197-37-1]	o-Phenylenediamine, 4-ethoxy-
22	[2687-25-4]	1,2-Diamino-3-methylbenzene
23	[3663-23-8]	o-Phenylenediamine, 4-butyl-
24	[5042-55-7]	m-Phenylenediamine, 5-nitro-
25	[5131-58-8]	m-Phenylenediamine, 4-nitro-
26	[5131-60-2]	m-Phenylenediamine, 4-chloro-
27	[5307-02-8]	p-Diaminoanisole
28	[5307-14-2]	p-Phenylenediamine, 2-nitro-
29	[6219-67-6]	m-Phenylenediamine, 4-methoxy, sulfate
30	[6219-71-2]	p-Phenylenediamine, 2-chloro-, sulfate
31	[6219-77-8]	o-Phenylenediamine, 4-nitro-, dihydrochloride
32	[6369-59-1]	1,4-Benzenediamine, 2-methyl-, sulfate
33	[15872-73-8]	4,6-Diamino-o-cresol
34	[16245-77-6]	p-Phenylenediamine sulfate
35	[18266-52-9]	p-Phenylenediamine, 2-nitro-, dihydrochloride
36	[20103-09-7]	p-Phenylenediamine, 2,5-dichloro-
37	[25376-45-8]	Diaminotoluene
38	[39156-41-7]	2,4-Diaminoanisole sulfate
39	[42389-30-0]	1,2-Benzenediamine, 5-chloro-3-nitro-
40	[62654-17-5]	1,4-Benzenediamine, ethanedioate (1:1)
41	[65879-44-9]	4,6-Diamino-2-methylphenol, hydrochloride
42	[66422-95-5]	Ethanol, 2-(2,4-diaminophenoxy)-, dihydrochloride
43	[67801-06-3]	1,3-Benzenediamine, 4-ethoxy-, dihydrochloride
44	[68015-98-5]	1,3-Benzenediamine, 4-ethoxy-, sulfate (1:1)
45	[68239-80-5]	m-Phenylenediamine, 4-chloro-, sulfate
46	[68239-82-7]	1,2-Benzenediamine, 4-nitro-, sulfate (1:1)
47	[68239-83-8]	1,4-Benzenediamine, 2-nitro-, sulfate (1:1)
48	[68459-98-3]	1,2-Benzenediamine, 4-chloro-, sulfate (1:1)
49	[68966-84-7]	1,3-Benzenediamine, ar-ethyl-ar-methyl

[a] The list published in the *Federal Register* has been edited and validated so that only those chemicals adhering to the ITC definition are included. One chemical, number 22, a,a'-Diamino-m-xylene (CAS No. 1477-55-0) has been deleted from the list. Number 22 is deleted because it does not adhere to the ITC's definition of a phenylenediamine given in the *Federal Register*. Note also that for CAS No. 6369-59-1 (*Federal Register* No. 33, no. 32 in Table 4) the correct name is 1,4-benzenediamine, 2-methyl-, sulfate, not 1,4-benzenediamine, ethanedioate (1:1) as listed in the *Federal Register*. Number 29 and 38 are the same chemical with two different CAS numbers. The names of the chemicals are the names listed in the Ninth Collective Index. There may be differences within the text of some of the CAS numbers. The names of the chemicals used in the text are the names for the chemicals that are found in the literature.

[62654-17-5] 1,4-Benzenediamine, ethane-dioate (1:1)
[65879-44-9] 4,6-Diamino-2-methyl-phenol, hydrochloride
[66422-95-5] Ethanol, 2-(2,4-diamino-phenoxy-, dihydrochloride
[67801-06-3] 1,3-Benzenediamine, 4-ethoxy-, dihydrochloride
[68015-98-5] 1,3-Benzenediamine, 4-ethoxy-, sulfate (1:1)
[68239-80-5] m-Phenylenediamine, 4-chloro-, sulfate
[68239-82-7] 1,2-Benzenediamine, 4-nitro-, sulfate (1:1)
[68239-83-8] 1,4-Benzenediamine, 2-nitro-, sulfate (1:1)
[68459-98-3] 1,2-Benzenediamine, 4-chloro-, sulfate (1:1)
[68966-84-7] 1,3-Benzenediamine, ar-ethyl-methyl

See individual phenylenediamines.
*TSCA 8(d) Unpublished Health and Safety Studies Reporting, 40 CFR 716, 47 FR 38800, September 2, 1982; 48 FR 13178, March 30, 1983
TSCA-CHIP available
TSCA 4(e), ITC
Sixth Report of the Interagency Testing Committee to the administrator, Environmental Protection Agency, April 1980; 45 FR 35897, May 28, 1980, responded to by the EPA administrator 47 FR 973, January 8, 1982

m-PHENYLENEDIAMINE, SULFATE (1:1). See Phenylenediamines. [541-70-8]

TSCA 4(e), ITC
Sixth Report of the Interagency Testing Committee to the administrator, Environmental Protection Agency, April 1980, 45 FR 35897, May 28, 1980, responded to by the EPA administrator 47 FR 973, January 8, 1982
*TSCA 8(d) Unpublished Health and Safety Studies Reporting, 40 CFR 716, 47 FR 38800, September 2, 1982; 48 FR 13178, March 30, 1983

p-PHENYLENEDIAMINE SULFATE. See Phenylenediamines. [16245-77-6]

TSCA 4(e), ITC
Sixth Report of the Interagency Testing Committee to the administrator, Environmental Protection Agency, April 1980, 45 FR 35897, May 28, 1980, responded to by the EPA administrator 47 FR 973, January 8, 1982
*TSCA 8(d) Unpublished Health and Safety Studies Reporting, 40 CFR 716, 47 FR 38800, September 2, 1982; 48 FR 13178, March 30, 1983

m-PHENYLENEDIAMMONIUM DICHLORIDE. See Phenylenediamines. [541-69-5]

TSCA 4(e), ITC
Sixth Report of the Interagency Testing Committee to the administrator, Environmental Protection Agency, April 1980; 45 FR 35897, May 28, 1980, responded to by the EPA Administrator 47 FR 973, January 8, 1982
*TSCA 8(d) Unpublished Health and Safety Studies Reporting, 40 CFR 716, 47 FR 38800, September 2, 1982; 48 FR 13178, March 30, 1983

1,10-(1,2-PHENYLENE)PYRENE.

*RCRA 40 CFR 261
EPA hazardous waste no. U137

PHENYLETHANOLAMINE. [122-98-5]

NFPA—hazardous chemical
health—2, flammability—1, reactivity—0

PHENYL ETHER (VAPOR). [101-84-8]

*OSHA 29 CFR 1910.1000 Table Z-1
PEL (TWA): 1 ppm, 7 mg/m^3
ACGIH-TLV
TWA: 1 ppm, 7 mg/m^3
STEL: 2 ppm, 14 mg/m^3

PHENYL ETHER-BIPHENYL MIXTURE (VAPOR).

*OSHA 29 CFR 1910.1000 Table Z-1

PHENYLETHYLENE. *See* Styrene, monomer. [100-42-5]

PHENYLFLUOROFORM. *See* Benzotrifluoride. [98-08-8]

PHENYL GLYCIDYL ETHER (PGE). [122-60-1]

*OSHA 29 CFR 1910.1000 Table Z-1
PEL (TWA): 10 ppm, 60 mg/m^3

ACGIH-TLV
TWA: 1 ppm, 6 mg/m^3

See Glycidyl ethers, NIOSH Criteria Document.

TSCA 4(e), ITC
Third report of the TSCA Interagency Testing Committee to the Administrator, Environmental Protection Agency, October 1978; 43 FR 50630, October 30, 1978

TSCA-CHIP available

PHENYLHYDRAZINE. [100-63-0]

*OSHA 29 CFR 1910.1000 Table Z-1

Skin

PEL (TWA): 5 ppm, 22 mg/m^3

ACGIH-TLV

Skin

TWA: 5 ppm, 20 mg/m^3
STEL: 10 ppm, 45 mg/m^3

See Hydrazine, NIOSH Criteria Document.

PHENYLHYDRAZINE HYDROCHLORIDE *See* Hydrazine, phenyl-, monohydrochloride. [59-88-1]

PHENYL MERCAPTAN. [108-98-5]

ACGIH-TLV
TWA: 0.5 ppm, 2 mg/m^3

PHENYLMERCURIC ACETATE. [62-38-4]

NFPA—hazardous chemical
when dry: health—3, flammability—1, reactivity—0
when wet: health—3, flammability—2, reactivity—0

*RCRA 40 CFR 261
EPA hazardous waste no. P092

N-PHENYL-2-NAPHTHYLAMINE. [135-88-6]

ACGIH-TLV
TWA
Industrial substance suspect of carcinogenic potential for humans—which is suspect of inducing cancer, based on either (1) limited epidemiologic evidence, exclusive of clinical reports of single cases, or (2) demonstration of carcinogenesis in one or more animal species by appropriate methods. Worker exposure by all routes should be carefully controlled to levels consistent with the animal and human experience data.

IARC—carcinogenic in animals

PHENYLPHOSPHINE. [638-21-1]

ACGIH-TLV
TWA: ceiling limit, 0.05 ppm, 0.25 mg/m^3

N-PHENYLTHIOUREA. [103-85-5]

*RCRA 40 CFR 261
EPA hazardous waste no. P093

PHENYTOIN. [57-41-0, 630-93-3]

IARC—carcinogenic in animals

PHORATE. [298-02-2]

ACGIH-TLV

Skin

TWA: 0.05 mg/m^3
STEL: 0.2 mg/m^3

FIFRA—Data Call In, 45 FR 75488, November 14, 1980
Status: agency decision reached

*RCRA 40 CFR 261
EPA hazardous waste no. P094

PHOSALONE. [2310-17-0]

FIFRA—Registration Standard—Issued September 1981 (insecticide)

PHOSDRIN (MEVINPHOS®). [7786-34-7]

*OSHA 29 CFR 1910.1000 Table Z-1

Skin
PEL(TWA): 0.1 mg/m³

PHOSGENE (CARBONYL CHLORIDE).
[75-44-5]

*OSHA 29 CFR 1910.1000 Table Z-1
PEL(TWA): 0.1 ppm, 0.4 mg/m³
ACGIH-TLV
TWA: 0.1 ppm, 0.4 mg/m³
NIOSH Criteria Document (Pub. No. 76-137), NTIS Stock No. PB 267514, February 23, 1976
"Phosgene" is defined as gaseous or liquefied phosgene. Synonyms for phosgene include carbonyl chloride, carbon oxychloride, chloroformyl chloride, and CG (designation used by military agencies).

"Occupational exposure to phosgene" is defined as exposure above half the recommended time-weighted average (TWA) environmental limit. Exposure at lower concentrations will require adherence to Sections 3, 4(a), 4(b), 4(c)(3), 4(c)(5), 4(c)(6), 5, 6, 7, and 8(a). "Overexposure" is defined as known or suspected exposure above either the TWA or ceiling concentrations, or any exposure that leads to development of pulmonary symptoms.

"Occupational exposure to phosgene shall be controlled so that no worker is exposed to phosgene at a concentration greater than one-tenth part phosgene per million parts of air (0.1 ppm) determined as a TWA concentration for up to a 10-hour workday, 40-hour workweek, or to more than two-tenths part phosgene per million parts of air (0.2 ppm) as a ceiling concentration for any 15-minute period. The standard is designed to protect the health and safety of workers for up to a 10-hour workday, 40-hour workweek, over a working lifetime.

NFPA—hazardous chemical
health—4, flammability—0, reactivity—0
TSCA-CHIP available
*CWA 311(b)(2)(A), 40 CFR 116, 117
Discharge RQ = 5000 pounds (2270 kilograms)
CAA 112

Chemical proposed to be assessed as a Hazardous Air Pollutant by a House of Representatives bill to amend the CAA
*RCRA 40 CFR 261
EPA hazardous waste no. P095 (hazardous substance)

PHOSPHAMIDON. [13171-21-6]
FIFRA—Data Call In, 45 FR 75488, November 14, 1980
Status: letter issued 9/30/81

PHOSPHINE. [3803-51-2]
*OSHA 29 CFR 1910.1000 Table Z-1
PEL(TWA): 0.3 ppm, 0.4 mg/m³
ACGIH-TLV
TWA: 0.3 ppm, 0.4 mg/m³
STEL: 1 ppm, 1 mg/m³
*RCRA 40 CFR 261
EPA hazardous waste no. P096

PHOSPHINE, TRIPHENYL-. [603-35-0]
*TSCA 8(a), 40 CFR 712, 47 FR 26992, June 22, 1982, Chemical Information Rule

PHOSPHORIC ACID. [7664-38-2]
*OSHA 29 CFR 1910.1000 Table Z-1
PEL(TWA): 1 mg/m³
ACGIH-TLV
TWA: 1 mg/m³
STEL: 3 mg/m³
NFPA—hazardous chemical
health—2, flammability—0, reactivity—0
*CWA 311(b)(2)(A), 40 CFR 116, 117
Discharge RQ = 5000 pounds (2270 kilograms)

PHOSPHORIC ACID, 2,2-BIS(BROMOMETHYL)-3-CHLOROPROPYL BIS[2-CHLORO-1-(CHLOROMETHYL)ETHYL ESTER. [66108-37-0]
*TSCA 8(a), 40 CFR 712, 47 FR 26992, June 22, 1982, Chemical Information Rule

PHOSPHORIC ACID, 2-CHLORO-1-(2,4,5-TRICHLOROPHENYL)VINYL DIMETHYL ESTER. Syn: Tetrachlorovinphos. [961-11-5]

OSHA Candidate List of Potential Occupational Carcinogens, 45 FR 53672, August 12, 1980

PHOSPHORIC ACID, DIBUTYL PHENYL ESTER. See Aryl phosphates. [2528-36-1]

*TSCA 8(a), 40 CFR 712, 47 FR 26992, June 22, 1982, Chemical Information Rule

*TSCA 8(d) Unpublished Health and Safety Studies Reporting, 40 CFR 716, 47 FR 387, September 2, 1982

PHOSPHORIC ACID, DIETHYL p-NITROPHENYL ESTER. [311-45-5]

*RCRA 40 CFR 261
EPA hazardous waste no. P041

PHOSPHORIC ACID, DIISODECYL PHENYL ESTER. See Aryl phosphates. [51363-64-5]

*TSCA 8(a), 40 CFR 712, 47 FR 26992, June 22, 1982, Chemical Information Rule

*TSCA 8(d) Unpublished Health and Safety Studies Reporting, 40 CFR 716, 47 FR 387, September 2, 1982

PHOSPHORIC ACID, (1,1-DIMETHYLETHYL)PHENYL DIPHENYL ESTER. See Aryl phosphates. [56803-37-3]

*TSCA 8(a), 40 CFR 712, 47 FR 26992, June 22, 1982, Chemical Information Rule

*TSCA 8(d) Unpublished Health and Safety Studies Reporting, 40 CFR 716, 47 FR 387, September 2, 1982

PHOSPHORIC ACID, 2-ETHYLHEXYL DIPHENYL ESTER. See Aryl phosphates. [1241-94-7]

*TSCA 8(a), 40 CFR 712, 47 FR 26992, June 22, 1982, Chemical Information Rule

*TSCA 8(d) Unpublished Health and Safety Studies Reporting, 40 CFR 716, 47 FR 387, September 2, 1982

PHOSPHORIC ACID, ISODECYL DIPHENYL ESTER. See Aryl phosphates. [29761-21-5]

*TSCA 8(a), 40 CFR 7112, 47 FR 26992, June 22, 1982, Chemical Information Rule

*TSCA 8(d), Unpublished Health and Safety Studies Reporting, 40 CFR 716, 47 FR 387, September 2, 1982

PHOSPHORIC ACID, LEAD SALT. [7446-27-7]

*RCRA 40 CFR 261
EPA hazardous waste no. U145

PHOSPHORIC ACID, (1-METHYLETHYL) PHENYL DIPHENYL ESTER. See Aryl phosphates. [28108-99-8]

*TSCA 8(d) Unpublished Health and Safety Studies Reporting, 40 CFR 716, 47 FR 387, September 2, 1982

PHOSPHORIC ACID, METHYLPHENYL DIPHENYL ESTER. See Aryl phosphates. [26444-49-5]

*TSCA 8(a), 40 CFR 712, 47 FR 26992, June 22, 1982, Chemical Information Rule

*TSCA 8(d) Unpublished Health and Safety Studies Reporting, 40 CFR 716, 47 FR 387, September 2, 1982

PHOSPHORIC ACID, TRIMETHYL ESTER. Syn: Methyl phosphate. [512-56-1]

OSHA Candidate List of Potential Occupational Carcinogens, 45 FR 53672, August 12, 1980

PHOSPHORIC ACID, TRIPHENYL ESTER. See Aryl phosphates. [115-86-6]

*TSCA 8(a), 40 CFR 712, 47 FR 26992, June 22, 1982, Chemical Information Rule

*TSCA 8(d) Unpublished Health and Safety Studies Reporting, 40 CFR 716, 47 FR 387, September 2, 1982

PHOSPHORIC ACID, TRIS(METHYLPHENYL) ESTER. See Aryl phosphates. [1330-78-5]

*TSCA 8(a), 40 CFR 712, 47 FR 26992, June 22, 1982, Chemical Information Rule

*TSCA 8(d) Unpublished Health and Safety

Studies Reporting, 40 CFR 716, 47 FR 387, September 2, 1982

PHOSPHORIC ACID, TRIS(2-METHYLPHENYL) ESTER. See Aryl phosphates. [78-30-8]

*TSCA 8(a), 40 CFR 712, 47 FR 26992, June 22, 1982, Chemical Information Rule

*TSCA 8(d) Unpublished Health and Safety Studies Reporting, 40 CFR 716, 47 FR 387, September 2, 1982

PHOSPHORIC ACID, TRIS(3-METHYLPHENYL) ESTER. See Aryl phosphates. [563-04-2]

*TSCA 8(d) Unpublished Health and Safety Studies Reporting, 40 CFR 716, 47 FR 387, September 2, 1982

PHOSPHORIC ACID, TRIS(4-METHYLPHENYL) ESTER. See Aryl phosphates. [78-32-0]

*TSCA 8(a), 40 CFR 712, 47 FR 26992, June 22, 1982, Chemical Information Rule

*TSCA 8(d) Unpublished Health and Safety Studies Reporting, 40 CFR 716, 47 FR 387, September 2, 1982

PHOSPHORIC SULFIDE. See Phosphorus pentasulfide. [1314-80-3]

PHOSPHORIC TRIAMIDE, HEXAMETHYL-. Syn: HMPA. [680-31-9]

OSHA Candidate List of Potential Occupational Carcinogens, 45 FR 53672, August 12, 1980

ACGIH-TLV

Skin

TWA: Industrial substance suspect of carcinogenic potential for humans—which is suspect of inducing cancer, based on either (1) limited epidemiologic evidence, exclusive of clinical reports of single cases, or (2) demonstration of carcinogenesis in one or more animal species by appropriate methods. Worker exposure by all routes should be carefully controlled to levels consistent with the animal and human experience data.

TSCA-CHIP available

NIOSH Current Intelligence Bulletin No. 6—cancers related bulletin

IARC—carcinogenic in animals

PHOSPHORODITHIOIC ACID, O,O-DIETHYL, S-METHYL ESTER. [3288-58-2]

*RCRA 40 CFR 261
EPA hazardous waste no. U087

PHOSPHOROFLUORIDIC ACID, BIS(1-METHYLETHYL) ESTER. [55-91-4]

*RCRA 40 CFR 261
EPA hazardous waste no. P043

PHOSPHOROTHIOIC ACID, O,O-DIETHYL, S-(ETHYLTHIO)METHYL ESTER. [2600-69-3]

*RCRA 40 CFR 261
EPA hazardous waste no. P094

PHOSPHOROTHIOIC ACID, O,O-DIETHYL O-(p-NITROPHENYL) ESTER. [56-38-2]

*RCRA 40 CFR 261
EPA hazardous waste no. P089 (hazardous substance)

PHOSPHOROTHIOIC ACID, O,O-DIETHYL O-PYRAZINYL ESTER. [297-97-2]

*RCRA 40 CFR 261
EPA hazardous waste no. P040

PHOSPHOROTHIOIC ACID, O,O-DIMETHYL O-[p-((DIMETHYLAMINO)SULFONYL)PHENYL] ESTER.

*RCRA 40 CFR 261
EPA hazardous waste no. P097

PHOSPHORODITHIOIC ACID, O,O-DIMETHYL S-[2-(METHYLAMINO)-2-OXOETHYL] ESTER. [1113-02-6]

*RCRA 40 CFR 261
EPA hazardous waste no. P044

PHOSPHOROUS ACID, TRIMETHYL ESTER. [121-45-9]

*TSCA 8(a), 40 CFR 712, 47 FR 26992, June 22, 1982, Chemical Information Rule

PHOSPHORUS (YELLOW). [7723-14-0]

*OSHA 29 CFR 1910.1000 Table Z-1
PEL(TWA): 0.1 mg/m^3
ACGIH-TLV
TWA: 0.1 mg/m^3
STEL: 0.3 mg/m^3
NFPA—hazardous chemical (red)
health—0, flammability—1, reactivity—1
NFPA—hazardous chemical (white or yellow)
health—3, flammability—3, reactivity—1
*CWA 311(b)(2)(A), 40 CFR 116, 117
Discharge RQ = 1 pound (0.454 kilogram)

PHOSPHORUS CHLORIDE. See
Phosphorus trichloride. [7719-12-2]

PHOSPHORUS OXYCHLORIDE.
[10026-13-8, 10025-87-3]

ACGIH-TLV (1982 Addition)
TWA: 0.1 ppm, 0.6 mg/m^3
STEL: 0.5 ppm, 3 mg/m^3
NFPA—hazardous chemical
health—3, flammability—0, reactivity—2,
W—water may be hazardous in fire fighting
*CWA 311(b)(2)(A), 40 CFR 116, 117
Discharge RQ = 5000 pounds (2270 kilograms)

PHOSPHORUS PENTACHLORIDE.
[10026-13-8]

*OSHA 29 CFR 1910.1000 Table Z-1
PEL(TWA): 1 mg/m^3
ACGIH-TLV
TWA: 0.1 ppm, 1 mg/m^3

PHOSPHORUS PENTASULFIDE.
[1314-80-3]

*OSHA 29 CFR 1910.1000 Table Z-1
PEL(TWA): 1 mg/m^3
ACGIH—TLV
TWA: 1 mg/m^3
STEL: 3 mg/m^3
NFPA—hazardous chemical
when nonfire: health—2, flammability—1, reactivity—2, W—water may be hazardous in fire fighting
when fire: health—3, flammability—1, reactivity—2, W—water may be hazardous in fire fighting
*CWA 311(b)(2)(A), 40 CFR 116, 117
Discharge RQ = 100 pounds (45.4 kilograms)

PHOSPHORUS PERSULFIDE. See
Phosphorus pentasulfide. [1314-80-3]

PHOSPHORUS SEQUISULFIDE.
[1314-85-8]

NFPA—hazardous chemical
when nonfire: health—0, flammability—1, reactivity—1
when fire: health—2, flammability—1, reactivity—1

PHOSPHORUS SULFIDE. [1314-80-3]

*RCRA 40 CFR 261
EPA hazardous waste no. U189 (reactive waste)

PHOSPHORUS TRICHLORIDE.
[7719-12-2]

*OSHA 29 CFR 1910.1000 Table Z-1
PEL(TWA): 0.5 ppm, 3 mg/m^3
ACGIH-TLV (1982 Addition)
TWA: 0.2 ppm, 1.5 mg/m^3
STEL: 0.5 ppm, 3 mg/m^3
*CWA 311(b)(2)(A), 40 CFR 116, 117
Discharge RQ = 5000 pounds (2270 kilograms)
NFPA—hazardous chemical
health—3, flammability—0, reactivity—2,
W—water may be hazardous in fire fighting

PHOSPHORYL CHLORIDE. See
Phosphorus oxychloride. [10025-87-3]

PHTHALATE ESTERS.

CWA 304(a)(1), 45 FR 79318, November 28, 1980, Water Qualitry Criteria

Freshwater Aquatic Life

The available data for phthalate esters indicate that acute and chronic toxicity to freshwater aquatic life occur at concentrations as low as 940 and 3 µg/liter, respectively, and would occur at lower concentrations among species that are more sensitive than those tested.

Saltwater Aquatic Life

The available data for phthalate esters indicate that acute toxicity to saltwater aquatic life occurs at concentrations as low as 2944 µg/liter and would occur at lower concentrations among species that are more sensitive than those tested. No data are available concerning the chronic toxicity of phthalate esters to sensitive saltwater aquatic life, but toxicity to one species of algae occurs at concentrations as low as 3.4 µg/liter.

Human Health

For the protection of human health from the toxic properties of dimethyl-phthalate ingested through water and contaminated aquatic organisms, the ambient water criterion is determined to be 313 mg/liter. For the protection of human health from the toxic properties of dimethyl-phthalate ingested through contaminated aquatic organisms alone, the ambient water criterion is determined to be 2.9 g/liter. For the protection of human health from the toxic properties of diethyl-phthalate ingested through water and contaminated aquatic organisms, the ambient water criterion is determined to be 350 mg/liter. For the protection of human health from the toxic properties of diethyl-phthalate ingested through contaminated aquatic organisms alone, the ambient water criterion is determined to be 1.8 g/liter. For the protection of human health from the toxic properties of dibutyl-phthalate ingested through water and contaminated aquatic organisms, the ambient water criterion is determined to be 34 mg/liter. For the protection of human health from the toxic properties of dibutyl-phthalate ingested through contaminated aquatic organisms alone, the ambient water criterion is determined to be 154 mg/liter. For the protection of human health from the toxic properties of di-2-ethylhexyl-phthalate ingested through water and contaminated aquatic organisms, the ambient water criterion is determined to be 15 mg/liter. For the protection of human health from the toxic properties of di-2-ethylhexyl-phthalate ingested through contaminated aquatic organisms alone, the ambient water criterion is determined to be 50 mg/liter.

PHTHALIC ANHYDRIDE. [85-44-9]
*OSHA 29 CFR 1910.1000 Table Z-1
PEL(TWA): 2 ppm, 12 mg/m^3
ACGIH-TLV
TWA: 1 ppm, 6 mg/m^3
STEL: 4 ppm, 24 mg/m^3
NFPA—hazardous chemical
health—2, flammability—1, reactivity—0
*RCRA 40 CFR 261
EPA hazardous waste no. U190

m-PHTHALODINITRILE. [626-17-5]
ACGIH-TLV
TWA: 5 mg/m^3

PICFUME. See Chloropicrin. [76-06-2]

PICLORAM. [1918-02-1]
OSHA Candidate List of Potential Occupational Carcinogens, 45 FR 53672, August 12, 1980
ACGIH-TLV
TWA: 10 mg/m^3
STEL: 20 mg/m^3

2-PICOLINE. [109-06-8]
*RCRA 40 CFR 261
EPA hazardous waste no. U191

PICOLINIC ACID, 4-AMINO-3,5,6-TRICHLORO-. See Picloram. [1918-02-1]

PICRIC ACID. [88-89-1]
*OSHA 29 CFR 1910.1000 Table Z-1
Skin
PEL(TWA): 0.1 mg/m^3
ACGIH-TLV

Skin
TWA: 0.1 mg/m³
STEL: 0.3 mg/m³
NFPA—hazardous chemical
health—2, flammability—4, reactivity—4

PINDONE. [83-26-1]
ACGIH-TLV
TWA: 0.1 mg/m³
STEL: 0.3 mg/m³

PIPERAZINE DIHYDROCHLORIDE.
[142-64-3]
ACGIH—TLV (1982 Addition)
TWA: 5 mg/m³

PIPERAZINE, 1,4-DINITROSO-. Syn:
N,N'-Dinitrosopiperazine. [140-79-4]
OSHA Candidate List of Potential Occupational Carcinogens, 45 FR 53672, August 12, 1980

PIPERONYL BUTOXIDE. [51-03-6]
40 CFR 162.11, FIFRA-RPAR
Notice of Intent to Register Issued
Current Status: Decision document in agency review. Returned to Registration Division on September 30, 1981.
Criteria Possibly Met or Exceeded: Co-carcinogenicity and Carcinogenicity

PIPERONYL SULFOXIDE. [120-62-7]
NCI—carcinogenic in animals

PIVAL® (2-PIVALYL-1,3-INDANDIONE). [83-26-1]
*OSHA 29 CFR 1910.1000 Table Z-1
PEL(TWA): 0.1 mg/m³

PIVALOLACTONE. [1955-45-9]
NCI—carcinogenic in animals

2-PIVALYL—1,3-INDANDIONE. See
Pindone. [83-26-1]

PLANTPIN® (BUTOXICARBOXIME).
[34681-23-7]
FIFRA—Registration Standard—Issued September 1981 (insecticide)

PLASTER OF PARIS.
ACGIH-TLV
TWA: nuisance particulate, 30 mppcf or 10 mg/m³ of total dust < 1% quartz, or 5 mg/m³ respirable dust
STEL: 20 mg/m³

PLATINUM (SOLUBLE SALTS).
[7440-06-4]
*OSHA 29 CFR 1910.1000 Table Z-1
PEL(TWA): 0.002 mg/m³ (as Pt)
ACGIH-TLV (metal)
TWA: 1 mg/m³
ACGIH-TLV (soluble salts)
TWA: 0.002 mg/m³

PLUMBANE, TETRAETHYL-. [78-00-2]
*RCRA 40 CFR 261
EPA hazardous waste no. P110 (hazardous substance)

POLYBROMINATED BIPHENYL MIXTURE (FIREMASTER FF-1).
[67774-32-7]
NCI—carcinogenic in animals

POLYCHLORINATED BIPHENYLS (PCBs). [1336-36-3]
NIOSH Criteria Document (Pub. No. 77-225), NTIS Stock No. PB 276849, September 27, 1977
"PCBs" are defined for this recommended standard as commercial preparations of chlorinated biphenyl compounds, including those preparations that may be described as single isomers or classes of isomers, such as decachlorodiphenyl. Biphenyl and its monochlorinated derivatives occurring in commercial preparations of PCBs shall be measured along with the polychlorinated derivatives, and shall be treated in this standard as the polychlorinated components of the preparations.
"Occupational exposure to PCBs" is defined as working with PCBs or with equipment containing PCBs that can become airborne or that can spill or splash on the skin or into the eyes, or the handling of any solid products that may result in exposure to

PCBs by skin contact or by inhalation. The term "PCB work area" is defined as an area where there is occupational exposure to PCBs. In areas where no occupational exposure to PCBs occurs, but where PCBs are present in equipment in the workplace, adherence is required only to Section 8(a).

Occupational exposure to polychlorinated biphenyls (PCBs) shall be controlled so that no worker is exposed to PCBs at a concentration greater than 1.0 $\mu g/m^3$, determined as a time-weighted average (TWA) concentration, for up to a 10-hour workday, 40-hour workweek. The standard is designed to protect the health and provide for the safety of employees for up to a 10-hour workday, 40-hour workweek, over a normal working lifetime.

NIOSH Current Intelligence Bulletin No. 7—cancer-related bulletin

OSHA Candidate List of Potential Occupational Carcinogens (EPA Carcinogen Assessment Group List), 45 FR 53672, August 12, 1980

IARC—carcinogenic in animals

*TSCA 40 CFR 761

I. *Disposal*
—43 FR 7150, February 17, 1978
Polychlorinated Biphenyls (PCBs), Disposal and Marking.

Action: Final Rule.
—48 FR 13181, March 30, 1983
Polychlorinated Biphenyls (PCBs); Procedural Amendment of the Approval Authority for PCB Disposal Facilities and Guidance for Obtaining Approval.

Action: Procedural Rule Amendment and Statement of Policy.

II. *Ban Regulation*
—44 FR 31514, May 31, 1979
Polychlorinated Biphenyls (PCBs) Manufacturing, Processing, Distribution in Commerce, and Use Prohibitions; Criteria Modification; Hearings.

Action: Final Rule.

III. *Electrical Equipment/Totally Enclosed*
—46 FR 16090, March 10, 1981
Polychlorinated Biphenyls (PCBs) Manufacturing, Processing, Distribution in Commerce, and Use Prohibitions; Use in Electrical Equipment; Court Order on Inspection and Maintenance.

Action: Rule-Related Court Order and Enforcement Notice.
—47 FR 37342, August 25, 1982
Polychlorinated Biphenyls (PCBs) Manufacturing, Processing, Distribution in Commerce, and Use Prohibitions; Use in Electrical Equipment.

Action: Final Rule.
—47 FR 54436, December 3, 1982
Polychlorinated Biphenyls (PCBs); Manufacturing, Processing, Distribution in Commerce, and Use Prohibitions; Use in Electrical Equipment; Correction.

Action: Rule; Correction.
—48 FR 5729, February 8, 1983
Polychlorinated Biphenyls (PCBs) Manufacturing, Processing, Distribution in Commerce, and Use Prohibitions; Incorporations by Reference Revisions [incorporates certain revised test methods of the American Society for Testing and Materials (ASTM).]

Action: Final Rule.
—48 FR 15125, April 7, 1983
Polychlorinated Biphenyls (PCBs); Manufacturing, Processing, Distribution in Commerce, and Use Prohibitions; Incorporations by Reference Revisions; Correction.

Action: Final Rule; Correction.
—48 FR 7172, February 18, 1983
Polychlorinated Biphenyls (PCBs) Manufacturing, Processing, Distribution in Commerce, and Use Prohibitions; Use in Electrical Equipment; Statement of Policy (statement of policy as to how the agency will determine whether electrical equipment

poses an exposure risk to food or feed).

Action: Rule-Related Notice; Statement of General Policy.

IV. A. *Railroads*
—48 FR 124, January 3, 1983 Polychlorinated Biphenyls (PCBs) Manufacturing, Processing, Distribution in Commerce, and Use Prohibitions; Amendment to Use Authorization for PCB Railroad Transformers.

Action: Final Rule.

V. *Closed and Controlled*
—47 FR 46980, October 21, 1982 Polychlorinated Biphenyls (PCBs); Manufacturing, Processing, Distribution in Commerce, and Use Prohibitions; Use in Closed and Controlled Waste Manufacturing Processes.

Action: Final Rule.

—48 FR 4467, February 1, 1983 Polychlorinated Biphenyls (PCBs) Manufacturing, Processing, Distribution in Commerce, and Use Prohibitions; Use in Closed and Controlled Waste Manufacturing Processes; Correction.

Action: Rule; Correction.

CWA 304(a)(1), 45 FR 79318, November 28, 1980, Water Quality Criteria

Freshwater Aquatic Life

For polychlorinated biphenyls the criterion to protect freshwater aquatic life as derived using the guidelines is 0.014 μg/liter as a 24-hour average. The available data indicate that acute toxicity to freshwater aquatic life probably will occur only at concentrations above 2.0 μg/liter and that the 24-hour average should provide adequate protection against the acute toxicity.

Saltwater Aquatic Life

For polychlorinated biphenyls the criterion to protect saltwater aquatic life as derived using the guidelines is 0.030 μg/liter as a 24-hour average. The available data indicate that acute toxicity to saltwater aquatic life probably will occur only at concentrations above 10 μg/liter and that the 24-hour average should provide adequate protection against acute toxicity.

Human Health

For the maximum protection of human health from the potential carcinogenic effects due to exposure of PCBs through ingestion of contaminated aquatic organisms, the ambient water concentration should be zero based on the nonthreshold assumption for this chemical. However, zero level may not be attainable at the present time. Therefore, the levels that may result in incremental increase of cancer risk over the lifetime are estimated at 10^{-5}, 10^{-6} and 10^{-7}. The corresponding criteria are 0.79 ng/liter, 0.079 ng/liter, and 0.0079 ng/liter, respectively. If the above estimates are made for consumption of aquatic organisms only, excluding consumption of water, the levels are 0.79 ng/liter, 0.079 ng/liter, and 0.0079 ng/liter, respectively. Other concentrations representing different risk levels may be calculated by use of the guidelines. The risk estimate range is presented for information purposes and does not represent an agency judgment on an "acceptable" risk level.

*CWA 311(b)(2)(A), 40 CFR 116, 117
Discharge RQ = 10 pounds (4.54 kilograms)
CAA 112
Chemical proposed to be assessed as a Hazardous Air Pollutant by a House of Representatives bill to amend the CAA

POLYCHLORINATED TERPHENYLS—POLYCHLORINATED *ORTHO-*, *META-*, AND *PARA*-TEROPHENYLS.
[11126-42-4, 12642-23-8, 61788-33-8]

*TSCA 8(d) Unpublished Health and Safety Studies Reporting, 40 CFR 716, 47 FR 387, September 2, 1982

TSCA 4(e), ITC
This category consists of the polychlorinated *ortho-*, *meta-*, and *para*-terphenyls. Second Report of the Interagency Testing Committee to the administrator, Environmental Pro-

tection Agency, April 1978; 43 FR 16684, April 19, 1978, responded to by the EPA administrator 44 FR 28095, May 14, 1979, EPA responded to the committee's recommendation for testing 46 FR 54482, November 2, 1981

POLYCHLOROBIPHENYLS. *See* Chlorodiphenyls. [1336-36-3]

POLYFORMALDEHYDE. *See* Paraformaldehyde. [30525-89-4]

POLYNUCLEAR AROMATIC HYDROCARBONS (PAHs).

CWA 304(a)(1), 45 FR 79318, November 28, 1980, Water Quality Criteria

Freshwater Aquatic Life

The limited freshwater data base available for polynuclear aromatic hydrocarbons, mostly from short-term bioconcentration studies with two compounds, does not permit a statement concerning acute or chronic toxicity.

Saltwater Aquatic Life

The available data for polynuclear aromatic hydrocarbons indicate that acute toxicity to saltwater aquatic life occurs at concentrations as low as 300 µg/liter and would occur at lower concentrations among species that are more sensitive than those tested. No data are available concerning the chronic toxicity of polynuclear aromatic hydrocarbons to sensitive saltwater aquatic life.

Human Health

For the maximum protection of human health from the potential carcinogenic effects due to exposure to PAHs through ingestion of contaminated water and contaminated aquatic organisms, the ambient water concentration should be zero based on the non-threshold assumption for this chemical. However, zero level may not be attainable at the present time. Therefore, the levels that may result in incremental increase of cancer risk over the lifetime are estimated at 10^{-5}, 10^{-6}, and 10^{-7}. The corresponding criteria are 28 ng/liter, 2.8 ng/liter, and 0.28 ng/liter, respectively. If the above estimates are made for consumption of aquatic organisms only, excluding consumption of water, the levels are 311 ng/liter, 31.1 ng/liter, and 3.11 ng/liter, respectively. Other concentrations representing different risk levels may be calculated by use of the guidelines. The risk estimate range is presented for information purposes and does not represent an agency judgment on an "acceptable" risk level.

POLYOXYMETHYLENE GLYCOL. *See* Paraformaldehyde. [30525-89-4]

POLYSORBATE 20. [9005-64-5]

TSCA-CHIP available

POLYTETRAFLUOROETHYLENE DECOMPOSITION PRODUCTS.

ACGIH-TLV
TWA: Thermal decomposition of the fluorocarbon chain in air leads to the formation of oxidized products containing carbon, fluorine, and oxygen. Because these products decompose in parts by hydrolysis in alkaline solution, they can be quantitatively determined in air as fluoride to provide an index of exposure. No TLV is recommended pending determination of the toxicity of the products, but air concentrations should be minimal.
Trade names: Algoflon, Fluon, Halon, Teflon, Tetran

PONCEAU MX. [3761-53-3]

IARC—carcinogenic in animals

PONCEAU 3R. [3564-09-8]

IARC—carcinogenic in animals

POTASSIUM. [7440-09-7]

NFPA—hazardous chemical
health—3, flammability—1, reactivity—2, W̶—water may be hazardous in fire fighting

POTASSIUM ARSENATE. [7784-41-0]

*CWA 311(b)(2)(A), 40 CFR 116, 117
Discharge RQ = 1000 pounds (454 kilograms)

POTASSIUM ARSENITE. [10124-50-2]
*CWA 311(b)(2)(A), 40 CFR 116, 117
Discharge RQ = 1000 pounds (454 kilograms)

POTASSIUM BICHROMATE.
See Dichromates. [7778-50-9]
*CWA 311(b)(2)(A), 40 CFR 116, 117
Discharge RQ = 1000 pounds (454 kilograms)
See Chromium (VI), NIOSH Criteria Document.

POTASSIUM BIS(2-HYDROXYETHYL) DITHIOCARBAMATE. See Carbamic acid (2-hydroxyethyl)dithiomonopotassium salt. [23746-34-1]
OSHA Candidate List of Potential Occupational Carcinogens, 45 FR 53672, August 12, 1980
IARC—carcinogenic in animals

POTASSIUM BROMATE. [7758-02-3]
NFPA—hazardous chemical
when nonfire: health—0, flammability—0, reactivity—0, oxy—oxidizing chemical
when fire: health—1, flammability—0, reactivity—0, oxy—oxidizing chemical

POTASSIUM CHLORATE. [3811-04-9]
NFPA—hazardous chemical
when nonfire: health—0, flammability—0, reactivity—0, oxy—oxidizing chemical
when nonfire: health—2, flammability—0, reactivity—0, oxy—oxidizing chemical

POTASSIUM CHROMATE. [7789-00-6]
See Chromium(VI), NIOSH Criteria Document.
*CWA 311(b)(2)(A), 40 CFR 116, 117
Discharge RQ = 1000 pounds (454 kilograms)

POTASSIUM CYANIDE. [151-50-8]
See Cyanide, hydrogen, and cyanide salts, NIOSH Criteria Document.
NFPA—hazardous chemical
health—3, flammability—0, reactivity—0
*CWA 311(b)(2)(A), 40 CFR 116, 117
Discharge RQ = 10 pounds (4.54 kilograms)
*RCRA 40 CFR 261
EPA hazardous waste no. P098

POTASSIUM DICHLOROISOCYANURATE. See Potassium dichloro-S-triazinetrione. [2244-21-5]

POTASSIUM DICHLORO-S-TRIAZINETRIONE. [2244-21-5]
NFPA—hazardous chemical
health—3, flammability—0, reactivity—2, oxy—oxidizing chemical

POTASSIUM DICHROMATE. See Potassium bichromate. [7778-50-9]
See Chromium(VI), NIOSH Criteria Document.

POTASSIUM HYDROXIDE (LYE). [1310-58-3]
ACGIH-TLV
TWA: ceiling limit, 2 mg/m^3
NFPA—hazardous chemical
health—3, flammability—0, reactivity—1
*CWA 311(b)(2)(A), 40 CFR 116, 117
Discharge RQ = 1000 pounds (454 kilograms)

POTASSIUM NITRATE. [7757-79-1]
NFPA—hazardous chemical
when nonfire: health—0, flammability—0, reactivity—0, oxy—oxidizing chemical
when fire: health—1, flammability—0, reactivity—0, oxy—oxidizing chemical

POTASSIUM PERCHLORATE. [7778-74-7]
NFPA—hazardous chemical
health—1, flammability—0, reactivity—2, oxy—oxidizing chemical

POTASSIUM PERMANGANATE. [7722-64-7]
NFPA—hazardous chemical
health—1, flammability—0, reactivity—0, oxy—oxidizing chemical
*CWA 311(b)(2)(A), 40 CFR 116, 117
Discharge RQ = 100 pounds (45.4 kilograms)

POTASSIUM PEROXIDE. [17014-71-0]
NFPA—hazardous chemical
health—3, flammability—0, reactivity—2, W—water may be hazardous in fire fighting, oxy—oxidizing chemical, reacts explosively with water or steam to produce toxic and corrosive fumes

POTASSIUM PERSULFATE.
[7727-21-1]
NFPA—hazardous chemical
health—1, flammability—0, reactivity—0, oxy—oxidizing chemical

POTASSIUM SILVER CYANIDE.
[506-61-6]
*RCRA 40 CFR 261
EPA hazardous waste no. P099

POTASSIUM SULFIDE. [1312-73-8]
NFPA—hazardous chemical
health—2, flammability—1, reactivity—0

PROCARBAZINE HYDROCHLORIDE.
See p-Toluamide, N-isopropyl-alpha-(2-methylhydrazino)-, monohydrochloride.
[366-70-1]

PROFLAVINE. [92-62-6]
NCI—carcinogenic in animals

PROFLURALIN. [26399-36-0]
FIFRA—Data Call In, 45 FR 75488, November 14, 1980
Status: letter being drafted

PROGESTERONE. [57-83-0]
IARC—carcinogenic in animals

PRONAMIDE. [23950-58-5]
OSHA Candidate List of Potential Occupational Carcinogens (EPA Carcinogen Assessment Group List), 45 FR 53672, August 12, 1980
FIFRA—Data Call In, 45 FR 75488, November 14, 1980
Status: letter being drafted
40 CFR 162.11, FIFRA-RPAR Completed
Current Status: PD 1 published, 42 FR 32302 (5/20/77); comment period closed 8/29/77: NTIS# PB81 109472.
PD 2/3 completed and Notice of Proposed Determination published, 44 FR 3083 (1/15/79); comment period closed 2/14/79: NTIS# PB80 213911.
PD 4 completed and Final Determination published, 44 FR 61640 (10/26/79): NTIS# PB81 112716.
Criteria Possibly Met or Exceeded: Oncogenicity
FIFRA—Registration Standard Development—Data Evaluation and Development of Regulatory Position (herbicide)
*RCRA 40 CFR 261
EPA hazardous waste no. U192

PRONETALOL (HYDROCHLORIDE).
[51-02-5]
IARC—carcinogenic in animals

PROPACHLOR. [1918-16-7]
FIFRA—Data Call In, 45 FR 75488, November 14, 1980
Status: letter being drafted

PROPANAL. See Propionaldehyde.
[123-38-6]

PROPANAL, 2-METHYL-2-(METHYLTHIO)-, O-[(METHYLAMINO)CARBONYL] OXIME. [116-06-3]
*RCRA 40 CFR 261
EPA hazardous waste no. P070

1-PROPANAMINE. [107-10-8]
*RCRA 40 CFR 261
EPA hazardous waste no. U194 (ignitable waste, toxic waste)

2-PROPANAMINE, 1-CHLORO-N,N-DIMETHYL-, HYDROCHLORIDE.
[17256-39-2]
*TSCA 8(a), 40 CFR 712, 47 FR 26992, June 22, 1982, Chemical Information Rule

1-PROPANAMINE, N-PROPYL-.
[142-84-7]
*RCRA 40 CFR 261
EPA hazardous waste no. U110 (ignitable waste)

PROPANE. [74-98-6]

*OSHA 29 CFR 1910.1000 Table Z-1
PEL(TWA): 1000 ppm, 1800 mg/m^3
ACGIH-TLV
TWA: asphyxiant—gas or vapors, when present in high concentrations in air, act primarily as simple asphyxiant without other significant physiologic effects
NFPA—hazardous chemical (flammable chemical)
health—1, flammability—4, reactivity—0

PROPANE, 2,2-BIS[P-(2,3-EPOXYPROPOXY)PHENYL]-. Syn: Bisphenol A-diglycidyl ether. [1675-54-3]
OSHA Candidate List of Potential Occupational Carcinogens, 45 FR 53672, August 12, 1980

PROPANE, 1,2-DIBROMO-3-CHLORO-. [96-12-8]
*TSCA 8(a), 40 CFR 712, 47 FR 26992, June 22, 1982, Chemical Information Rule
*RCRA 40 CFR 261
EPA hazardous waste no. U066

PROPANE, 1,2-DICHLORO-. [78-87-5]
*TSCA 8(a), 40 CFR 712, 47 FR 26992, June 22, 1982, Chemical Information Rule

PROPANEDINITRLE.
*RCRA 40 CFR 261
EPA hazardous waste no. U149

PROPANENITRILE. [107-12-0]
*RCRA 40 CFR 261
EPA hazardous waste no. P101

PROPANENITRILE, 3-CHLORO-. [542-76-7]
*RCRA 40 CFR 261
EPA hazardous waste no. P027

PROPANENITRILE, 2-HYDROXY-2-METHYL-. [75-86-5]
*RCRA 40 CFR 261
EPA hazardous waste no. P069 (hazardous substance)

PROPANE, 2-NITRO-. Syn: Isonitropropane. [79-46-9]

OSHA Candidate List of Potential Occupational Carcinogens, 45 FR 53672, August 12, 1980
*RCRA 40 CFR 261
EPA hazardous waste no. U171 (ignitable waste)

PROPANE, 2,2'-OXYBIS(2-CHLORO-.
*RCRA 40 CFR 261
EPA hazardous waste no. U027

PROPANE SULTONE. [1120-71-4]
ACGIH-TLV
TWA: Industrial substance suspect of carcinogenic potential for humans—which is suspect of inducing cancer, based on either (1) limited epidemiologic evidence, exclusive of clinical reports of single cases, or (2) demonstration of carcinogenesis in one or more animal species by appropriate methods. Worker exposure by all routes should be carefully controlled to levels consistent with the animal and human experience data.
OSHA Candidate List of Potential Occupational Carcinogens (EPA Carcinogen Assessment Group List), 45 FR 53672, August 12, 1980
IARC—carcinogenic in animals
*RCRA 40 CFR 261
EPA hazardous waste no. U193

1-PROPANETHIOL. [107-03-9]
See Thiols, NIOSH Criteria Document.

1,2,3-PROPANETRIOL, TRINITRATE. [55-63-0]
*RCRA 40 CFR 261
EPA hazardous waste no. P081 (reactive waste)

PROPANOIC ACID. See Propionic acid. [79-09-4]

1-PROPANOL, 2,3-DIBROMO-, PHOSPHATE (3:1) [126-72-7]
*RCRA 40 CFR 261
EPA hazardous waste no. U235

1-PROPANOL, 2,3-EPOXY-. [556-52-5]

*RCRA 40 CFR 261
EPA hazardous waste no. U126

1-PROPANOL, 2-METHYL-. [78-83-1]
*RCRA 40 CFR 261
EPA hazardous waste no. U140 (ignitable waste, toxic waste)

2-PROPANONE. [67-64-1]
*RCRA 40 CFR 261
EPA hazardous waste no. U002 (ignitable waste)

2-PROPANONE, 1-BROMO-. [598-31-2]
*RCRA 40 CFR 261
EPA hazardous waste no. P017

PROPARGITE. [2312-35-8]
*CWA 311(b)(2)(A), 40 CFR 116, 117
Discharge RQ = 10 pounds (4.54 kilograms)

PROPARGYL ALCOHOL. [107-19-7]
ACGIH-TLV

Skin

TWA: 1 ppm, 2 mg/m^3
STEL: 3 ppm, 6 mg/m^3
*RCRA 40 CFR 261
EPA hazardous waste no. P102

PROPARGYL BROMIDE.
See 3-Bromopropyne. [106-96-7]

PROPENAL. See Acrolein. [107-02-8]

2-PROPENAMIDE. [79-06-1]

*TSCA 8(a), 40 CFR 712, 47 FR 26992, June 22, 1982, Chemical Information Rule
*RCRA 40 CFR 261
EPA hazardous waste no. U007

PROPENE. See Propylene. [115-07-1]

PROPENE, 1,3-DICHLORO-. [542-75-6]

*RCRA 40 CFR 261
EPA hazardous waste no. U084 (hazardous substance)

1-PROPENE, 1,1,2,3,3,3-HEXACHLORO-. [1888-71-7]
*RCRA 40 CFR 261
EPA hazardous waste no. U243

1-PROPENE, 1,1,2,3,3,3,-HEXAFLUORO-. [116-15-4]
*TSCA 8(a), 40 CFR 712, 47 FR 26992, June 22, 1982, Chemical Information Rule

2-PROPENENITRILE. [107-13-1]
*RCRA 40 CFR 261
EPA hazardous waste no. U009 (hazardous substance)

2-PROPENENITRILE, 2-METHYL-. [126-98-7]
*RCRA 40 CFR 261
EPA hazardous waste no. U152 (ignitable waste, toxic waste)

2-PROPENOIC ACID. [79-10-7]
*RCRA 40 CFR 261
EPA hazardous waste no. U008 (ignitable waste)

2-PROPENOIC ACID, BICYCLO [2.2.1]HEPT-5-EN-2-YL METHYL ESTER. [95-39-6]
*TSCA 8(a), 40 CFR 712, 47 FR 26992, June 22, 1982, Chemical Information Rule

2-PROPENOIC ACID, BUTYL ESTER. [141-32-2]
*TSCA 8(a), 40 CFR 712, 47 FR 26992, June 22, 1982, Chemical Information Rule

2-PROPENOIC ACID, 2-CYANOETHYL ESTER. [106-71-8]
*TSCA 8(a), 40 CFR 712, 47 FR 26992, June 22, 1982, Chemical Information Rule

2-PROPENOIC ACID, DECYL ESTER. [2156-96-9]
*TSCA 8(a), 40 CFR 712, 47 FR 26992, June 22, 1982, Chemical Information Rule

2-PROPENOIC ACID, 2-(DIETHYLAMINO) ETHYL ESTER. [2426-54-2]

*TSCA 8(a), 40 CFR 712, 47 FR 26992, June 22, 1982, Chemical Information Rule

2-PROPENOIC ACID, 1,1-DIMETHYLETHYL ESTER. [1663-39-4]

*TSCA 8(a), 40 CFR 712, 47 FR 26992, June 22, 1982, Chemical Information Rule

2-PROPENOIC ACID, 2,2-DIMETHYL-1,3-PROPANEDIYL ESTER. [2223-82-7]

*TSCA 8(a), 40 CFR 712, 47 FR 26992, June 22, 1982, Chemical Information Rule

2-PROPENOIC ACID, 2,2-DINITROPROPYL ESTER. [17997-09-2]

*TSCA 8(a), 40 CFR 712, 47 FR 26992, June 22, 1982, Chemical Information Rule

2-PROPENOIC ACID, 2-ETHOXYETHYL ESTER. [106-74-1]

*TSCA 8(a), 40 CFR 712, 47 FR 26992, June 22, 1982, Chemical Information Rule

2-PROPENOIC ACID, 2-ETHYLBUTYL ESTER. [3953-10-4]

*TSCA 8(a), 40 CFR 712, 47 FR 26992, June 22, 1982, Chemical Information Rule

2-PROPENOIC ACID, ETHYL ESTER. [140-88-5]

*TSCA 8(a), 40 CFR 712, 47 FR 26992, June 22, 1982, Chemical Information Rule

*RCRA 40 CFR 261
EPA hazardous waste no. U113 (ignitable waste)

2-PROPENOIC ACID, 2-ETHYLHEXYL ESTER. [103-11-7]

*TSCA 8(a), 40 CFR 712, 47 FR 26992, June 22, 1982, Chemical Information Rule

2-PROPENOIC ACID, 3a,4,5,6,7,7a-HEXAHYDRO-4,7-METHANO-1H-INDENYL ESTER. [33791-58-1]

*TSCA 8(a), 40 CFR 712, 47 FR 26992, June 22, 1982, Chemical Information Rule

2-PROPENOIC ACID, 1,6-HEXANEDIYL ESTER. [13048-33-4]

*TSCA 8(a), 40 CFR 712, 47 FR 26992, June 22, 1982, Chemical Information Rule

2-PROPENOIC ACID, HEXYL ESTER. [2499-95-8]

*TSCA 8(a), 40 CFR 712, 47 FR 26992, June 22, 1982, Chemical Information Rule

2-PROPENOIC ACID, 2-HYDROXYETHYL ESTER. [818-61-1]

*TSCA 8(a), 40 CFR 712, 47 FR 26992, June 22, 1982, Chemical Information Rule

2-PROPENOIC ACID, ISODECYL ESTER. [1330-61-6]

*TSCA 8(a), 40 CFR 712, 47 FR 26992, June 22, 1982, Chemical Information Rule

2-PROPENOIC ACID, 2-METHYL-, 2-[BIS(1-METHYLETHYL)AMINO]ETHYL ESTER. [16715-83-6]

*TSCA 8(a), 40 CFR 712, 47 FR 26992, June 22, 1982, Chemical Information Rule

2-PROPENOIC ACID, 2-METHYL-, 1,4-BUTANEDIYL ESTER. [2082-81-7]

*TSCA 8(a), 40 CFR 712, 47 FR 26992, June 22, 1982, Chemical Information Rule

2-PROPENOIC ACID, 2-METHYL-, BUTYL ESTER. [97-88-1]

*TSCA 8(a), 40 CFR 712, 47 FR 26992, June 22, 1982, Chemical Information Rule

2-PROPENOIC ACID, 2-METHYL-, CYCLOHEXYL ESTER. [101-43-9]

*TSCA 8(a), 40 CFR 712, 47 FR 26992, June 22, 1982, Chemical Information Rule

2-PROPENOIC ACID, 2-METHYL-, 2-(DIETHYLAMINO) ETHYL ESTER.

*TSCA 8(a), 40 CFR 712, 47 FR 26992, June 22, 1982, Chemical Information Rule

2-PROPENOIC ACID, 2-METHYL-, 2-(DIMETHYLAMINO) ETHYL ESTER. [2867-47-2]

*TSCA 8(a), 40 CFR 712, 47 FR 26992, June 22, 1982, Chemical Information Rule

2-PROPENOIC ACID, 2-METHYL-, 2-[(1,1-DIMETHYLETHYL)AMINO] ETHYL ESTER. [3775-90-4]

*TSCA 8(a), 40 CFR 712, 47 FR 26992, June 22, 1982, Chemical Information Rule

2-PROPENOIC ACID, 2-METHYL-, DODECYL ESTER. [142-90-5]

*TSCA 8(a), 40 CFR 712, 47 FR 26992, June 22, 1982, Chemical Information Rule

2-PROPENOIC ACID, METHYL ESTER. [96-33-3]

*TSCA 8(a), 40 CFR 712, 47 FR 26992, June 22, 1982, Chemical Information Rule

2-PROPENOIC ACID, 2-METHYL-, 1,2-ETHANEDIYLBIS(OXY-2,1-ETHANEDIYL) ESTER. [109-16-0]

*TSCA 8(a), 40 CFR 712, 47 FR 26992, June 22, 1982, Chemical Information Rule

2-PROPENOIC ACID, 2-METHYL-, ETHYL ESTER. [97-86-9]

*TSCA 8(a), 40 CFR 712, 47 FR 26992, June 22, 1982, Chemical Information Rule
*RCRA 40 CFR 261
EPA hazardous waste no. U118

2-PROPENOIC ACID, 1-METHYLETHYL ESTER. [689-12-3]

*TSCA 8(a), 40 CFR 712, 47 FR 26992, June 22, 1982, Chemical Information Rule

2-PROPENOIC ACID, 2-METHYL-, 2-ETHYLHEXYL ESTER. [688-84-6]

*TSCA 8(a), 40 CFR 712, 47 FR 26992, June 22, 1982, Chemical Information Rule

2-PROPENOIC ACID, 2-METHYL-, HEXYL ESTER. [142-09-6]

*TSCA 8(a), 40 CFR 712, 47 FR 26992, June 22, 1982, Chemical Information Rule

2-PROPENOIC ACID, 2-METHYL-, 2-HYDROXYETHYL ESTER. [868-77-9]

*TSCA 8(a), 40 CFR 712, 47 FR 26992, June 22, 1982, Chemical Information Rule

2-PROPENOIC ACID, 2-METHYL-, METHYL ESTER. [80-62-6]

*TSCA 8(a), 40 CFR 712, 47 FR 26992, June 22, 1982, Chemical Information Rule
*RCRA 40 CFR 261
EPA hazardous waste no. U162 (hazardous substance, ignitable waste, toxic waste)

2-PROPENOIC ACID, 2-METHYL-, OCTADECYL ESTER. [32360-05-7]

*TSCA 8(a), 40 CFR 712, 47 FR 26992, June 22, 1982, Chemical Information Rule

2-PROPENOIC ACID, 2-METHYL-, OXIRANYLMETHYL ESTER. [106-91-2]

*TSCA 8(a), 40 CFR 712, 47 FR 26992, June 22, 1982, Chemical Information Rule

2-PROPENOIC ACID, 2-METHYL-, OXYDI-2, 1-ETHANEDIYL ESTER. [2425-79-8]

*TSCA 8(a), 40 CFR 712, 47 FR 26992, June 22, 1982, Chemical Information Rule

2-PROPENOIC ACID, 2-METHYL-, 2-PROPENYL ESTER. [96-05-9]

*TSCA 8(a), 40 CFR 712, 47 FR 26992, June 22, 1982, Chemical Information Rule

2-PROPENOIC ACID, 2-METHYL-, PROPYL ESTER. [2210-28-8]

*TSCA 8(a), 40 CFR 712, 47 FR 26992, June 22, 1982, Chemical Information Rule

2-PROPENOIC ACID, 2-METHYLPROPYL ESTER. [106-63-8]

*TSCA 8(a), 40 CFR 712, 47 FR 26992, June 22, 1982, Chemical Information Rule

2-PROPENOIC ACID, 2-METHYL-, SODIUM SALT. [5536-61-8]

*TSCA 8(a), 40 CFR 712, 47 FR 26992, June 22, 1982, Chemical Information Rule

2-PROPENOIC ACID, 2-METHYL-, (TETRAHYDRO-2-FURANYL) METHYL ESTER. [2455-24-5]

*TSCA 8(a), 40 CFR 712, 47 FR 26992, June 22, 1982, Chemical Information Rule

2-PROPENOIC ACID, 2-METHYL-, 3-(TRIMETHOXYSILYL) PROPYL ESTER. [2530-85-0]

*TSCA 8(a), 40 CFR 712, 47 FR 26992, June 22, 1982, Chemical Information Rule

2-PROPENOIC ACID, 2-METHYL-, 1,7,7-TRIMETHYLBICYCLO [2.2.1] HEPT-2-YL ESTER, *exo*-. [7534-94-3]

*TSCA 8(a), 40 CFR 712, 47 FR 26992, June 22, 1982, Chemical Information Rule

2-PROPENOIC ACID, MONOESTER with 1,2-PROPANEDIOL. [25584-83-2]

*TSCA 8(a), 40 CFR 712, 47 FR 26992, June 22, 1982, Chemical Information Rule

2-PROPENOIC ACID, OXIRANYL-METHYL ESTER. [106-90-1]

*TSCA 8(a), 40 CFR 712, 47 FR 26992, June 22, 1982, Chemical Information Rule

2-PROPENOIC ACID, PROPYL ESTER. [925-60-0]

*TSCA 8(a), 40 CFR 712, 47 FR 26992, June 22, 1982, Chemical Information Rule

2-PROPENOIC ACID, SODIUM SALT. [7446-81-3]

*TSCA 8(a), 40 CFR 712, 47 FR 26992, June 22, 1982, Chemical Information Rule

2-PROPENOIC ACID, TRIDECYL ESTER. [3076-04-8]

*TSCA 8(a), 40 CFR 712, 47 FR 26992, June 22, 1982, Chemical Information Rule

2-PROPEN-1-OL. [107-18-6]

*RCRA 40 CFR 261
EPA hazardous waste no. P005 (hazardous substance)

beta-PROPIOLACTONE. [57-57-8]

*OSHA Standard 29 CFR 1910.1013
Cancer-Suspect Agent
Shall not apply to solid or liquid mixtures containing less than 1.0% by weight or volume of beta-propiolactone

ACGIH-TLV
TWA: 0.5 ppm, 1.5 mg/m^3
STEL: 1 ppm, 3 mg/m^3
Industrial substance suspect of carcinogenic potential for humans—which is suspect of inducing cancer, based on either (1) limited epidemiologic evidence, exclusive of clinical reports of single cases, or (2) demonstration of carcinogenesis in one or more animal species by appropriate methods. Worker exposure by all routes should be carefully controlled to levels consistent with the animal and human experience data.
IARC—carcinogenic in animals

PROPIONALDEHYDE. [123-38-6]

NFPA—hazardous chemical
health—2, flammability—3, reactivity—1

PROPIONIC ACID. [79-09-4]

ACGIH-TLV
TWA: 10 ppm, 30 mg/m^3
STEL: 15 ppm, 45 mg/m^3
NFPA—hazardous chemical
health—2, flammability—2, reactivity—0
*CWA 311(b)(2)(A), 40 CFr 116, 117
Discharge RQ = 5000 pounds (2270 kilograms)

PROPIONIC ACID, 2-(2,4,5-TRICHLOROPHENOXY)-. [93-72-1]

*RCRA 40 CFR 261
EPA hazardous waste no. U233 (hazardous substance)

PROPIONIC ANHYDRIDE. [123-62-6]

*CWA 311(b)(2)(A), 40 CFR 116, 117
Discharge RQ = 5000 pounds (2270 kilograms)
NFPA—hazardous chemical
health—2, flammability—2, reactivity—1

PROPIONITRILE. [123-38-6]

See Nitriles, NIOSH Criteria Document.

PROPOXUR. [114-26-1]

ACGIH-TLV
TWA: 0.5 mg/m^3
STEL: 2 mg/m^3

n-PROPYL ACETATE. [109-60-4]
*OSHA 29 CFR 1910.1000 Table Z-1
PEL(TWA): 200 ppm, 840 mg/m^3
ACGIH-TLV
TWA: 200 ppm, 840 mg/m^3
STEL: 250 ppm, 1050 mg/m^3
NFPA—hazardous chemical (flammable chemical)
health—1, flammability—3, reactivity—0

PROPYL ALCOHOL. [71-23-8]
*OSHA 29 CFR 1910.1000 Table Z-1
PEL(TWA): 200 ppm, 500 mg/m^3
ACGIH-TLV

Skin

TWA: 200 ppm, 500 mg/m^3
STEL: 250 ppm, 625 mg/m^3

n-PROPYLAMINE. [107-10-8]
NFPA—hazardous chemical
health—3, flammability—3, reactivitiy—0
*RCRA 40 CFR 261
EPA hazardous waste no. U194 (ignitable waste, toxic waste)

n-PROPYL CARBAMATE. [627-12-3]
IARC—carcinogenic in animals

PROPYLENE. [115-07-1]
ACGIH-TLV
TWA: Asphyxiant—gas or vapors, when present in high concentrations in air, act primarily as simple asphyxiant without other significant physiologic effects.
NFPA—hazardous chemical
health—1, flammability—4, reactivity—1

PROPYLENE DICHLORIDE. [78-87-5]
*OSHA 29 CFR 1910.1000 Table Z-1
PEL(TWA): 75 ppm, 350 mg/m^3
NCI—carcinogenic in animals
NFPA—hazardous chemical
health—2, flammability—3, reactivity—0
TSCA 4(e), ITC
Third Report of the TSCA Interagency Testing Committee to the adminstrator, Environmental Protection Agency, October 1978; 43 FR 50630, October 30, 1978
*TSCA 8(d), Unpublished Health and Safety Studies Reporting, 40 CFR 716, 47 FR 387, September 2, 1982
CWA 307(a)(1), 40 CFR 125, Priority Pollutant
*RCRA 40 CFR 261
EPA hazardous waste no. U083 (hazardous substance)

PROPYLENE GLYCOL. [57-55-6]
NFPA—hazardous chemical (flammable chemical)
health—0, flammability—1, reactivity—0

PROPYLENE GLYCOL DINITRATE (PGDN). [6423-43-4]
ACGIH-TLV

Skin

TWA: 0.02 ppm, 0.1 mg/m^3
STEL: 0.05 ppm, 0.3 mg/m^3

PROPYLENE GLYCOL MONOMETHYL ETHER. [107-98-2]
ACGIH-TLV
TWA: 100 ppm, 360 mg/m^3
STEL: 150 ppm, 540 mg/m^3

PROPYLENE IMINE. [75-55-8]
*OSHA 29 CFR 1910.1000 Table Z-1

Skin

PEL(TWA): 2 ppm, 5 mg/m^3
ACGIH-TLV

Skin

TWA: 2 ppm, 5 mg/m^3
Intended changes for 1982:
TWA: 2 ppm, 5 mg/m^3
Industrial substance suspect of carcinogenic potenital for humans—which is suspect of inducing cancer, based on either (1) limited epidemiologic evidence, exclusive of clinical reports of single cases, or (2) demonstration of carcinogenesis in one or more animal species by appropriate methods. Worker expo-

sure by all routes should be carefully controlled to levels consistent with the animal and human experience data.

*RCRA 40 CFR 261
EPA hazardous waste no. P067

PROPYLENE OXIDE. [75-56-9]

*OSHA 29 CFR 1910.1000 Table Z-1
PEL(TWA): 100 ppm, 240 mg/m^3

ACGIH-TLV
TWA: 20 ppm, 50 mg/m^3

IARC—carcinogenic in animals

NCI—carcinogenic in animals

NFPA—hazardous chemical
health—2, flammability—4, reactivity—2

CAA 112
Chemical proposed to be assessed as a Hazardous Air Pollutant by a House of Representatives bill to amend the CAA

*CWA 311(b)(2)(A), 40 CFR 116, 117
Discharge RQ = 5000 pounds (2270 kilograms)

PROPYL GALLATE. [121-79-9]

NCI—may be carcinogenic in animals

n-PROPYL NITRATE. [627-13-4]

*OSHA 29 CFR 1910.1000 Table Z-1
PEL(TWA): 25 ppm, 110 mg/m^3

ACGIH-TLV
TWA: 25 ppm, 105 mg/m^3
STEL: 40 ppm, 470 mg/m^3

NFPA—hazardous chemical
health—2, flammability—3, reactivity—3, oxy—oxidizing agent.

PROPYLTHIOURACIL. [51-52-5]

OSHA Candidate List of Potential Occupational Carcinogens (EPA Assessment Group List), 45 FR 53672, August 12, 1980

IARC—carcinogenic in animals

PROPYLTRICHLOROSILANE (NORMAL). [141-57-1]

NFPA—hazardous chemical
health—3, flammability—3, reactivity—1

PROPYNE. See Methyl acetylene.
[74-99-7]

2-PROPYN-1-OL. [107-19-7]

*RCRA 40 CFR 261
EPA hazardous waste no. P102

PRUSSIC ACID. See Hydrogen cyanide.
[74-90-8]

PYRENE. See Coal tar pitch
volatiles. [129-00-0]

CWA 307(a)(1). 40 CFR 125, Priority Pollutant

PYRETHRINS. [121-29-9, 121-21-1]

*CWA 311(b)(2)(A), 40 CFR 116, 117
Discharge RQ = 1000 pounds (454 kilograms)

PYRETHRUM. [8003-34-7]

*OSHA 29 CFR 1910.1000 Table Z-1
PEL(TWA): 5 mg/m^3

4-PYRIDINAMINE. [504-24-5]

*RCRA 40 CFR 261
EPA hazardous waste no. P008

PYRIDINE. [110-86-1]

*OSHA 29 CFR 1910.1000 Table Z-1
PEL(TWA): 5 ppm, 15 mg/m^3

ACGIH-TLV
TWA: 5 ppm, 15 mg/m^3
STEL: 10 ppm, 30 mg/m^3

NFPA—hazardous chemical
health—2, flammability—3, reactivity—0

TSCA4(e), ITC
Second Report of the Interagency Testing Committee to the administrator, Environmental Protection Agency, April 1978; 43 FR 16684, April 19, 1978, responded to by the EPA administrator 44 FR 28095, May 14, 1979; 47 FR 58031, December 29, 1982

*TSCA 8(a), 40 CFR 712, 47 FR 26992, June 22, 1982, Chemical Information Rule

*TSCA 8(d) Unpublished Health and Safety Studies Reporting, 40 CFR 716, 47 FR 387, September 2, 1982

*RCRA 40 CFR 261
EPA hazardous waste no. U196

PYRIDINE, 2-6-DIAMINO-3-(PHENYLAZO)-, MONOHYDROCHLORIDE. Syn: Phenazopyridine hydrochloride. [136-40-3]

OSHA Candidate List of Potential Occupational Carcinogens, 45 FR 53672, August 12, 1980

PYRIDINE, 2-[(2-DIMETHYLAMINO)ETHYL)-2-THENYLAMINO]-. [91-80-5]

*RCRA 40 CFR 261
EPA hazardous waste no. U155

PYRIDINE, HEXAHYDRO-N-NITROSO-.

*RCRA 40 CFR 261
EPA hazardous waste no. U179

PYRIDINE, 2-METHYL-. [109-06-8]

*RCRA 40 CFR 261
EPA hazardous waste no. U191

PYRIDINE, (S)-3-(1-METHYL-2-PYRROLIDINYL)-, and SALTS.

*RCRA 40 CFR 261
EPA hazardous waste no. P075

PYRIMETHANINE.

IARC—carcinogencic in animals

4(1H)-PYRIMIDIONE, 2,3-DIHYDRO-6-METHYL-2-THIOXO-.

*RCRA 40 CFR 261
EPA hazardous waste no. U164

PYROPHOSPHORIC ACID, TETRAETHYL ESTER. [690-49-3]

*RCRA 40 CFR 261
EPA hazardous waste no. P111 (hazardous substance)

PYROXYLIN. See Cellulose nitrate. [9004-70-0]

PYRROLE, TETRAHYDRO-N-NITROSO-.

*RCRA 40 CFR 261
EPA hazardous waste no. U180

Q

QUARTZ. *See* Silica.

QUENCHING OIL.
NFPA—hazardous chemical (flammable chemical)
health—0, flammability—1, reactivity—0

QUICKLIME. *See* Calcium oxide.
[1305-78-8]

QUINOLINE. Syn: 1-azanaphthalene.
[91-22-5]
OSHA Candidate List of Potential Occupational Carcinogens, 45 FR 53672, August 12, 1980
*CWA 311(b)(2)(A), 40 CFR 116, 117
Discharge RQ = 1000 pounds (454 kilograms)

QUINOLINE, 8-NITRO-. [607-35-2]
OSHA Candidate List of Potential Occupational Carcinogens, 45 FR 53672, August 12, 1980

QUINONE. Syn: *p*-BENZOQUINONE.
[106-51-4]
*OSHA 29 CFR 1910.1000 Table Z-1
PEL(TWA): 0.1 ppm, 0.4 mg/m^3
ACGIH-TLV
TWA: 0.1 ppm, 0.4 mg/m^3
STEL: 0.3 ppm, 1 mg/m^3
TSCA 4(e), ITC
Fifth report of the TSCA Interagency Testing Committee to the administrator, Environmental Protection Agency, November 1979; 44 FR 70664, December 7, 1979
*TSCA 8(d) Unpublished Health and Safety Studies Reporting, 40 CFR 716, 47 FR 387, September 2, 1982
*RCRA 40 CFR 261
EPA hazardous waste no. U197

***p*-QUINONE DIOXIME.**
See p-Benzoquinone dioxime.

QUINTOZENE.
IARC—carcinogenic in animals

R

RADIOFREQUENCY (RF) SEALERS and HEATERS.
Potential Health Hazards and Their Prevention
NIOSH Current Intelligence Bulletin No. 33

RANDON DAUGHTERS.
NIOSH Current Intelligence Bulletin No. 10

RANEY NICKEL. See Nickel Catalyst. [7440-02-0]

RDX. See Cyclonite. [121-82-4]

RESERPINE. [50-55-5]
OSHA Candidate List of Potential Occupational Carcinogens (EPA Carcinogen Assessment Group List), 45 FR 53672, August 12, 1980
NCI—carcinogenic in animals
*RCRA 40 CFR 261
EPA hazardous waste no. U200

RESORCINOL. [108-46-3]
ACGIH-TLV
TWA: 10 ppm, 45 mg/m^3
STEL: 20 ppm, 90 mg/m^3
*CWA 311(b)(2)(A), 40 CFR 116, 117
Discharge RQ = 1000 pounds (454 kilograms)
*RCRA 40 CFR 261
EPA hazardous waste no. U201 (hazardous substance)

RETRORSINE. [480-54-6]
IARC—carcinogenic in animals

RHODAMINE B. [81-88-9]
IARC—carcinogenic in animals
TSCA—CHIP available

RHODAMINE 6G. [989-38-8]
IARC carcinogenic in animals

RHODIUM. [7440-16-6]
*OSHA 29 CFR 1910.1000 Table Z-1
PEL(TWA):
Metal fume and dusts 0.1 mg/m^3 (as Rh)
Soluble salts 0.001 mg/m^3
Metal
ACGIH-TLV (1982 Addition)
TWA: 1 mg/m^3
Soluble salts, as Rh
ACGIH-TLV
TWA: 0.001 mg/m^3
STEL: 0.003 mg/m^3
Intended changes for 1982 (1982 Revision or Addition):
Insoluble compounds, as Rh
TWA: 1 mg/m^3
Soluble compounds, as Rh
TWA: 0.01 mg/m^3

RONNEL. [299-84-3]
O,O-dimethyl-O-(2,4,5-trichlorophenyl) phosphorothioate (related to 2,4,5-T)
*OSHA 29 CFR 1910.1000 Table Z-1
PEL(TWA): 15 mg/m^3
ACGIH-TLV
TWA: 10 mg/m^3
40 CFR 162.11, FIFRA Pre-RPAR review
Current Status: Schedule to be developed
Criteria Possibly Met or Exceeded: Oncogenicity, Teratogenicity, and Fetotoxicity

ROSIN CORE SOLDER PYROLYSIS PRODUCTS, as FORMALDEHYDE.

ACGIH-TLV
TWA: 0.1 mg/m^3
STEL: 0.3 mg/m^3

ROTENONE (COMMERCIAL).
[83-79-4]

*OSHA 29 CFR 1910.1000 Table Z-1
PEL(TWA): 5 mg/m^3

ACGIH-TLV
TWA: 5 mg/m^3
STEL: 10 mg/m^3

40 CFR 162.11, FIFRA-RPAR
Notice of Intent to Register Issued
Current Status: Notice of Availability of Decision document published, 46 FR 36745 (July 15, 1981)
Criteria Possibly Met or Exceeded: Oncogenicity, Mutagenicity, Teratogenicity, Reproductive Effects, Chronic Toxicity, Significant Wildlife Populations Reductions, and Acute Toxicity to Aquatic Wildlife

ROUGE. [1309-37-1]

ACGIH-TLV
TWA: nuisance particulate—30 mppcf or 100 mg/m^3 of total dust < 1% quartz, or 5 mg/m^3 respirable dust
STEL: 20 mg/m^3

RUBBER SOLVENT (NAPHTHA).

ACGIH-TLV
TWA: 400 ppm, 1600 mg/m^3

RUBIDIUM CHROMATE.

See Chromium (VI), NIOSH Criteria Document.

RUBIDIUM DICHROMATE.
[13446-73-6]

See Chromium (VI), NIOSH Criteria Document.

S

SACCHARATED IRON OXIDE.
[8047-67-4]

IARC—carcinogenic in animals

SACCHARIN. [81-07-2]

OSHA Candidate List of Potential Occupational Carcinogens (EPA Carcinogen Assessment Group List), 45 FR 53672, August 12, 1980

IARC—carcinogenic in animals

SACCHARIN AND SALTS. [81-07-2]

*RCRA 40 CFR 261
EPA hazardous waste no. U202

SAFROLE. [94-59-7]

OSHA Candidate List of Potential Occupational Carcinogens (EPA Carcinogen Assessment Group List), 45 FR 53672, August 12, 1980

IARC—carcinogenic in animals

40 CFR 162.11, FIFRA—RPAR, Final Action, Voluntary Cancellation
Current Status: 42 FR 11039, February 25, 1977; 42 FR 16844, March 30, 1977; and 42 FR 29957, June 10, 1977
Criteria Possibly Met or Exceeded: Oncogenicity and Mutagenicity

*RCRA 40 CFR 261
EPA hazardous waste no. U203

SALTPETER. See Potassium nitrate.
[7757-79-1]

SELENIOUS ACID. [7783-00-8]

*RCRA 40 CFR 261
EPA hazardous waste no. U204

SELENIUM. [7782-49-2]

CWA 304(a)(1), 45 FR 79318, November 28, 1980, Water Quality Criteria

Freshwater Aquatic Life

For total recoverable inorganic selenite the criterion to protect freshwater aquatic life as derived using the guidelines is 35 μg/liter as a 24-hour average and the concentration should not exceed 260 μg/liter at any time.

The available data for inorganic selenate indicate that acute toxicity to freshwater aquatic life occurs at concentrations as low as 760 μg/liter and would occur at lower concentrations among species that are more sensitive than those tested. No data are available concerning the chronic toxicity of inorganic selenate to sensitive freshwater aquatic life.

Saltwater Aquatic Life

For total recoverable inorganic selenite the criterion to protect saltwater aquatic life as derived using the guidelines is 54 μg/liter as a 24-hour average and the concentration should not exceed 410 μg/liter at any time.

No data are available concerning the toxicity of inorganic selenate to saltwater aquatic life.

Human Health

The ambient water quality criterion for selenium is recommended to be identical to the existing drinking water standard which is 10 μg/liter. Analysis of the toxic effects data resulted in a calculated level that is protective of human health against the ingestion of contaminated water and contaminated aquatic organisms. The calculated value is comparable to the present standard. For this reason a selective criterion based on exposure solely from consumption of 6.5 g of aquatic organisms was not derived.

CWA 307(a)(1), Priority Pollutant

*National Interim Primary Drinking Water Regulations, 40 CFR 141; 40 FR 59565,

December 24, 1975; amended by 41 FR 28402, July 9, 1976; 44 FR 68641, November 29, 1979; corrected by 45 FR 15542, March 11, 1980; 45 FR 57342, August 27, 1980; 47 FR 18998, March 3, 1982; corrected by 47 FR 10998, March 12, 1982
Maximum contaminant level—0.01 mg/liter

SELENIUM COMPOUNDS. [7782-49-2]

*OSHA 29 CFR 1910.1000 Table Z-1
PEL(TWA): 0.2 mg/m^3 (as Se)
ACGIH-TLV
TWA: 0.2 mg/m^3

SELENIUM DIOXIDE. [7446-08-4]

*RCRA 40 CFR 261
EPA hazardous waste no. U204 (hazardous substance)

SELENIUM DISULFIDE. [7488-56-4]

*RCRA 40 CFR 261
EPA hazardous waste no. U205 (reactive waste, toxic waste)

SELENIUM HEXAFLUORIDE. [7783-79-1]

*OSHA 29 CFR 1910.1000 Table Z-1
PEL(TWA): 0.05 ppm, 0.4 mg/m^3
ACGIH-TLV
TWA: 0.05 ppm, 0.2 mg/m^3

SELENIUM OXIDE. [7446-08-4]

*CWA 311(b)(2)(A), 40 CFR 116, 117
Discharge RQ = 1000 pounds (454 kilograms)

SELENIUM SULFIDE (R). [7488-56-4]

OSHA Candidate List of Potential Occupational Carcinogens (EPA Carcinogen Assessment List), 45 FR 53672, August 12, 1980
NCI—carcinogenic in animals

SELENOUREA. [630-10-4]

*RCRA 40 CFR 261
EPA hazardous waste no. P103

SEMICARBAZIDE. Syn: Carbamylhydrazine. [57-56-7]

OSHA Candidate List of Potential Occupational Carcinogens, 45 FR 53672, August 12, 1980
TSCA-CHIP available

SEMICARBAZIDE MONOHYDROCHLORIDE.

Syn: Carbamylhdrazine hydrochloride. [563-41-7]
IARC—carcinogenic in animals
OSHA Candidate List of Potential Occupational Carcinogens, 45 FR 53672, August 12, 1980

SEMICARBAZIDE, 1-PHENYL-. Syn: Cryogenine. [103-03-7]

OSHA Candidate List of Potential Occupational Carcinogens, 45 FR 53672, August 12, 1980

L-SERINE, DIAZOACETATE (ESTER). [115-02-6]

*RCRA 40 CFR 261
EPA hazardous waste no. U015

SESONE. [136-78-7]

ACGIH-TLV
TWA: 10 mg/m^3
STEL: 20 mg/m^3

SILANE. See Silicon tetrahydride. [7803-62-5]

SILANE A-186. See TMOHS. [3388-04-3]

TSCA-CHIP available

SILICA (SiO$_2$) [60676-86-0]. See Silica, quartz, crystalline; Silica, cristobalite, crystalline; Tridymite, silica, crystalline; silica, fused, crystalline; Tripoli, silica, crystalline; and Silica, amorphous, crystalline.

SILICA, AMORPHOUS, CRYSTALLINE. [7631-86-9]

ACGIH-TLV
20 mppcf
Intended changes for 1982:
6 mg/m^3, Total dust (all sampled sizes)
3 mg/m^3, Respirable dust (< 5 μm)

SILICA, CRISTOBALITE, CRYSTALLINE. [14464-46-1]

ACGIH-TLV
Use one-half the value calculated from the count or mass formulae for quartz

SILICA, CRYSTALLINE

NIOSH Criteria Document (Pub. No. 75-120), NTIS Stock No. PB 246697, 1974

Crystalline silica, hereafter referred to in this document as free silica, is defined as silicon dioxide (SiO_2). "Crystalline" refers to the orientation of SiO_2 molecules in a fixed pattern as opposed to a nonperiodic, random molecular arrangement defined as "amorphous." The three most common crystalline forms of free silica encountered in industry are quartz, tridymite, and cristobalite. Micro- and crypto-crystalline varieties of free silica, also included in the recommended standard, are composed of minute grains of free silica cemented together with amorphous silica and include tripoli, flint, chalcedony, agate, onyx, and silica flour. Other forms of free silica which, upon analysis, are found to have a crystalline structure as part of their composition are also subject to the recommended standard.

"Exposure to free silica" means exposure of the worker to an airborne concentration of free silica greater than half of the recommended environmental level in the workplace. Worker exposure at lower environmental concentrations will not require adherence to the standard.

Occupational exposure shall be controlled so that no worker is exposed to a time-weighted average (TWA) concentration of free silica greater than 50 μg per cubic meter of air (50 $\mu g/m^3$; 0.050 mg/m^3) as determined by a full-shift sample for up to a 10-hour workday, 40-hour workweek.

The standard is designed to protect the health and safety of workers for up to a 10-hour workday, 40-hour workweek, over a working lifetime.

SILICA, FUSED, CRYSTALLINE.
[60676-86-0]

ACGIH-TLV
Use quartz formulae

SILICA, QUARTZ, CRYSTALLINE.
[1480-60-7]

ACGIH-TLV
TLV, mppcf: 300/(% quartz + 10)
TLV for respirable dust, mg/m^3: 10 mg/m^3/(% respirable quartz + 2)
TLV for "total dust" respirable and nonrespirable: 30 mg/m^3/(% quartz + 3)
TSCA-CHIP available

SILICA FLOUR. [14808-60-7]

Silicosis
NIOSH Current Intelligence Bulletin No. 36

SILICATES (< 1% QUARTZ). *See* Asbestos, Graphite, Mica, Mineral wool fiber, Perlite, Soapstone, Talc (nonasbestiform or fibrous).

SILICOCHLOROFORM. *See* Trichlorosilane. [10025-78-2]

SILICON. [7440-21-3]

ACGIH-TLV
TWA: nuisance particulate—30 mppcf or 10 mg/m^3 of total dust < 1% quartz, or 5 mg/m^3 respirable dust
STEL: 20 mg/m^3

SILICON CARBIDE. [409-21-2]

ACGIH-TLV
TWA: nuisance particulate—30 mppcf or 10 mg/m^3 of total dust < 1% quartz, or 5 mg/m^3 respirable dust
STEL: 20 mg/m^3

SILICON DIOXIDE. *See* Silica, crystalline, NIOSH Criteria Document. [60676-86-0]

SILICON TETRAHYDRIDE. [7803-62-5]

ACGIH-TLV
TWA: 0.5 ppm, 0.7 mg/m^3
STEL: 1 ppm, 1.5 mg/m^3
Intended changes for 1982:
TWA: 5 ppm, 7 mg/m^3

SILVER. [7440-22-4]

CWA 304(a)(1), 45 FR 79318, November 28, 1980, Water Quality Criteria

Freshwater Aquatic Life

For freshwater aquatic life the concentration (in micrograms per liter) of total recoverable silver should not exceed the numerical value given by $e[1.72(\ln(\text{hardness})-6.52)]$ at any time. For example, at hardnesses of 50, 100, and 200 mg/liter as $CaCO_3$ the concentration of total recoverable silver should not exceed 1.2, 4.1, and 13 µg/liter, respectively, at any time. The available data indicate that chronic toxicity to freshwater aquatic life may occur at concentration as low as 0.12 µg/liter.

Saltwater Aquatic Life

For saltwater aquatic life the concentration of total recoverable silver should not exceed 2.3 µg/liter at any time. No data are available concerning the chronic toxicity of silver to sensitive saltwater aquatic life.

Human Health

The ambient water quality criterion for silver is recommended to be identical to the existing drinking water standard which is 50 µg/liter. Analysis of the toxic effects data resulted in a calculated level that is protective of human health against the ingestion of contaminated water and contaminated aquatic organisms. The calculated value is comparable to the present standard. For this reason a selective criterion based on exposure solely from consumption of 6.5 g aquatic organisms was not derived.

CWA 307(a)(1), Priority Pollutant

*National Interim Primary Drinking Water Regulations, 40 CFR 141; 40 FR 59565, December 24, 1975; amended by 41 FR 28402, July 9, 1976; 44 FR 68641, November 29, 1979; corrected by 45 FR 15542, March 11, 1980; 45 FR 57342, August 27, 1980; 47 FR 18998, March 3, 1982; corrected by 47 FR 10998, March 12, 1982
Maximum contaminant level—0.05 mg/liter

SILVER, METAL and SOLUBLE COMPOUNDS. [7440-22-4]

*OSHA 29 CFR 1910.1000 Table Z-1
PEL(TWA): 0.01 mg/m^3

ACGIH-TLV (metal)
TWA: 0.1 mg/m^3

ACGIH-TLV (soluble compounds)
TWA: 0.01 mg/m^3

SILVER CYANIDE. [506-64-9]

*RCRA 40 CFR 261
EPA hazardous waste no. P104

SILVER NITRATE. [7761-88-8]

NFPA—hazardous chemical
when nonfire: health—0, flammability—0, reactivity—0, oxy—oxidizing chemical
when fire: health—1, flammability—0, reactivity—0, oxy—oxidizing chemical

*CWA 311(b)(2)(A), 40 CFR 116, 117
Discharge RQ = 1 pound (0.454 kilogram)

SILVEX, 2,4,5-TP(2,4,5-TRICHLOROPHENOXYPROPIONIC ACID). [93-72-1]

40 CFR 162.11, FIFRA—RPAR completed
Current Status: PD 4 completed and Final Determination published, 44 FR 72316, December 13, 1979; applies to uses not included in suspension actions described on prior page under "Notice of Intent to Cancel/Suspend Issued."
Criteria Possibly Met or Exceeded: Oncogenicity, Teratogenicity, and Fetotoxicity

40 CFR 162.11, FIFRA—RPAR, Notice of Intent to Cancel/Suspend Issued
Current Status: PD 1 published, 43 FR 17116, April 21, 1978: 2,4,5-T—NTIS# PB80 212665. Emergency Suspension Order and Notice of Intent to Cancel published, 44 FR 15874, March 15, 1979: 2,4,5-T—NTIS# PB80 225923. Silvex—NTIS# PB80 226376.
PD 4 published, 44 FR 72316, December 13, 1979, Cancellation Hearings began March 14, 1980.
Criteria Possibly Met or Exceeded: Oncogenicity, Teratogenicity, and Fetotoxicity

*RCRA 40 CFR 261
EPA hazardous waste no. U233 (hazardous substance)

*National Interim Primary Drinking Water Regulations, 40 CFR 141; 40 FR 59565, December 24, 1975; amended by 41 FR 28402, July 9, 1976; 44 FR 68641, November 29, 1979; corrected by 45 FR 15542, March 11, 1980; 45 FR 57342, August 27, 1980; 47

FR 18998, March 3, 1982; corrected by 47 FR 10998, March 1982
Maximum contaminant level—0.01 mg/liter

SIMAZINE. [122-34-9]
FIFRA—Registration Standard Development—Data Collection (herbicide)
FIFRA—Data Call In, 45 FR 75488, November 14, 1980
Status: letter issued December 2, 1981

SMOKING and the occupational environment, adverse health effects of.
NIOSH Current Intelligence Bulletin No. 31—cancers-related bulletin

SOAPSTONE.
ACGIH-TLV
20 mppcf

SODIUM. [7440-23-5]
NFPA—hazardous chemical
health—3, flammability—1, reactivity—2, W—water may be hazardous in fire fighting
*CWA 311(b)(2)(A), 40 CFR 116, 117
Discharge RQ = 1000 pounds (454 kilograms)

SODIUM ARSENATE. [7631-89-2]
*CWA 311(b)(2)(A), 40 CFR 116, 117
Discharge RQ = 1000 pounds (454 kilograms)

SODIUM ARSENITE. [7784-46-5]
*CWA 311(b)(2)(A), 40 CFR 116, 117
Discharge RQ = 1000 pounds (454 kilograms)

SODIUM ARSENITE (two products).
40 CFR 162.11, FIFRA—RPAR, Final Action, Voluntary Cancellation
Current Status: 43 FR 48267, October 18, 1978
Criteria Possibly Met or Exceeded: Oncogenicity, Mutagenicity, and Teratogenicity

SODIUM AZIDE. [26628-22-8]
ACGIH-TLV
TWA: ceiling limit, 0.1 ppm, 0.3 mg/m^3
TSCA-CHIP available

*RCRA 40 CFR 261
EPA hazardous waste no. P105

SODIUM BICHROMATE. See Dichromates. [10588-01-9]
*CWA 311(b)(2)(A), 40 CFR 116, 117
Discharge RQ = 1000 pounds (454 kilograms)
See Chromium (VI), NIOSH Criteria Document.

SODIUM BIFLUORIDE. [1333-83-1]
*CWA 311(b)(2)(A), 40 CFR 116, 117
Discharge RQ = 5000 pounds (2270 kilograms)

SODIUM BISULFITE. [7631-90-5]
ACGIH-TLV
TWA: 5 mg/m^3
*CWA 311(b)(2)(A), 40 CFR 116, 117
Discharge RQ = 5000 pounds (2270 kilograms)

SODIUM CHLORATE. [7775-09-9]
NFPA—hazardous chemical
when nonfire: health—0, flammability—0, reactivity—2, oxy—oxidizing chemical
when fire: health—1, flammability—0, reactivity—2, oxy—oxidizing chemical

SODIUM CHLORITE. [7758-19-2]
NFPA—hazardous chemical
health—1, flammability—1, reactivity—2, oxy—oxidizing chemical

SODIUM CHROMATE. [7775-11-3]
*CWA 311(b)(2)(A), 40 CFR 116, 117
Discharge RQ = 1000 pounds (454 kilograms)

SODIUM CHROMATE, TETRAHYDRATE.
See Chromium (VI), NIOSH Criteria Document.

SODIUM CYANIDE. [143-33-9]
See Cyanide, hydrogen and cyanide salts, NIOSH Criteria Document.
NFPA—hazardous chemical
health—3, flammability—0, reactivity—0

*CWA 311(b)(2)(A), 40 CFR 116, 117
Discharge RQ = 10 pounds (4.54 kilograms)
*RCRA 40 CFR 261
EPA hazardous waste no. P106 (hazardous substance)

SODIUM DICHLOROISOCYANURATE.
See Sodium dichloro-s-triazinetrione. [2893-78-9]

SODIUM DICHLOROISOCYANURATE DIHYDRATE. See Sodium dichloro-s-triazinetrione dihydrate.

SODIUM 2,4-DICHLOROPHENOXYETHYL SULFATE. See Sesone. [136-78-7]

SODIUM DICHLORO-S-TRIAZINETRIONE. [2893-78-9]
NFPA—hazardous chemical
health—3, flammability—0, reactivity—2, oxy—oxidizing chemical

SODIUM DICHLORO-S-TRIAZINETRIONE DIHYDRATE.
NFPA—hazardous chemical
health—3, flammability—0, reactivity—1, oxy—oxidizing chemical

SODIUM DICHROMATE. See Sodium bichromate. [10588-01-9]

SODIUM DODECYLBENZENESULFONATE. [25155-30-0]
*CWA 311(b)(2)(A), 40 CFR 116, 117
Discharge RQ = 1000 pounds (454 kilograms)

SODIUM FLUORIDE. [7681-49-4]
NFPA—hazardous chemical
health—2, flammability—0, reactivity—0
*CWA 311(b)(2)(A), 40 CFR 116, 117
Discharge RQ = 5000 pounds (2270 kilograms)

SODIUM FLUOROACETATE (1080). [62-74-8]
*OSHA 29 CFR 1910.1000 Table Z-1

Skin
PEL(TWA): 0.05 mg/m^3
ACGIH-TLV

Skin
TWA: 0.05 mg/m^3
STEL: 0.15 mg/m^3
40 CFR 162.11, FIFRA—RPAR issued
Current Status: PD 1 published, 41 FR 52792, December 1, 1976; comment period closed March 15, 1977: NTIS# PB80 216823. PD 2/3 in agency review
Criteria Possibly Met or Exceeded: Reduction in Nontarget and Endangered Species and No Antidote

SODIUM HYDRIDE. [7646-69-7]
NFPA—hazardous chemical
health—3, flammability—3, reactivity—2, W—water may be hazardous in fire fighting

SODIUM HYDROSULFIDE. [16721-80-5]
*CWA 311(b)(2)(A), 40 CFR 116, 117
Discharge RQ = 5000 pounds (2270 kilograms)

SODIUM HYDROSULFITE. [7775-14-6]
NFPA—hazardous chemical
health—3, flammability—1, reactivity—2

SODIUM HYDROXIDE. [1310-73-2]
*OSHA 29 CFR 1910.1000 Table Z-1
PEL(TWA): 2 mg/m^3
ACGIH-TLV
TWA: ceiling limit, 2 mg/m^3
NIOSH Criteria Document (Pub. No. 76-105), NTIS Stock No. PB 246694, September 16, 1975
Synonyms for sodium hydroxide include caustic soda, lye, and white caustic.
"Occupational exposure to sodium hydroxide" is defined as exposure to airborne concentrations of sodium hydroxide exceeding one-half of the recommended workplace environmental limit. Adherence only to Sections 4(a), 4(b), 6(a) (1, 2, 7, 8, and 9), and 7(a) is required when workplace environmental concentrations of sodium hydroxide

are not greater than one-half of the recommended workplace environmental limit.

Occupational exposure to sodium hydroxide shall be controlled so that no worker is exposed to sodium hydroxide at a concentration greater than 2.0 mg/m^3 of air for any 15-minute sampling period.

The standard is designed to protect the health and safety of workers over a working lifetime.

NFPA—hazardous chemical
health—3, flammability—0, reactivity—1
*CWA 311(b)(2)(A), 40 CFR 116, 117
Discharge RQ = 1000 pounds (454 kilograms)

SODIUM HYPOCHLORITE. [7681-52-9, 10022-70-5]

*CWA 311(b)(2)(A), 40 CFR 116, 117
Discharge RQ = 100 pounds (45.4 kilograms)

SODIUM METABISULFITE. [7681-57-4]

ACGIH-TLV
TWA: 5 mg/m^3

SODIUM METHYLATE. [124-41-4]

*CWA 311(b)(2)(A), 40 CFR 116, 117
Discharge RQ = 1000 pounds (454 kilograms)

SODIUM NITRATE. [7631-99-4]

NFPA—hazardous chemical
when nonfire: health—0, flammability—0, reactivity—0, oxy—oxidizing chemical
when fire: health—1, flammability—0, reactivity—0, oxy—oxidizing chemical

SODIUM NITRITE. [7632-00-0]

*CWA 311(b)(2)(A), 40 CFR 116, 117
Discharge RQ = 100 pounds (45.4 kilograms)

SODIUM PERCHLORATE. [7601-89-0]

NFPA—hazardous chemical
health—2, flammability—0, reactivity—2, oxy—oxidizing chemical

SODIUM PEROXIDE. [1313-60-6]

NFPA—hazardous chemical
health—3, flammability—0, reactivity—2,
W—water may be hazardous in fire fighting, oxy—oxidizing chemical, reacts explosively with water or steam to produce toxic and corrosive fumes

SODIUM PHOSPHATE, DIBASIC. [7558-79-4, 10039-32-4, 10140-65-5]

*CWA 311(b)(2)(A), 40 CFR 116, 117
Discharge RQ = 5000 pounds (2270 kilograms)

SODIUM PHOSPHATE, TRIBASIC. [7785-84-4, 7601-54-9, 10101-89-0, 10361-89-4, 7758-29-4, 10124-56-8]

*CWA 311(b)(2)(A), 40 CFR 116, 117
Discharge RQ = 5000 pounds (2270 kilograms)

SODIUM-POTASSIUM ALLOYS. [11135-81-2]

NFPA—hazardous chemical
health—3, flammability—3, reactivity—2,
W—water may be hazardous in fire fighting

SODIUM SELENITE. [10102-18-8, 7782-82-3]

*CWA 311(b)(2)(A), 40 CFR 116, 117
Discharge RQ = 1000 pounds (454 kilograms)

SODIUM SULFIDE. [16721-80-5]

NFPA—hazardous chemical
health—2, flammability—1, reactivity—0

SODIUM SUPEROXIDE. See Sodium peroxide. [1313-60-6]

SOOTS and TARS.

OSHA Candidate List of Potential Occupational Carcinogens (EPA Carcinogen Assessment Group List), 45 FR 53672, August 12, 1980

SOOT, TARS, and OILS.

IARC—carcinogenic in humans

SOY BEAN OIL.

NFPA—hazardous chemical (flammable chemical)
health—0, flammability—1, reactivity—0

STANNIC CHLORIDE. [7646-78-8]
NFPA—hazardous chemical
health—3, flammability—0, reactivity—1

STARCH. [9005-84-9]
ACGIH-TLV
TWA: nuisance particulate—30 mppcf or 10 mg/m^3 of total dust < 1% quartz, or 5 mg/m^3 respirable dust
STEL: 20 mg/m^3

STERIGMATOCYSTIN. [10048-13-2]
IARC—carcinogenic in animals

STIBINE. [7803-52-3]
*OSHA 29 CFR 1910.1000 Table Z-1
PEL(TWA): 0.1 ppm, 0.5 mg/m^3
ACGIH-TLV
TWA: 0.1 ppm, 0.5 mg/m^3
STEL: 0.3 ppm, 1.5 mg/m^3

4,4-STILBENEDIOL, alpha, alpha'-DIETHYL-. [56-53-1]
*RCRA 40 CFR 261
EPA hazardous waste no. U089

4,4-STIBENEDIOL, alpha, alpha'-DIETHYL-, DIPROPIONATE, (E)-.
Syn: Diethylstilbesterol dipropionate. [130-80-3]
OSHA Candidate List of Potential Occupational Carcinogens, 45 FR 53672, August 12, 1980

STODDARD SOLVENT. [8052-41-3]
*OSHA 29 CFR 1910.1000 Table Z-1
PEL(TWA): 500 ppm, 2900 mg/m^3
ACGIH-TLV (1982 Addition)
TWA: 100 ppm, 525 mg/m^3
STEL: 200 ppm, 1050 mg/m^3

STREPTOZATOCIN. [18883-66-4]
OSHA Candidate List of Potential Occupational Carcinogens (EPA Carcinogen Assessment Group List), 45 FR 53672, August 12, 1980
IARC—carcinogenic in animals (used as positive carcinogenic control by NCI)
*RCRA 40 CFR 261
EPA hazardous waste no. U206

STREUNEX. *See* Lindane. [58-89-9]

STROBANE. Syn: Terpene polychlorinate. [8001-50-1]
OSHA Candidate List of Potential Occupational Carcinogens, 45 FR 53672, August 12, 1980
40 CFR 162.11, FIFRA—RPAR, Final Action, Voluntary Cancellation
Current Status: 41 FR 26607, June 28, 1976
Criteria Possibly Met or Exceeded: Oncogenicity

STRONTIUM CHROMATE. [7789-06-2]
IARC—carcinogenic in animals
*CWA 311(b)(2)(A), 40 CFR 116, 117
Discharge RQ = 1000 pounds (454 kilograms)

STRONTIUM DICHROMATE. *See* Dichromates.

STRONTIUM NITRATE. [10042-76-9]
NFPA—hazardous chemical
when nonfire: health—0, flammability—0, reactivity—0, oxy—oxidizing chemical
when fire: health—1, flammability—0, reactivity—0, oxy—oxidizing chemical

STRONTIUM PEROXIDE. [1314-18-7]
NFPA—hazardous chemical
health—1, flammability—0, reactivity—0, oxy—oxidizing chemical

STRONTIUM SULFIDE. [1314-96-1]
*RCRA 40 CFR 261
EPA hazardous waste no. P107

STRYCHNINE. [57-24-9]
*OSHA 29 CFR 1910.1000 Table Z-1
PEL(TWA): 0.15 mg/m^3
ACGIH-TLV
TWA: 0.15 mg/m^3
STEL: 0.45 mg/m^3
*CWA 311(b)(2)(A), 40 CFR 116, 117
Discharge RQ = 10 pounds (4.54 kilograms)

STRYCHNINE and SALTS.
*RCRA 40 CFR 261
EPA hazardous waste no. P108

STRYCHNINE/STRYCHNINE SULFATE. [57-24-9, 60-41-3]

40 CFR 162.11, FIFRA—RPAR issued
Current Status: PD 1 published, 42 FR 2713, January 13, 1977; comment period closed March 15, 1977: NTIS# PB80 216807.
PD 2/3 completed and Notice of Determination published, 45 FR 73602, November 5, 1980; comment period closed September 30, 1981: NTIS# PB81 123960.
PD 4 being drafted.
Criteria Possibly Met or Exceeded: Reduction in Nontarget and Endangered Species

STYRENE, MONOMER. [100-42-5]

*OSHA 1910.1000 Table Z-2
100 ppm, 8-hour TWA
200 ppm, acceptable ceiling concentration
600 ppm, 5 minutes in any 3 hours; acceptable maximum peak
ACGIH-TLV
TWA: 50 ppm, 215 mg/m^3
STEL: 100 ppm, 425 mg/m^3

NCI—may be carcinogenic in animals
NFPA—hazardous chemical
health—2, flammability—3, reactivity—2
*CWA 311(b)(2)(A), 40 CFR 116, 117
Discharge RQ = 1000 pounds (454 kilograms)

STYRENE OXIDE. [96-09-3]

TSCA—CHIP available

SUBTILISINS (proteolytic enzymes as 100% pure crystalline enzyme). [1395-21-7]

ACGIH-TLV
TWA: ceiling limit, 0.00006 mg/m^3, based on "highs volume" sampling

SUCCINIC ACID, MONO(2,2-DIMETHYLHYDRAZIDE). Syn: Daminozide. [1596-84-5]

OSHA Candidate List of Potential Occupational Carcinogens, 45 FR 53672, August 12, 1980

SUCCINIC ANHYDRIDE. [108-30-5]

IARC—carcinogenic animals

SUCCINONITRILE. [110-6102]

See Nitriles, NIOSH Criteria Document.

SUCROSE. [57-50-1]

ACGIH-TLV
TWA: nuisance particulate—30 mppcf or 10 mg/m^3 of total dust < 1% quartz, or 5 mg/m^3 respirable dust
STEL: 20 mg/m^3

SUDAN I. [842-07-9]

IARC—carcinogenic in animals

SUDAN II. [3118-97-6]

IARC—carcinogenic in animals

SULFALLATE. [95-06-7]

NCI—carcinogenic in animals

SULFAMETHOXAZOLE. See Sulfanilamide, N'-(5-methyl-3-isoxazolyl)-. [723-46-6]

SULFANILAMIDE, N'-(5-METHYL-3-ISOXAZOLYL)-. Syn: Sulfamethoxazole. [723-46-6]

OSHA Candidate List of Potential Occupational Carcinogens, 45 FR 53672, August 12, 1982

SULFATE.

*National Secondary Drinking Water Regulations, 40 CFR 143; 44 FR 42198, July 19, 1979, effective January 19, 1981
Maximum contaminant level—250 mg/liter

SULFOTEP. [3689-24-5]

ACGIH-TLV

Skin

TWA: 0.2 mg/m^3
STEL: 0.6 mg/m^3

SULFUR. [7704-34-9]

FIFRA—Registration Standard Development—Data Evaluation and Development of Regulatory Position (fungicide)
NFPA—hazardous chemical
when nonfire: health—1, flammability—1, reactivity—0

when fire: health—2, flammability—1, reactivity—0

SULFUR DIOXIDE. [7446-09-5]

*OSHA 29 CFR 1910.1000 Table Z-1
PEL(TWA): 5 ppm, 13 mg/m^3
ACGIH-TLV
TWA: 2 ppm, 5 mg/m^3
STEL: 5 ppm, 10 mg/m^3
NIOSH Criteria Document (Pub. No. 74-111), NTIS Stock No. PB 228152, February 11, 1974
"Exposure to sulfur dioxide" means exposure to a concentration of sulfur dioxide equal to or above one-half the recommended workroom environmental standard. Exposures at lower environmental concentrations will not require adherence to the recommended standard.
Occupational exposure to sulfur dioxide shall be controlled so that workers shall not be exposed to sulfur dioxide at a concentration greater than 2 parts per million parts of air (5 mg/m^3 of air) determined as a time-weighted average exposure for an 8-hour workday.
The standard is designed to protect the health and safety of workers for an 8-hour day, 40-hour workweek, over a working lifetime.
NFPA—hazardous chemical
health—2, flammability—0, reactivity—0

SULFUR HEXAFLUORIDE. [2551-62-4]

*OSHA 29 CFR 1910.1000 Table Z-1
PEL(TWA): 1000 ppm, 6000 mg/m^3
ACGIH-TLV
TWA: 1000 ppm, 6000 mg/m^3
STEL: 1250 ppm, 7500 mg/m^3
TSCA-CHIP available

SULFUR HYDRIDE. [7783-06-4]

*RCRA 40 CFR 261
EPA hazardous waste no. U135

SULFURIC ACID. [7864-93-9]

*OSHA 29 CFR 1910.1000 Table Z-1
PEL(TWA): 1 mg/m^3
ACGIH-TLV
TWA: 1 mg/m^3
NIOSH Criteria Document (Pub. No. 74-128), NTIS Stock No. PB 233098, June 6, 1974
"Exposure to sulfuric acid" means exposure to a concentration of liquid, mist, or special dry powder of sulfuric acid, or to sulfur trioxide associated with oleum (fuming sulfuric acid) equal to or above one-half the recommended environmental standard. Exposures at lower environmental concentrations will not require adherence to all sections except for work practices, equipment, and clothing which may be necessary to guard against the occurrence of forseeable accidents such as from spray or splash.
Occupational exposure to sulfuric acid mist shall be controlled so that workers shall not be exposed to a concentration greater than 1 mg/m^3 of air determined as a time-weighted average (TWA) exposure for up to a 10-hour workday, 40-hour workweek.
The standard is designed to protect the health and safety of workers for up to a 40-hour workweek over a working lifetime.
NFPA—hazardous chemical
health—3, flammability—0, reactivity—0,
W—water may be hazardous in fire fighting
*CWA 311(b)(2)(A), 40 CFR 116, 117
Discharge RQ = 1000 pounds (454 kilograms)

SULFURIC ACID, DIETHYL ESTER.
Syn: Ethyl sulfate. [64-67-5]

OSHA Candidate List of Potential Occupational Carcinogens, 45 FR 53672, August 12, 1980
*TSCA 8(a), 40 CFR 712, 47 FR 26992, June 22, 1982, Chemical Information Rule

SULFURIC ACID, DIMETHYL ESTER.
[77-78-1]

*TSCA 8(a), 40 CFR 712, 47 FR 26992, June 22, 1982, Chemical Information Rule
*RCRA 40 CFR 261
EPA hazardous waste no. U103

SULFURIC ACID MIST. [7864-93-9]

SULFURIC ACID, THALLIUM (I) SALT.
[7446-18-6]

*RCRA 40 CFR 261
EPA hazardous waste no. P115 (hazardous substance)

SULFURIC OXYCHLORIDE. See
Sulfuryl chloride. [7791-25-5]

SULFUR MONOCHLORIDE.
[10025-67-9, 12771-08-3]

*OSHA 29 CFR 1910.1000 Table Z-1
PEL(TWA): 1 ppm, 6 mg/m^3

ACGIH-TLV
TWA: 1 ppm, 6 mg/m^3
STEL: 3 ppm, 18 mg/m^3

NFPA—hazardous chemical
health—2, flammability—1, reactivity—1

*CWA 311(b)(2)(A), 40 CFR 116, 117
Discharge RQ = 1000 pounds (454 kilograms)

SULFUROUS OXYCHLORIDE. See
Thionyl chloride. [7719-09-7]

SULFUR OXIDES.

*CAA § 109 (b); Part C § 160–178; 40 CFR 50, National Ambient Air Quality Standard (NAAQS) (not to be exceeded more than once per year)

	µg/m^3	
	Primary	Secondary
Annual	80 (0.03 ppm)	—
24 hr	365 (0.14 ppm)	—
3 hr	—	1300 (0.5 ppm)

PSD Air Quality Increments Maximum Allowable Increases, µg/m^3

	Class I	Class II	Class III	vs.	Primary NAAQS
Annual	2	20	40		80
24 hr	5	91	182		365
3 hr	25	512	700		—

SULFUR PENTAFLUORIDE.
[5714-22-7]

*OSHA 29 CFR 1910.1000 Table Z-1
PEL(TWA): 0.025 ppm, 0.25 mg/m^3

ACGIH-TLV
TWA: 0.025 ppm, 0.25 mg/m^3
STEL: 0.075 ppm, 0.75 mg/m^3

SULFUR PHOSPHIDE.

*RCRA 40 CFR 261
EPA hazardous waste no. U189 (reactive waste)

SULFUR SELENIDE.

*RCRA 40 CFR 261
EPA hazardous waste no. U205 (reactive waste, toxic waste)

SULFUR TETRAFLUORIDE.
[7783-60-0]

ACGIH-TLV
TWA: 0.1 ppm, 0.4 mg/m^3
STEL: 0.3 ppm, 1 mg/m^3

SULFURYL CHLORIDE. [7791-25-5]

NFPA—hazardous chemical
health—3, flammability—0, reactivity—2,
W—water may be hazardous in fire fighting

SULFURYL FLUORIDE. [2699-79-8]

*OSHA 29 CFR 1910.1000 Table Z-1
PEL(TWA): 5 ppm, 20 mg/m^3

ACGIH-TLV
TWA: 5 ppm, 20 mg/m^3
STEL: 10 ppm, 40 mg/m^3

SYSTOX. See Dementon®. [8065-48-3]

T

2,4,5-T. [93-76-5]
*OSHA 29 CFR 1910.1000 Table Z-1
PEL(TWA): 10 mg/m³
ACGIH-TLV
TWA: 10 mg/m³
STEL: 20 mg/m³
*CWA 311(b)(2)(A), 40 CFR 116, 117
Discharge RQ = 100 pounds (45.4 kilograms)
*RCRA 40 CFR 261
EPA hazardous waste no. U232 (hazardous substance)

2,4,5-T AMINES. [6369-96-6, 6369-97-7, 1319-72-8, 3813-14-7]
*CWA 311(b)(2)(A), 40 CFR 116, 117
Discharge RQ = 100 pounds (45.4 kilograms)

2,4,5-T ESTERS. [2545-59-7, 93-79-8, 61792-07-2, 1928-47-8, 25168-15-4]
*CWA 311(b)(2)(A), 40 CFR 116, 117
Discharge RQ = 100 pounds (45.4 kilograms)

2,4,5-T SALTS. [13560-99-1]
*CWA 311(b)(2)(A), 40 CFR 116, 117
Discharge RQ = 100 pounds (45.4 kilograms)

TALC (fibrous).
ACGIH-TLV
Use asbestos limit
Intended changes for 1982:
2 fibers/cc, >5 µm in length

TALC (NONASBESTIFORM).
[14807-96-6]
ACGIH-TLV
20 mppcf
Intended changes for 1982:
(containing no fiber)
15 mppcf or 2 mg/m³, respirable dust

TALLOW OIL.
NFPA—hazardous chemical (flammable chemical)
health—0, flammability—1, reactivity—0

TANNIC ACID. [1401-55-4]
IARC—carcinogenic in animals

TANTALUM. [7440-25-7]
*OSHA 29 CFR 1910.1000 Table Z-1
PEL(TWA): 5 mg/m³
ACGIH-TLV
TWA: 5 mg/m³
STEL: 10 mg/m³

TCBC. [1344-32-7]
FIFRA—Registration Standard Development—Data Collection (herbicide)
FIFRA—Data Call In, 45 FR 75488, November 14, 1980
Status: letter issued November 30, 1981

TDE. *See* Ethane, 1,1-dichloro-2,2-bis(*p*-chlorophenyl)-; UDT and metabolites. [72-54-8]

TDI. *See* Toluene diisocyanate (2,4- and 2,6-isomers; TDI). [584-84-9]

TEBUTHIURON. [34014-18-1]
FIFRA—Registration Standard Development—Data Evaluation and Development of Regulatory Position (herbicide)

TEDP. *See* Sulfotep. [3689-24-5]
*OSHA 29 CFR 1910.1000 Table Z-1

Skin
PEL(TWA): 0.2 mg/m³

TEL COMPOUNDS. See Motor fuel antiknock compounds (containing lead). [78-00-2]

TELLURIUM. [13494-80-9]
*OSHA 29 CFR 1910.1000 Table Z-1
PEL(TWA): 0.1 mg/m³

TELLURIUM and COMPOUNDS, as Te. [13494-80-9]
ACGIH-TLV
TWA: 0.1 mg/m³
TSCA-CHIP available

TELLURIUM HEXAFLUORIDE, as Te. [7783-80-4]
*OSHA 29 CFR 1910.1000 Table Z-1
PEL(TWA): 0.02 ppm, 0.2 mg/m³
ACGIH-TLV
TWA: 0.02 ppm, 0.2 mg/m³

TELLURIUM, TETRABIS(DIETHYL-DITHIOCARBAMATO)-. Syn: Ethyl tellurac. [20941-65-5]
OSHA Candidate List of Potential Occupational Carcinogens, 45 FR 53672, August 12, 1980

TEMEPHOS (ABATE®). [3383-96-8]
ACGIH-TLV
TWA: 10 mg/m³
FIFRA—Registration Standard—Issued August 1981 (insecticide)

TEPP. [107-49-3]
*OSHA 29 CFR 1910.1000 Table Z-1

Skin
PEL(TWA): 0.05 mg/m³
ACGIH-TLV

Skin
TWA: 0.004 ppm, 0.05 mg/m³
STEL: 0.01 ppm, 0.2 mg/m³

FIFRA—Data Call In, 45 FR 75488, November 14, 1980
Status: agency decision reached

TERBACIL. [5902-51-2]
FIFRA—Registration Standard Development—Data Evaluation and Development of Regulatory Position (herbicide)

TERBUFOS. [13071-79-9]
FIFRA—Data Call In, 45 FR 75488, November 14, 1980
Status: letter issued April 1, 1981

TEREPHTHALATE, BIS(2-ETHYLHEXYL). [6422-86-2]
*TSCA 8(d) Unpublished Health and Safety Studies Reporting, 40 CFR 716, 47 FR 54624, December 3, 1982

TEREPHTHALIC ACID. [100-21-0]
TSCA-CHIP available

TERPENE POLYCHLORINATE. See Strobane. [8001-50-1]
IARC—carcinogenic in animals

TERPHENYLS. [92-94-4]
*OSHA 29 CFR 1910.1000 Table Z-1
PEL(Ceiling): 1 ppm, 9 mg/m³
ACGIH-TLV
TWA: ceiling limit, 0.5 ppm, 5 mg/m³

TERPHENYLS, CHLORINATED. Syn: PCTs. [61788-33-8]
OSHA Candidate List of Potential Occupational Carcinogens, 45 FR 53672, August 12, 1980

TERRAZOLE®. [2593-15-9]
FIFRA—Registration Standard—Issued September 1980, NTIS# PB81 126716 (fungicide)

TESTOSTERONE. [58-22-0]
IARC—carcinogenic in animals

1,1,2,2-TETRABROMOETHANE. [79-27-6]
OSHA 29 CFR 1910.1000 Table Z-1
PEL(TWA): 1 ppm, 14 mg/m³

ACGIH-TLV
TWA: 1 ppm, 15 mg/m³
STEL: 1.5 ppm, 20 mg/m³

NFPA—hazardous chemical
health—3, flammability—0, reactivity—1

TSCA-CHIP available

1,2,3,4-TETRACHLOROBENZENE.
[634-66-2]

TSCA 4(e), ITC

Third Report of the TSCA Interagency Testing Committee to the administrator, Environmental Protection Agency, October 1978; 43 FR 50630, October 30, 1978, responded to by the EPA administrator 45 FR 48524, July 18, 1980

1,2,3,5-TETRACHLOROBENZENE.
[634-90-2]

TSCA 4(e), ITC

Third Report of the TSCA Interagency Testing Committee to the administrator, Environmental Protection Agency, October 1978; 43 FR 50630, October 30, 1978, responded to by the EPA administrator 45 FR 48524, July 18, 1980

1,2,4,5-TETRACHLOROBENZENE.
[95-94-3]

TSCA 4(e), ITC

Third Report of the TSCA Interagency Testing Committee to the administrator, Environmental Protection Agency, October 1978; 43 FR 50630, October 30, 1978, responded to by the EPA administrator 45 FR 48524, July 18, 1980

*RCRA 40 CFR 261
EPA hazardous waste no. U207

2,3,7,8-TETRACHLORODIBENZO-*p*-DIOXIN (TCDD). [1746-01-6]

OSHA Candidate List of Potential Occupational Carcinogens (EPA Carcinogen Assessment Group List), 45 FR 53672, August 12, 1980

NCI—carcinogenic in animals

CWA 307(a)(1), Priority Pollutant

CAA 112

Chemical proposed to be assessed as a Hazardous Air Pollutant by a House of Representatives bill to amend the CAA

1,1,1,2-TETRACHLORO-2,2-DIFLUOROETHANE. [76-11-9]

*OSHA 29 CFR 1910.1000 Table Z-1
PEL(TWA): 500 ppm, 4170 mg/m³

ACGIH-TLV
TWA: 500 ppm, 4170 mg/m³
STEL: 625 ppm, 5210 mg/m³

1,1,2,2-TETRACHLORO-1,2-DIFLUOROETHANE. [76-12-0]

*OSHA 29 CFR 1910.1000 Table Z-1
PEL(TWA): 500 ppm, 4170 mg/m³

ACGIH-TLV
TWA: 500 ppm, 4170 mg/m³
STEL: 625 ppm, 5210 mg/m³

TETRACHLORODIPHENYLETHANE (TDE). [72-54-8]

NCI—may be carcinogenic in animals

TETRACHLOROETHANE. See 1,1,2,2-tetrachloroethane. [79-34-5]

1,1,1,2-TETRACHLOROETHANE.
[630-20-6]

OSHA Candidate List of Potential Occupational Carcinogens (EPA Carcinogen Assessment Group List), 45 FR 53672, August 12, 1980

NCI—carcinogenic in animals

*RCRA 40 CFR 261
EPA hazardous waste no. U208

1,1,2,2-TETRACHLOROETHANE.
[79-34-5]

*OSHA 29 CFR 1910.1000 Table Z-1

Skin

PEL(TWA): 5 ppm, 35 mg/m³

ACGIH-TLV (1982 addition)

Skin

TWA: 1 ppm, 7 mg/m³
STEL: 5 ppm, 35 mg/m³

NIOSH Criteria Document (Pub. No. 77-121), NTIS Stock No. PB 273802, December 17, 1976

These criteria and the recommended standard apply to exposure of workers to the symmetrical isomer of the chlorinated hydrocarbon compound, CHCl$_2$-CHCl$_2$ referred to as 1,1,2,2-tetrachloroethane. Acetylene tetrachloride and sym-tetrachloroethane are synonyms. "Tetrachloroethane" will be used throughout this document to mean the symmetrical isomer unless otherwise stated. The "action level" is defined as one-half the recommended time-weighted average (TWA) environmental limit. "Occupational exposure to tetrachloroethane," because of systemic effects and dermal irritation produced by contact of tetrachloroethane with the skin, is defined as work in an area where tetrachloroethane is stored, produced, processed, or otherwise used. If an employee is occupationally exposed to airborne concentrations of tetrachloroethane in excess of the action level, then all sections of the recommended standard shall be complied with; if the employee is occupationally exposed at or below the action level, then all sections of the recommended standard shall be complied with except section 8.

When skin exposure is prevented, occupational exposure to tetrachloroethane shall be controlled so that no employee is exposed to tetrachloroethane at a concentration greater than 1.0 part per million parts of air by volume (6.87 mg/m^3 of air) determined as a TWA concentration for up to a 10-hour workday, 40-hour workweek.

The standard is designed to protect the health and safety of employees for up to a 10-hour work shift, 40-hour workweek, over a working lifetime.

NCI—carcinogenic in animals

TSCA-CHIP available

CWA 307(a)(1), Priority Pollutant

*RCRA 40 CFR 261
EPA hazardous waste no. U209

TETRACHLOROETHENE. See Tetrachloroethylene. [127-18-4]

TETRACHLOROETHYLENE. [127-18-4]

*OSHA 1910.1000 Table Z-2
100 ppm, 8-hour TWA
200 ppm, acceptable ceiling concentration
300 ppm, 5 minutes in any 3 hours; acceptable maximum peak

ACGIH-TLV (1982 Revision or Addition)
TWA: 50 ppm, 335 mg/m^3
STEL: 200 ppm, 1340 mg/m^3

NIOSH Criteria Document (Pub. No. 76-185), NTIS Stock No. PB 266583, July 1, 1976

"Occupational exposure to tetrachloroethylene" is defined as exposure at a concentration greater than the action level (TWA) environmental limit. An "action level" is defined as half the time-weighted average (TWA) environmental limit. Occupational exposure to tetrachloroethylene will require adherence to all sections. Exposure at lower environmental concentrations will not require adherence to Sections 3(a), 4(a)(4), 4(b), 4(c), 5, 6, 7(a).

Occupational exposure shall be controlled so that no workers are exposed to tetrachloroethylene in excess of 50 ppm (339 mg/m^3) determined as a time-weighted average (TWA) concentration for up to a 10-hour workday, 40-hour workweek, or at greater than a ceiling concentration of 100 ppm (678 mg/m^3) determined by 15-minute samples, twice daily.

The standard is designed to protect the health and safety of workers for up to a 10-hour workday, 40-hour workweek, over a working lifetime.

NIOSH Current Intelligence Bulletin No. 20—cancer-related bulletin

OSHA Candidate List of Potential Occupational Carcinogens (EPA Carcinogen Assessment Group List), 45 FR 53672, August 12, 1980

NCI—carcinogenic in animals

NFPA—hazardous chemical
health—2, flammability—0, reactivity—0

CWA 304(a)(1), 45 FR 79318, November 28, 1980, Water Quality Criteria

Freshwater Aquatic Life

The available data for tetrachloroethylene indicate that acute and chronic toxicity to freshwater aquatic life occur at concentrations as low as 5280 and 840 µg/liter, respectively, and would occur at lower concentrations among species that are more sensitive than those tested.

Saltwater Aquatic Life

The available data for tetrachloroethylene indicate that acute and chronic toxicity to saltwater aquatic life occur at concentrations low as 10,200 and 450 µg/liter, respectively, and would occur at lower concentrations among species that are more sensitive than those tested.

Human Health

For the maximum protection of human health from the potential carcinogenic effects due to exposure of tetrachloroethylene through ingestion of contaminated water and contaminated aquatic organisms, the ambient water concentration should be zero based on the nonthreshold assumption for this chemical. However, zero level may not be attainable at the present time. Therefore, the levels that may result in incremental increase of cancer risk over the lifetime are estimated at 10^{-5}, 10^{-6}, and 10^{-7}. The corresponding criteria are 8 µg/liter, 0.8 µg/liter, and 0.08 µg/liter, respectively. If the above estimates are made for consumption of aquatic organisms only, excluding consumption of water, the levels are 88.5 µg/liter, 8.85 µg/liter, and 0.88 µg/liter, respectively. Other concentrations representing different risk levels may be calculated by use of the guidelines. The risk estimate range is presented for information purposes and does not represent an agency judgment on an "acceptable" risk level.

CWA 307(a)(1), Priority Pollutant

CAA 112

Chemical proposed to be assessed as a Hazardous Air Pollutant by a House of Representatives bill to amend the CAA

*RCRA 49 CFR 172.101

EPA hazardous waste no. U210

TETRACHLOROMETHANE. See Carbon tetrachloride. [56-23-5]

TETRACHLORONAPHTHALENE. [1335-88-2]

*OSHA 29 CFR 1910.1000 Table Z-1

Skin

PEL(TWA): 2 mg/m^3

ACGIH—TLV

TWA: 2 mg/m^3

STEL: 4 mg/m^3

2,3,4,6-TETRACHLOROPHENOL. [58-90-2]

*RCRA 40 CFR 261

EPA hazardous waste no. U212

m-**TETRACHLOROPHTHALONITRILE.**
See Isophthalonitrile, tetrachloro-.
[1897-45-6]

TETRACHLORVINPHOS. See Phosphoric acid, 2-chloro-1-(2,4,5-trichlorophenyl) vinyl dimethyl ester. [961-11-5]

NCI—carcinogenic in animals

TETRADIFON. [116-29-0]

FIFRA—Data Call In, 45 FR 75488, November 14, 1980

Status: letter issued March 30, 1981

FIFRA—Registration Standard Development—Data Collection (insecticide)

TETRAETHYLDITHIOPYROPHOSPHATE. [3689-24-5]

*RCRA 40 CFR 261

EPA hazardous waste no. P109

TETRAETHYL LEAD (TEL COMPOUND).
See Motor fuel antiknock compounds (containing lead). [78-00-2]

TETRAETHYL LEAD, as Pb. [78-00-2]

*OSHA 29 CFR 1910.1000 Table Z-1

Skin

PEL(TWA): 0.075 mg/m^3 (as Pb)

ACGIH-TLV

Skin

TWA: 0.100 mg/m^3, for control of general room air, biologic monitoring is essential for personnel control
STEL: 0.3 mg/m^3

*CWA 311(b)(2)(A), 40 CFR 116, 117
Discharge RQ = 100 pounds (45.4 kilograms)

*RCRA 40 CFR 261
EPA hazardous waste no. P110 (hazardous substance)

TETRAETHYL PYROPHOSPHATE.
[107-49-3]

*CWA 311(b)(2)(A), 40 CFR 116, 117
Discharge RQ = 100 pounds (45.4 kilograms)

*RCRA 40 CFR 261
EPA hazardous waste no. P111 (hazardous substance)

TETRAFLUOROETHENE. See Fluoroalkenes. [116-14-3]

TSCA 4(e), ITC
Seventh Report of the TSCA Interagency Testing Committee to the administrator, Environmental Protection Agency, October 1980; 45FR 78432, November 25, 1980

*TSCA 8(d) Unpublished Health and Safety Studies Reporting, 40 CFR 716, 47 FR 38800, September 2, 1982; 48 FR 13178, March 30, 1983

NFPA—hazardous chemical
when nonfire: health—2, flammability—4, reactivity—3
when fire: health—3, flammability—4, reactivity—3

TETRAHYDROFURAN. [109-99-9]

*OSHA 29 CFR 1910.1000 Table Z-1
PEL(TWA): 200 ppm, 590 mg/m^3

ACGIH-TLV
TWA: 200 ppm, 590 mg/m^3
STEL: 250 ppm, 735 mg/m^3

NFPA—hazardous chemical
health—2, flammability—3, reactivity—1

TSCA-CHIP available

*RCRA 40 CFR 261
EPA hazardous waste no. U213 (ignitable waste)

TETRAHYDRONAPHTHALENE.
[119-64-2]

NFPA—hazardous chemical (flammable chemical)
health—1, flammability—2, reactivity—0

TETRAHYDRO-p-OXAZINE. See Morpholine. [110-91-8]

4-(1,3,3-TETRAMETHYLBUTYL)PHENOL.
[140-66-9]

TSCA 4(e), ITC
Eleventh Report of the Interagency Testing Committee to the administrator, Environmental Protection Agency, November 1982; 47 FR 54626, December 3, 1982

*TSCA 8(d) Unpublished Health and Safety Studies Reporting, 40 CFR 716, 47 FR 54624, December 3, 1982

TETRAMETHYLENE OXIDE. See Tetrahydrofuran. [109-99-9]

TETRAMETHYL LEAD. See Motor fuel antiknock compounds (containing lead). [75-74-1]

*OSHA 29 CFR 1910.1000 Table Z-1

Skin

PEL(TWA): 0.075 mg/m^3 (as Pb)
ACGIH-TLV

Skin

TWA: 0.150 mg/m^3, for control of general room air, biologic monitoring is essential for personnel control
STEL: 0.5 mg/m^3

TETRAMETHYL SUCCINONITRILE.
[3333-52-6]

*OSHA 29 CFR 1910.1000 Table Z-1

Skin

PEL(TWA): 0.5 ppm, 3 mg/m^3

ACGIH-TLV

TWA: 0.5 ppm, 3 mg/m^3

STEL: 2 ppm, 9 mg/m^3

See Nitriles, NIOSH Criteria Document.

TSCA-CHIP available

TETRANITROMETHANE. [509-14-8]

*OSHA 29 CFR 1910.1000 Table Z-1

PEL(TWA): 1 ppm, 8 mg/m^3

ACGIH-TLV

TWA: 1 ppm, 8 mg/m^3

*RCRA 40 CFR 261

EPA hazardous waste no. P112 (reactive waste)

TETRAPHOSPHORIC ACID, HEXAETHYL ESTER. [757-58-4]

*RCRA 40 CFR 261

EPA hazardous waste no. P062

TETRASODIUM PYROPHOSPHATE. [7722-88-5]

ACGIH-TLV

TWA: 5 mg/m^3

TETRYL(2,4,6-TRINITROPHENYL-METHYLNITRAMINE). [479-45-8]

*OSHA 29 CFR 1910.1000 Table Z-1

Skin

PEL(TWA): 1.5 mg/m^3

ACGIH-TLV

Skin

TWA: 1.5 mg/m^3

STEL: 3.0 mg/m^3

THALLIC OXIDE. [1314-32-5]

*RCRA 40 CFR 261

EPA hazardous waste no. P113

THALLIUM. [7440-28-0]

CWA 304(a)(1), 45 FR 79318, November 28, 1980, Water Quality Criteria

Freshwater Aquatic Life

The available data for thallium indicate that acute and chronic toxicity to freshwater aquatic life occur at concentrations as low as 1400 and 40 µg/liter, respectively, and would occur at lower concentrations among species that are more sensitive than those tested. Toxicity to one species of fish occurs at concentrations as low as 20 µg/liter after 2600 hours of exposure.

Saltwater Aquatic Life

The available data for thallium indicate that acute toxicity to saltwater aquatic life occurs at concentrations as low as 2130 µg/liter and would occur at lower concentrations among species that are more sensitive than those tested. No data are available concerning the chronic toxicity of thallium to sensitive saltwater aquatic life.

Human Health

For the protection of human health from the toxic properties of thallium ingested through water and contaminated aquatic organisms, the ambient water criterion is determined to be 13 µg/liter.

For the protection of human health from the toxic properties of thallium ingested through contaminated aquatic organisms alone, the ambient water criterion is determined to be 48 µg/liter.

CWA 307(a)(1), Priority Pollutant

THALLIUM (SOLUBLE COMPOUNDS). [7440-28-0]

*OSHA 29 CFR 1910.1000 Table Z-1

Skin

PEL(TWA): 0.1 mg/m^3 (as Tl)

ACGIH-TLV

Skin

TWA: 0.1 mg/m^3

THALLIUM (I) ACETATE. [563-68-8]

*RCRA 40 CFR 261

EPA hazardous waste no. U214

THALLIUM (I) CARBONATE.
[6533-73-9]
*RCRA 40 CFR 261
EPA hazardous waste no. U215

THALLIUM (I) CHLORIDE.
[7791-12-0]
*RCRA 40 CFR 261
EPA hazardous waste no. U216

THALLIUM (I) NITRATE.
[10102-45-1]
*RCRA 40 CFR 261
EPA hazardous waste no. U217

THALLIUM (III) OXIDE.
*RCRA 40 CFR 261
EPA hazardous waste no. P113

THALLIUM (I) SELENITE.
[12039-52-0]
*RCRA 40 CFR 261
EPA hazardous waste no. P114

THALLIUM SULFATE.
[10031-59-1, 7446-18-6]
*CWA 311(b)(2)(A), 40 CFR 116, 117
Discharge RQ = 1000 pounds (454 kilograms)
*RCRA 40 CFR 261
EPA hazardous waste no. P115 (hazardous substance)

THF. See Tetrahydrofuran. [109-99-9]

THIAZOLE, 2-AMINO-5-NITRO-.
Syn: Entramin. [121-66-4]
OSHA Candidate List of Potential Occupational Carcinogens, 45 LFR 53672, August 12, 1980

THIOACETAMIDE. [62-55-5]
IARC—carcinogenic in animals
*RCRA 40 CFR 261
EPA hazardous waste no. U218

4,4'-THIOANILINE.
IARC—carcinogenic in animals
NCI—carcinogenic in animals

4,4'-THIOBIS(6-tert-BUTYL-m-CRESOL). [96-69-5]
ACGIH—TLV
TWA: 10 mg/m^3
STEL: 20 mg/m^3

THIOFANOX. [39196-18-4]
*RCRA 40 CFR 261
EPA hazardous waste no. P045

THIOGLYCOLIC ACID. [68-11-1]
ACGIH-TLV
TWA: 1 ppm, 5 mg/m^3

THIOMIDODICARBONIC DIAMIDE.
*RCRA 40 CFR 261
EPA hazardous waste no. P049

THIOLS.
NIOSH Criteria Document (Pub. No. 78-213), NTIS# PB 81 225609, September 26, 1978

These criteria and the recommended standard apply to exposure of employees to selected monofunctional organic sulfhydryl compounds, specifically, the 14 n-alkane thiols having the general molecular formula $C_nH_{2n} + 1SH$ (where $n = 1, 2 \ldots 12, 16,$ and 18), the aliphatic cyclic thiol, cyclohexanethiol, and the aromatic thiol, benzenethiol; hereinafter they may be referred to as "thiols." Synonyms for thiols include mercaptans, thioalcohols, and sulfyhdrates.

Because of systemic effects, absorption through the skin on contact and possible dermal irritation, "occupational exposure to thiols" is defined as work in any area where thiols are produced, processed, stored, or otherwise used. If thiols are handled or stored in intact, sealed containers, for example, during shipment, NIOSH recommends that only Sections 3, 5(a), and 6(g) of this proposed standard apply.

1. Occupational exposure shall be controlled so that no employee is exposed to benzenethiol at concentrations in excess of 0.5 mg/m^3 of air (0.1 ppm in air by volume) determined as a ceiling concentration for any 15-minute period.

2. Occupational exposure to aliphatic thiols shall be controlled so that employees are not exposed at concentrations greater than the limits, in milligrams per cubic meter of air, shown in Table 5 as a ceiling concentration for any 15-minute period.

3. Occupational exposure to mixtures of thiols shall be controlled so that no employee is exposed at an equivalent concentration for the mixture greater than that calculated by the formula given in 29 CFR 1910.1000 (d)(2)(i). The formula is given below.

In case of a mixture of air contaminants an employer shall compute the equivalent exposure as follows:

$$E_m = \frac{C_1}{L_1} + \frac{C_2}{L_2} + \cdots \frac{C_n}{L_n}$$

where:

E_m is the equivalent exposure for the mixture.
C is the concentration of a particular contaminant.
L is the exposure limit in parts per million for that contaminant, from Table Z-1. Values from table Z-1 are important in Table 5.

The value of E_m shall not exceed unity (1).

Only 3 of the 14 thiols in the NIOSH Criteria Document are listed in the Table Z-1 of the OSHA Standards for Toxic and Hazardous substances.

The recommended standard is designed to protect the health and provide for the safety of employees for up to a 10-hour work shift, in a 40-hour workweek, during a working lifetime.

THIOMETHANOL.

*RCRA 40 CFR 261
EPA hazardous waste no. U153 (ignitable waste, toxic waste)

THIONYL CHLORIDE. [7719-09-7]

NFPA—hazardous chemical
health—3, flammability—0, reactivity—2,
W—water may be hazardous in fire fighting

THIOPHANATE METHYL. [23564-05-8]

40 CFR 162.11, FIFRA-RPAR issued
Current Status: PD 1 published, 42 FR 61970, December 7, 1977; comment period closed March 27, 1978: NTIS# PB80 216856. PD 2 completed and Notice of Proposed Determination published, 44 FR 58798, October 11, 1979.
Criteria Possibly Met or Exceeded: Mutagenicity and Reduction in Nontarget Species (Rebutted)

TABLE 5. Recommended Exposure Limits for Aliphatic Thiols.

Thiol	OSHA Exposure Limits, ppm From Table Z-1	Ceiling Concentration Limits	
		mg/m³	Approximate ppm Equivalents
1-Methanethiol	10	1.0	0.5
1-Ethanethiol	10	1.3	0.5
1-Propanethiol	Not listed	1.6	0.5
1-Butanethiol	10	1.8	0.5
1-Pentanethiol	Not listed	2.1	0.5
1-Hexanethiol	Not listed	2.4	0.5
1-Heptanethiol	Not listed	2.7	0.5
1-Octanethiol	Not listed	3.0	0.5
1-Nonanethiol	Not listed	3.3	0.5
1-Decanethiol	Not listed	3.6	0.5
1-Undecanethiol	Not listed	3.9	0.5
1-Dodecanethiol	Not listed	4.1	0.5
1-Hexadecanethiol	Not listed	5.3	0.5
1-Octadecanethiol	Not listed	5.9	0.5
Cyclohexanethiol	Not listed	2.4	0.5

THIOPHENOL. [108-98-5]
*RCRA 40 CFR 261
EPA hazardous waste no. P014

THIOSEMICARBAZIDE. [79-19-6]
*RCRA 40 CFR 261
EPA hazardous waste no. P116

THIOURACIL. [141-90-2]
IARC—carcinogenic in animals

THIOUREA. [62-56-6]
IARC—carcinogenic in animals
TSCA-CHIP available
*RCRA 40 CFR 261
EPA hazardous waste no. U219

THIOUREA, (2-CHLOROPHENYL)-.
*RCRA 40 CFR 261
EPA hazardous waste no. P026

THIOUREA, 1-NAPHTHALENYL-.
*RCRA 40 CFR 261
EPA hazardous waste no. P072

THIOUREA, PHENYL-.
*RCRA 40 CFR 261
EPA hazardous waste no. P093

THIRAM. [137-26-8]
*OSHA 29 CFR 1910.1000 Table Z-1
PEL(TWA): 5 mg/m^3
ACGIH-TLV
TWA: 5 mg/m^3
STEL: 10 mg/m^3
FIFRA—Data Call In, 45 FR 75488, November 14, 1980
Status: letter issued August 7, 1981
*RCRA 40 CFR 261
EPA hazardous waste no. U244

THORIUM DIOXIDE. [1314-20-1]
TSCA-CHIP available

THORIUM NITRATE. [13823-29-5]
NFPA—hazardous chemical
when nonfire: health—0, flammability—0, reactivity—0, oxy—oxidizing chemical
when fire: health—1, flammability—0; reactivity—0, oxy—oxidizing chemical

TIN, METAL. [7440-31-5]
ACGIH-TLV (1982 addition)
TWA: 2 mg/m^3
STEL: 4 mg/m^3

TIN (INORGANIC COMPOUNDS, except oxides).
*OSHA 29 CFR 1910.1000 Table Z-1
PEL(TWA): 2 mg/m^3

TIN (ORGANIC COMPOUNDS),
as Sn. [7440-31-5]
*OSHA 29 CFR 1910.1000 Table Z-1
PEL(TWA): 0.1 mg/m^3
ACGIH-TLV

Skin
TWA: 0.1 mg/m^3
STEL: 0.2 mg/m^3

TIN, OXIDE and INORGANIC COMPOUNDS, except SnO$_4$, as Sn.
[7440-31-5]
ACGIH-TLV (1982 addition)
TWA: 2 mg/m^3
STEL: 4 mg/m^3

TIN TETRACHLORIDE. See Stannic chloride. [7646-78-8]

TITANIUM DIOXIDE, as Ti.
[13463-67-7]
*OSHA 29 CFR 1910.1000 Table Z-1
PEL(TWA): 15 mg/m^3
ACGIH-TLV
TWA: nuisance particulate—30 mppcf or 10 mg/m^3 of total dust < 1% quartz, or 5 mg/m^3 respirable dust.
STEL: 20 mg/m^3

TITANIUM TETRACHLORIDE.
[7550-45-0]
NFPA—hazardous chemical
health—3, flammability—0, reactivity—1

TML COMPOUND. See Motor fuel antiknock compounds (containing lead). [75-74-1]

TMOHS (SILANE A-186). [3388-04-3]
TSCA-CHIP available

TMTU. *See* Urea, 1,1,3,3-tetramethyl-2-thio-. [2782-91-4]

TNB. *See* Trinitrobenzene. [99-35-4]

TNT. *See* Trinitrotoluene. [118-96-7]

o-TOLIDINE. [119-93-7]
ACGIH-TLV
TWA: Industrial substance suspect of carcinogenic potential for humans—which is suspect of inducing cancer, based on either (1) limited epidemiologic evidence, exclusive of clinical reports of single cases, or (2) demonstration of carcinogenesis in one or more animal species by appropriate methods. Worker exposure by all routes should be carefully controlled to levels consistent with the animal and human experience data.

NIOSH Criteria Document (Pub. No. 78-179), NTIS# PB81 227084, August 11, 1978
The term "o-tolidine" refers to various physical forms of the compound and its salts. Synonyms for o-tolidine includes 3,3-dimethylbenzidine, 4,4-diamino-3,3-dimethylbiphenyl, diorthotoluidine, diaminoditolyl, azoic diazo reagent, gold diazo reagent, nitro coupling reagent, direct blue 63, fast blue R base, and benzo fast blue R. "Occupational exposure to o-tolidine" is defined as work in any place in which o-tolidine is produced, stored, used, packaged, or distributed. If o-tolidine is handled or stored only in intact, sealed containers, (e.g., during shipment), adherence to Sections 3, 5(a), 6(g), and 8(a) only is required.

This recommended standard does not apply to users of test tapes or test kits containing o-tolidine. Employees shall avoid skin contact with o-tolidine, since skin absorption can be a significant source of exposure.

Occupational exposure to o-tolidine shall be controlled so that employees are not exposed at a concentration greater than 20 $\mu g/m^3$ of air, determined as a ceiling concentration in a 1-hour sampling period. Skin contact with o-tolidine shall be avoided.

The recommended standard is designed to protect the health and provide for the safety of employees for up to a 10-hour work shift, 40-hour workweek, over a working lifetime.

o-TOLIDINE(3,3'-DIMETHYLBENZIDINE)-BASED DYES. [119-93-7]
TSCA 4(e), ITC
Fifth Report of the TSCA Interagency Testing Committee to the administrator, Environmental Protection Agency, November 1979; 44 FR 70664, December 7, 1979, EPA responded to the committee's recommendations for testing 46 FR 55005, November 5, 1981

p-TOLUAMIDE, N-ISOPROPYL-alpha-(2-METHYLHYDRAZINO)-, MONOHYDROCHLORIDE. Syn: Procarbazine hydrochloride. [366-70-1]
OSHA Candidate List of Potential Occupational Carcinogens, 45 FR 53672, August 12, 1980
NCI—carcinogenic in animals

TOLUENE (TOLUOL). [108-88-3]
*OSHA 1910.1000 Table Z-2
200 ppm, 8-hour TWA
300 ppm, acceptable ceiling concentration
500 ppm/10 minutes, acceptable maximum peak
ACGIH-TLV

Skin
TWA: 100 ppm, 375 mg/m^3
STEL: 150 ppm, 560 mg/m^3
NIOSH Criteria Document (Pub. No. 73-10023), NTIS Stock No. PB 222219, January 7, 1974
"Exposure to toluene" means exposure to a concentration of toluene equal to or above one-half the recommended workroom environmental standard. Exposures at lower environmental concentrations will not require adherence to the recommended standard.

Occupational exposure to toluene shall be controlled so that workers shall not be exposed to toluene at a concentration greater than 100 parts per million parts of air (375 mg/m^3 of air) determined as a time-weighted average (TWA) exposure for an 8-hour workday with a ceiling of 200 parts per million parts of air (750 mg/m^3 of air) as determined by a sampling time of 10 minutes.

The standard is designed to protect the health and safety of workers for an 8-hour workday, 40-hour workweek, over a working lifetime.

NFPA—hazardous chemical
health—2, flammability—3, reactivity—0

TSCA 4(e), ITC
Initial Report to the administrator, Environmental Protection Agency, TSCA Interagency Testing Committee, October 1, 1977; 42 FR 55026, October 12, 1977; responded to by the EPA administrator 43 FR 50134, October 26, 1978; 47 FR 56391, December 16, 1982

*TSCA 8(a), 40 CFR 712, 47 FR 26992, June 22, 1982, Chemical Information Rule

*TSCA 8(d) Unpublished Health and Safety Studies Reporting, 40 CFR 716, 47 FR 387, September 2, 1982

TSCA-CHIP available

CWA 304(a)(1), 45 FR 79318, November 28, 1980, Water Quality Criteria

Freshwater Aquatic Life

The available data for toluene indicate that acute toxicity to freshwater aquatic life occurs at concentrations as low as 17,500 μg/liter and would occur at lower concentrations among species that are more sensitive than those tested. No data are available concerning the chronic toxicity of toluene to sensitive freshwater aquatic life.

Saltwater Aquatic Life

The available data for toluene indicate that acute and chronic toxicity to saltwater aquatic life occur at concentrations as low as 6300 and 5000 μg/liter, respectively, and would occur at lower concentrations among species that are more sensitive than those tested.

Human Health

For the protection of human health from the toxic properties of toluene ingested through water and contaminated aquatic organisms, the ambient water criterion is determined to be 14.3 mg/liter.

For the protection of human health from the toxic properties of toluene ingested through contaminated aquatic organisms alone, the ambient water criterion is determined to be 424 mg/liter.

CWA 307(a)(1), Priority Pollutant
*CWA 311(b)(2)(A), 40 CFR 116, 117
Discharge RQ = 1000 pounds (454 kilograms)

CAA 112
Chemical proposed to be assessed as a Hazardous Air Pollutant by a House of Representatives bill to amend the CAA

*RCRA 40 CFR 261
EPA hazardous waste no. U220

TOLUENE-2,4-DIAMINE. Syn: m-Toluenediamine. [95-80-7]

OSHA Candidate List of Potential Occupational Carcinogens, 45 FR 53672, August 12, 1980

TSCA-CHIP available
*RCRA 40 CFR 261
EPA hazardous waste no. U221

TOLUENE DIISOCYANATE. [584-84-9]
*OSHA 29 CFR 1910.1000 Table Z-1
PEL(Ceiling): 0.02 ppm, 0.14 mg/m^3
ACGIH-TLV
TWA: ceiling limit, 0.02 ppm, 0.14 mg/m^3
Intended changes for 1982:
TWA: 0.005 ppm, 0.04 mg/m^3
NIOSH Criteria Document (Pub. No. 73-11022), NTIS Stock No. PB 222220, July 13, 1973

"Exposure to toluene diisocyanate" includes work in any area where toluene diisocyanate is stored, transported, or used.

Occupational exposure to toluene diisocyanate (TDI) shall be controlled so that no worker shall be exposed to a time-weighted average (TWA) of more than 0.005 ppm (0.036 mg/m^3) for any 8-hour workday, or for any 20-minute period to more than 0.02 ppm (0.14 mg/m^3).

Workers already sensitized to toluene diisocyanate should not be exposed to any amount at all.

The standard is designed to protect the health and safety of workers for an 8-hour workday, 40-hour workweek, over a working lifetime.

See Diisocyanates, NIOSH Criteria Document.

NFPA—hazardous chemical
health—3, flammability—1, reactivity—1

*RCRA 40 CFR 261
EPA hazardous waste no. U223 (reactive waste, toxic waste)

TOLUENE DIISOCYANATE (2,4- and 2,6- isomers; TDI). [584-84-9, 91-08-7]

NCI—carcinogenic in animals

o-**TOLUIDINE.** Syn: 1-Amino-2-methylbenzene. [95-53-4]

*OSHA 29 CFR 1910.1000 Table Z-1

Skin

PEL(TWA): 5 ppm, 22 mg/m^3

OSHA Candidate List of Potential Occupational Carcinogens, 45 FR 53672, August 12, 1980

ACGIH-TLV

Skin

TWA: 2 ppm, 9 mg/m^3
Intended changes for 1982 (1982 Revision or Addition):
TWA: 2 ppm, 9 mg/m^3
Industrial substance suspect of carcinogenic potential for humans—which is suspect of inducing cancer, based on either (1) limited epidemiologic evidence, exclusive of clinical reports of single cases, or (2) demonstration of carcinogenesis in one or more animal species by appropriate methods. Worker exposure by all routes should be carefully controlled to levels consistent with the animal and human experience data.

o-**TOLUIDINE, 5-CHLORO-.** Syn: 1-Amino-3-chloro-6-methylbenzene. [95-79-4]

OSHA Candidate List of Potential Occupational Carcinogens, 45 FR 53672, August 12, 1980

o-**TOLUIDINE, 4-CHLORO-, HYDROCHLORIDE.** *See* C.I. azoicdiazo component 11. [3165-93-3]

o-**TOLUIDINE, alpha,alpha, alpha-TRIFLUORO-2,6-DINITRO-*N,N*-DIPROPYL-.** Syn: Trifluralin. [1582-09-8]

OSHA Candidate List of Potential Occupational Carcinogens, 45 FR 53672, August 12, 1980

o-**TOLUIDINE HYDROCHLORIDE.** [636-21-5]

NCI—carcinogenic in animals

*RCRA 40 CFR 261
EPA hazardous waste no. U222

o-**TOLUIDINE, 4-(*o*-TOLYLAZO)-.** *See* C.I. solvent yellow 3. [97-56-3]

TOLUIDINES (*ortho* and *para*). [95-53-4, 106-49-0]

NFPA—hazardous chemical
health—3, flammability—2, reactivity—0

TOLUOL. *See* Toluene. [108-88-3]

1-(*o*-TOLYLAZO)-beta-NAPHTHOL. *See* C.I. solvent orange 2. [2646-17-5]

2,4-TOLYLENE DIISOCYANATE. *See* Toluene diisocyanate (2,4- and 2,6- isomers; TDI). [584-84-9]

TOTAL DISSOLVED SOLIDS (TDS).

*National Secondary Drinking Water Regulations, 40 CFR 143; 44 FR 42198, July 19, 1979, effective January 19, 1981
Maximum contaminant level—500 mg/liter

TOXAPHENE. *See* Chlorinated camphene. [8001-35-2]

NCI—carcinogenic in animals
40 CFR 162.11, FIFRA-RPAR issued
Current Status: PD 1 published, 42 FR 26860 (5/27/77); comment period closed 9/13/77: NTIS# PB80 216732.
PD 2/3 in agency review.
Criteria Possibly Met or Exceeded: Oncogenicity and Reduction in Nontarget species.

CWA 304 (a)(1), 45 FR 79318, November 28, 1980, Water Quality Criteria

Freshwater Aquatic Life

For toxaphene the criterion to protect freshwater aquatic life as derived using the guidelines is 0.013 µg/liter as a 24-hour average and the concentration should not exceed 1.6 µg/liter at any time.

Saltwater Aquatic Life

For saltwater aquatic life the concentration of toxaphene should not exceed 0.070 µg/liter at any time. No data are available concerning the chronic toxicity of toxaphene to sensitive saltwater aquatic life.

Human Health

For the maximum protection of human health from the potential carcinogenic effects due to exposure of toxaphene through ingestion of contaminated water and contaminated aquatic organisms, the ambient water concentration should be zero based on the nonthreshold assumption for this chemical. However, zero level may not be attainable at the present time. Therefore, the levels that may result in incremental increase of cancer risk over the lifetime are estimated at 10^{-5}, 10^{-6}, and 10^{-7}. The corresponding criteria are 7.1 ng/liter, 0.71 ng/liter, and 0.07 ng/liter, respectively. If the above estimates are made for consumption of aquatic organisms only, excluding the consumption of water, the levels are 7.3 ng/liter, 0.73 ng/liter, and 0.07 ng/liter, respectively. Other concentrations representing different risk levels may be calculated by use of the guidelines. The risk estimate range is presented for information purposes and does not represent an agency judgment on an "acceptable" risk level.

CWA 307(a)(1), Priority Pollutant
*CWA 311(b)(2)(A), 40 CFR 116, 117
Discharge RQ = 1 pound (0.454 kilogram)
*RCRA 40 CFR 261
EPA hazardous waste no. P123 (hazardous substance)

TOXAPHENE ($C_{10}H_{10}Cl_5$—technical chlorinated camphene, 67-69% chlorine). [8001-35-2]

*National Interim Primary Drinking Water Regulations, 40 CFR 141; 40 FR 59565, December 24, 1975; amended by 41 FR 28402, July 9, 1976; 44 FR 68641, November 29, 1979; corrected by 45 FR 15542, March 11, 1980; 45 FR 57342, August 27, 1980; 47 FR 18998, March 3, 1982; corrected by 47 FR 10998, March 12, 1982
Maximum contaminant level—0.005 mg/liter

2,4,5-TP ACID. [93-72-1]

*CWA 311(b)(2)(A), 40 CFR 116, 117
Discharge RQ = 100 pounds (45.4 kilograms)

2,4,5-TP ESTERS. [32534-95-5]

*CWA 311(b)(2)(A), 40 CFR 116, 117
Discharge RQ = 100 pounds (45.4 kilograms)

TRANSFORMER OIL.

NFPA—hazardous chemical (flammable chemical)
health—0, flammability—1, reactivity—0

TREMOLITE. See Asbestos. [60649-53-8]

*TSCA 8(d) Unpublished Health and Safety Studies Reporting, 40 CFR 716, 47 FR 387, September 2, 1982

TRIALLATE. [2303-17-5]

40 CFR 162.11, FIFRA-RPAR, Notice of Intent to Register Issued
Current Status: 45 FR 82349 (12/15/80). The agency determined that, in light of the low level of exposure, it should be returned to the registration process stipulating that a FIFRA 3(c)(2)(B) action be initiated to obtain necessary toxicity data to evaluate its hazards.
Criteria Possibly Met or Exceeded: Oncogenicity and Mutagenicity

TRIAMYLAMINE. See Tripentylamine.

1H-1,2,4-TRIAZOL-3-AMINE. [61-82-5]

*RCRA 40 CFR 261
EPA hazardous waste no. U011

2,4,6-TRIBROMOBENZENAMINE. See Aniline and chloro-, bromo-, and/or nitro-anilines. [147-82-0]

2,4,6-TRIBROMOPHENOL. [118-79-6]
TSCA-CHIP available

TRIBUTYLALUMINIUM. See alkylaluminums.

TRIBUTYLAMINE.
NFPA—hazardous chemical
health—2, flammability—2, reactivity—0

TRIBUTYL PHOSPHATE. [126-73-8]
*OSHA 29 CFR 1910.1000 Table Z-1
PEL(TWA): 5 mg/m^3
ACGIH-TLV
TWA: 0.2 ppm, 2.5 mg/m^3
STEL: 0.4 ppm, 5 mg/m^3

TRIBUTYL PHOSPHOROTRITHIOITE (MERPHOS, DEF). [150-50-5]
40 CFR 162.11, FIFRA-RPAR, Notice of Intent to Register Issued
Current Status: Pre-RPAR evaluation completed. Returned to Registration Division, 11/30/81, based on finding no potential unreasonable adverse effects.
Criteria Possibly Met or Exceeded: Neurotoxicity

TRICHLORFON. [52-68-6]
FIFRA—Data Call In, 45 FR 75488, November 14, 1980
Status: agency decision reached
40 CFR 162.11, FIFRA Pre-RPAR review
Current Status: PRE-RPAR evaluation under way. Review limited to trichlorfon as it degrades to DDVP. No direct review of trichlorfon is being conducted.
Criteria Possibly Met or Exceeded: Oncogenicity, Mutagenicity, Teratogenicity, Fetotoxicity, and Reproductive Effects
*CWA 311(b)(2)(A), 40 CFR 116, 117
Discharge RQ = 1000 pounds (454 kilograms)

TRICHLOROACETALDEHYDE.
[75-87-6; 302-17-0]
TSCA-CHIP available

TRICHLOROACETIC ACID. [76-03-9]
ACGIH-TLV
TWA: 1 ppm, 5 mg/m^3

2,4,6-TRICHLOROANILINE. See Aniline, 2,4,6-trichloro-. [634-93-5]

2,4,6-TRICHLOROBENZENAMINE.
See Aniline, 2,4,6-trichloro-. [634-93-5]

1,2,3-TRICHLOROBENZENE.
[87-61-6]
TSCA 4(e), ITC
Third Report of the TSCA Interagency Testing Committee to the administrator, Environmental Protection Agency, October 1978; 43 FR 50630, October 30, 1978, responded to by the EPA administrator 45 FR 48524, July 18, 1980

1,2,4-TRICHLOROBENZENE.
[120-82-1]
ACGIH-TLV
TWA: 5 ppm, 40 mg/m^3
TSCA 4(e), ITC
Third Report of the TSCA Interagency Testing Committee to the administrator, Environmental Protection Agency, October 1978; 43 FR 50630, October 30, 1978, responded to by the EPA administrator 45 FR 48524, July 18, 1980
CWA 307(a)(1), 40 CFR 125, Priority Pollutant

1,3,5-TRICHLOROBENZENE.
[108-70-3]
TSCA 4(e), ITC
Third Report of the TSCA Interagency Testing Committee to the administrator, Environmental Protection Agency, October 1978; 43 FR 50630, October 30, 1978; responded to by the EPA administrator 45 FR 48524, July 18, 1980

TRICHLOROBUTYLENE OXIDE.
[3083-25-8]
TSCA-CHIP available

1,1,1-TRICHLOROETHANE. [71-55-6]
*OSHA 29 CFR 1910.1000 Table Z-1
PEL(TWA): 350 ppm, 1900 mg/m^3
ACGIH-TLV
TWA: 350 ppm, 1900 mg/m^3
STEL: 450 ppm, 2450 mg/m^3
NIOSH Criteria Document (Pub. No. 76-

184), NTIS Stock No. PB 267069, July 1, 1976

1,1,1-Trichloroethane is also known as methyl chloroform. "Occupational exposure to 1,1,1-trichloroethane" is defined as exposure above 200 ppm measured as a time-weighted average (TWA) for up to a 10-hour workday, 40-hour workweek.

Occupational exposure to 1,1,1-trichloroethane requires adherence to all sections of the recommended standard. Exposure at lower concentrations will not require adherence to Sections 1, 2, 7(b), and 4(a), except 4(a)(5).

Occupational exposure shall be controlled so that workers are not exposed to 1,1,1-trichloroethane at greater than a ceiling concentration of 350 ppm (1910 mg/m^3) as determined by a 15-minute sample.

The standard is designed to protect the health and safety of workers for up to a 10-hour workday, 40-hour workweek, over a working lifetime.

NFPA—hazardous chemical
health—2, flammability—1, reactivity—0

TSCA 4(e), ITC
Second Report of the TSCA Interagency Testing Committee to the administrator, Environmental Protection Agency, April 1978; 43 FR 16684, April 19, 1978, responded to by the EPA administrator 46 FR 30300, June 5, 1981; removed from the priority List in the Ninth Report of the TSCA Interagency Testing Committee to the administrator, Environmental Protection Agency, October 30, 1981, 47 FR 5456, February 5, 1982

*TSCA 8(d) Unpublished Health and Safety Studies Reporting, 40 CFR 716, 47 FR 387, September 2, 1982

CAA 112
Chemical proposed to be assessed as a Hazardous Air Pollutant by a House of Representatives bill to amend the CAA.

*RCRA 40 CFR 261
EPA hazardous waste no. U226

1,1,2-TRICHLOROETHANE. [79-00-5]

*OSHA 29 CFR 1910.1000 Table Z-1

Skin

PEL(TWA): 10 ppm, 45 mg/m^3

ACGIH-TLV

Skin

TWA: 10 ppm, 45 mg/m^3
STEL: 20 ppm, 90 mg/m^3

NCI—carcinogenic in animals

NFPA—hazardous chemical
health—3, flammability—1, reactivity—0

TSCA-CHIP available

CWA 307(a)(1), Priority Pollutant

*RCRA 40 CFR 261
EPA hazardous waste no. U227

beta-TRICHLOROETHANE. See 1,1,2-Trichloroethane. [79-00-5]

TRICHLOROETHENE. See Trichloroethylene. [79-01-6]

TRICHLOROETHYLENE (TCE). [79-01-6]

*OSHA 1910.1000 Table Z-2
100 ppm, 8-hour TWA
200 ppm, acceptable ceiling concentration
300 ppm/5 minutes in any 2 hours, acceptable maximum peak

ACGIH-TLV
TWA: 50 ppm, 270 mg/m^3
STEL: 150 ppm, 805 mg/m^3
Intended changes for 1982 (1982 Revision or Addition):
TWA: 50 ppm, 270 mg/m^3
STEL: 200 ppm, 1080 mg/m^3

NIOSH Criteria Document (Pub. No. 73-11025), NTIS Stock No. PB 222222, July 23, 1973

"Occupational exposure to trichloroethylene" is defined as exposure to half of the time-weighted average limit of trichloroethylene in air or greater.

Occupational exposure to trichloroethylene shall be controlled so that workers will not be exposed to trichloroethylene at a concentration in excess of 100 ppm determined as a time-weighted average (TWA) exposure for an 8-hour workday, as measured by a minimum sampling time of 10 minutes.

No worker shall be exposed to a peak concentration of trichloroethylene in excess of 150 ppm, as measured by a maximum sampling time of 10 minutes.

The standard is designed to protect the health and safety of workers for an 8-hour workday, 40-hour workweek, over a working lifetime.

NIOSH Current Intelligence Bulletin No. 2—cancer-related bulletin

IARC—carcinogenic in animals

NCI—carcinogenic in animals

NFPA—hazardous chemical
health—2, flammability—1, reactivity—0

CWA 304(a)(1), 45 FR 79318, November 28, 1980, Water Quality Criteria

Freshwater Aquatic Life

The available data for trichloroethylene indicate that acute toxicity to freshwater aquatic life occurs at concentrations as low as 45,000 μg/liter and would occur at lower concentrations among species that are more sensitive than those tested. No data are available concerning the chronic toxicity of trichloroethylene to sensitive freshwater aquatic life but adverse behavioral effects occur to one species at concentrations as low as 21,900 μg/liter.

Saltwater Aquatic Life

The available data for trichloroethylene indicate that acute toxicity to saltwater aquatic life occurs at concentrations as low as 2000 μg/liter and would occur at lower concentrations among species that are more sensitive than those tested. No data are available concerning the chronic toxicity of trichloroethylene to sensitive saltwater aquatic life.

Human Health

For the maximum protection of human health from the potential carcinogenic effects due to exposure of trichloroethylene through ingestion of contaminated water and contaminated aquatic organisms, the ambient water concentration should be zero based on the nonthreshold assumption for this chemical. However, zero level may not be attainable at the present time. Therefore, the levels that may result in incremental increase of cancer risk over the lifetime are estimated at 10^{-5}, 10^{-6}, and 10^{-7}. The corresponding criteria are 27 μg/liter, 2.7 μg/liter, and 0.27 μg/liter, respectively. If the above estimates are made for consumption of aquatic organisms only, excluding consumption of water, the levels are 807 μg/liter, 80.7 μg/liter, and 8.07 μg/liter, respectively. Other concentrations representing different risk levels may be calculated by use of the guidelines. The risk estimate range is presented for information purposes and does not represent an agency judgment on an "acceptable" risk level.

CWA 307(a)(1), Priority Pollutant

*CWA 311(b)(2)(A), 40 CFR 116, 117
Discharge RQ = 1000 pounds (454 kilograms)

CAA 112
Chemical proposed to be assessed as a Hazardous Air Pollutant by a House of Representatives bill to amend the CAA

*RCRA 40 CFR 261
EPA hazardous waste no. U228 (hazardous substance)

TRICHLOROETHYLSILANE. [115-21-9]
NFPA—hazardous chemical
health—3, flammability—3, reactivity—0

TRICHLOROFLUOROMETHANE.
[75-69-4]
ACGIH-TLV
TWA: ceiling limit, 1000 ppm, 5600 mg/m^3
CWA 307(a)(1), 40 CFR 125, Priority Pollutant
*OSHA 29 CFR 1910.1000 Table Z-1
PEL(TWA): 1000 ppm, 5600 mg/m^3

TRICHLOROISOCYANURIC ACID. See Trichloro-s-triazinetrione. [87-90-1]

TRICHLOROMETHANE. See Chloroform. [67-66-3]

TRICHLOROMETHANETHIOL.
*RCRA 40 CFR 261
EPA hazardous waste no. P118

TRICHLOROMONOFLUORO-
METHANE. [75-69-4]
*RCRA 40 CFR 261
EPA hazardous waste no. U121

TRICHLOROMONOSILANE. See
Trichlorosilane. [10025-78-2]

TRICHLORONAPHTHALENE.
[1321-65-9]
*OSHA 29 CFR 1910.1000 Table Z-1

Skin
PEL(TWA): 5 mg/m^3
ACGIH-TLV
TWA: 5 mg/m^3
STEL: 10 mg/m^3

TRICHLORONITROMEHANE. See
Chloropicrin. [76-06-2]

2,4,5-TRICHLOROPHENOL. [95-95-4]
40 CFR 162.11, FIFRA-RPAR issued Current Status: PD 1 published, 42 FR 41268 (9/15/78); comment period closed 11/17/78. Possibility of TCDD contamination under investigation. 3(c)(2)(B) letters issued requesting additional data from registrants. Some voluntary cancellations and reformations being processed.
Criteria Possibly Met or Exceeded: Oncogenicity and Fetotoxicity
*RCRA 40 CFR 261
EPA hazardous waste no. U230 (hazardous substance)

2,4,5-TRICHLOROPHENOL
IN SOLVENT.
NFPA—hazardous chemical (flammable chemical)
health—1, flammability—1, reactivity—0

2,4,6-TRICHLOROPHENOL.
[25167-82-2]
OSHA Candidate List of Potential Occupational Carcinogens (EPA Carcinogen Assessment Group List), 45 FR 53672, August 12, 1980
NCI—carcinogenic in animals
CWA 307(a)(1), 40 CFR 125, Priority Pollutant
*CWA 311(b)(2)(A), 40 CFR 116, 117
Discharge RQ = 10 pounds (4.54 kilograms)
*RCRA 40 CFR 261
EPA hazardous waste no. U231 (hazardous substance)

2,4,6-TRICHLOROPHENOL
IN SOLVENT.
NFPA—hazardous chemical (flammable chemical)
health—1, flammability—1, reactivity—0

2,4,5-TRICHLOROPHENOXYACETIC
ACID. See 2,4,5-T. [93-76-5]

1,2,3-TRICHLOROPROPANE.
[96-18-4]
*OSHA 29 CFR 1910.1000 Table Z-1
PEL(TWA): 50 ppm, 300 mg/m^3
ACGIH-TLV
TWA: 50 ppm, 300 mg/m^3
STEL: 75 ppm, 450 mg/m^3

TRICHLOROSILANE. [10025-78-2]
NFPA—hazardous chemical
health—3, flammability—4, reactivity—2,
W—water may be hazardous in fire fighting

TRICHLORO-s-TRIAZINETRIONE.
[87-90-1]
NFPA—hazardous chemical
health—3, flammability—0, reactivity—2,
oxy—oxidizing chemical

1,1,2-TRICHLORO-1,2,2-
TRIFLUOROETHANE. [76-13-1]
*OSHA 29 CFR 1910.1000 Table Z-1
PEL(TWA): 1000 ppm, 7600 mg/m^3
ACGIH-TLV
TWA: 1000 ppm, 7600 mg/m^3
STEL: 1250 ppm, 9500 mg/m^3

TRICYCLOHEXYLTIN HYDROXIDE.
See Cyhexatin. [13121-70-5]

TRIDYMITE, SILICA, CRYSTALLINE.
[15468-32-3]

ACGIH-TLV
Use one-half the value calculated from formulae for quartz.

TRIETHANOLAMINE. [102-71-6]

NFPA—hazardous chemical
when nonfire: health—1, flammability—1, reactivity—1
when fire: health—2, flammability—1, reactivity—1

TSCA-CHIP available

TRIETHANOLAMINE DODECYLBENZENESULFONATE. [2732-34-1]

*CWA 311(b)(2)(A), 40 CFR 116, 117
Discharge RQ = 1000 pounds (454 kilograms)

TRIETHYLALUMINIUM. See Alkylaluminums.

TRIETHYLAMINE. [121-44-8]

*OSHA 29 CFR 1910.1000 Table Z-1
PEL(TWA): 25 ppm, 100 mg/m^3

ACGIH-TLV
TWA: 25 ppm, 100 mg/m^3
STEL: 40 ppm, 160 mg/m^3
Intended changes for 1982:
TWA: 10 ppm, 40 mg/m^3
STEL: 15 ppm, 60 mg/m^3

NFPA—hazardous chemical
health—2, flammability—3, reactivity—0

*CWA 311(b)(2)(A), 40 CFR 116, 117
Discharge RQ = 5000 pounds (2270 kilograms)

TRIETHYLENE GLYCOL. [112-27-6]

NFPA-hazardous chemical (flammable chemical)
health—1, flammability—1, reactivity—0

TRIETHYLENE GLYCOL DIGLYCIDYL ETHER. [1954-28-5]

IARC—carcinogenic in animals

TRIETHYLENE TETRAMINE.
[112-24-3]

TSCA-CHIP available

TRIFLUOROETHENE. See
Fluoroalkenes. [359-11-5]

TSCA 4(e), ITC
Seventh Report of the TSCA Interagency Testing Committee to the administrator, Environmental Protection Agency, October 1980; 45 FR 78432, November 25, 1980

*TSCA 8(d) Unpublished Health and Safety Studies Reporting, 40 CFR 716, 47 FR 38800, September 2, 1982; 48 FR 13178, March 30, 1983

TRIFLUOROMETHYLBENZENE. See
Benzotrifluoride. [677-21-4]

TRIFLUOROMETHYLETHENE. See
Fluoroalkenes. [677-21-4]

TSCA 4(e), ITC
Seventh Report of the TSCA Interagency Testing Committee to the administrator, Environmental Protection Agency, October 1980; 45 FR 78432, November 25, 1980

*TSCA 8(d) Unpublished Health and Safety Studies Reporting, 40 CFR 716, 47 FR 38800, September 2, 1982; 48 FR 13178, March 30, 1983

TRIFLUOROMONOBROMOMETHANE. [75-63-8]

*OSHA 29 CFR 1910.1000 Table Z-1
PEL(TWA): 1000 ppm, 6100 mg/m^3

ACGIH-TLV
TWA: 1000 ppm, 6100 mg/m^3
STEL: 1200 ppm, 7300 mg/m^3

TRIFLURALIN (TREFLAN). See p-Toluidine, alpha,alpha,alpha-trifluoro-2,6-dinitro-N,N-Dipropyl-. [1582-09-8]

NCI—carcinogenic in animals

40 CFR 162.11, FIFRA—RPAR issued
Current Status: PD 1/2/3 published, 44 FR 50911 (8/30/79); comment period closed 11/13/79: NTIS# PB80 213937.
PD 4 in agency review.
Criteria Possible Met or Exceeded: Oncogenicity and Mutagenicity

TRIHALOMETHANES.

*National Interim Primary Drinking Water Regulations, 40 CFR 141; 40 FR 59565,

December 24, 1975; amended by 41 FR 28402, July 9, 1976; 44 FR 68641, November 29, 1979; corrected by 45 FR 15542, March 11, 1980; 45 FR 57342, August 27, 1980; 47 FR 18998, March 3, 1982; corrected by 47 FR 10998, March 12, 1982

The sum of the concentrations of bromodichloromethane, dibromochloromethane, tribromomethane (bromoform), and trichloromethane (chloroform)—0.10 mg/liter

TRIISOBUTYLALUMINIUM. See Alkylaluminums.

TRIMELLITIC ANHYDRIDE (TMA).
[552-30-7]
ACGIH-TLV
TWA: 0.005 ppm, 0.04 mg/m^3
NIOSH Current Intelligence Bulletin No. 21
TSCA-CHIP available

TRIMETHYLAMINE. See Methylamines.
[75-50-3]
*CWA 311(b)(2)(A), 40 CFR 116, 117
Discharge RQ = 1000 pounds (454 kilograms)

2,4,5-TRIMETHYLANILINE. [137-17-7]
NCI—carcinogenic in animals

TRIMETHYLBENZENE. [25551-13-7]
ACGIH-TLV
TWA: 25 ppm, 125 mg/m^3
STEL: 35 ppm, 170 mg/m^3
TSCA 4(e), ITC
Tenth Report of the TSCA Interagency Testing Committee to the administrator, Environmental Protection Agency; 47 FR 22585, May 25, 1982
TSCA 8(a), 40 CFR 712, 48 FR 22697, May 19, 1983, Chemical Information Rule
TSCA 8(d) Unpublished Health and Safety Studies Reporting, 40 CFR 716, 48 FR 28483, June 22, 1983

1,2,3-TRIMETHYLBENZENE.
[526-73-8]
TSCA 4(e), ITC
Tenth Report of the TSCA Interagency Testing Committee to the administrator, Environmental Protection Agency; 47 FR 22585, May 25, 1982
TSCA 8(a), 40 CFR 712, 48 FR 22697, May 19, 1983, Chemical Information Rule
TSCA 8(d) Unpublished Health and Safety Studies Reporting, 40 CFR 716, 489 FR 28483, June 22, 1983

1,2,4-TRIMETHYLBENZENE.
[95-63-6]
TSCA 4(e), ITC
Tenth Report of the Interagency Testing Committee to the administrator, Environmental Protection Agency, May 1982; 47 FR 22585, May 25, 1982
*TSCA 8(d) Unpublished Health and Safety Studies Reporting, 40 CFR 716, 47 FR 38800, September 2, 1982; 48 FR 13178, March 30, 1983

1,3,5-TRIMETHYLBENZENE.
[106-67-8]
TSCA 4(e), ITC
Tenth Report of the TSCA Interagency Testing Committee to the administrator, Environmental Protection Agency; 47 FR 22585, May 25, 1982
TSCA 8(a), 40 CFR 712, 48 FR 22697, May 19, 1983, Chemical Information Rule
TSCA 8(d) Unpublished Health and Safety Studies Reporting, 40 CFR 716, 48 FR 28483, June 22, 1983

TRIMETHYL PHOSPHATE. [512-56-1]
NCI—carcinogenic in animals

TRIMETHYL PHOSPHITE. [121-45-9]
ACGIH-TLV (1982 Addition)
TWA: 2 ppm, 10 mg/m^3
STEL: 5 ppm, 25 mg/m^3
TSCA-CHIP available
Has received AD disposition

1,1,3-TRIMETHYL-2-THIOUREA. See Urea, 1,1,3-trimethyl-2-thio-. [2489-77-2]

N,N,N'-TRIMETHYLTHIOUREA. See Urea, 1,1,3-trimethyl-2-thio-. [2489-77-2]

TRINITROBENZENE. [99-35-4]
NFPA—hazardous chemical
health—2, flammability—4, reactivity—4
*RCRA 40 CFR 261
EPA hazardous waste no. U234 (reactive waste, toxic waste)

2,4,7-TRINITROFLUORENONE. See Fluoren-9-one, 2,4,7-trinitro-. [129-79-3]

2,4,7-TRINITRO-9h-FLUOREN-9-ONE. See Fluoren-9-one, 2,4,7-trinitro-. [129-79-3]

2,4,6-TRINITROPHENOL. See Picric acid. [88-89-1]

2,4,6-TRINITROPHENYLMETHYLNITRAMINE. See Tetryl. [479-45-8]

1,2,3-TRINITROPROPANETRIOL. [5437-63-8]
See Nitroglycerin and ethylene glycol dinitrate, NIOSH Criteria Document.

TRINITROTOLUENE (TNT). [118-96-7]
*OSHA 29 CFR 1910.1000 Table Z-1

Skin
PEL(TWA): 1.5 mg/m^3
ACGIH-TLV (1982 Addition)

Skin
TWA: 0.5 mg/m^3
STEL: 3 mg/m^3
NFPA—hazardous chemical
health—2, flammability—4, reactivity—4
TSCA-CHIP available

TRIOCTYLAMINE. [1116-76-3]
TSCA-CHIP available

TRIORTHOCRESYL PHOSPHATE. [78-30-8]
*OSHA 29 CFR 1910.1000 Table Z-1
PEL(TWA): 0.1 mg/m^3
ACGIH-TLV
TWA: 0.1 mg/m^3
STEL: 0.3 mg/m^3

1,3,5-TRIOXANE, 2,4,6-TRIMETHYL-. [123-63-7]
*RCRA 40 CFR 261
EPA hazardous waste no. U182

TRIPENTYLAMINE.
NFPA—hazardous chemical
health—2, flammability—1, reactivity—0

TRIPHENYL AMINE. [603-34-9]
ACGIH-TLV
TWA: 5 mg/m^3

TRIPHENYL PHOSPHATE. [115-86-6]
*OSHA 29 CFR 1910.1000 Table Z-1
*PEL(TWA): 3 mg/m^3
ACGIH-TLV
TWA: 3 mg/m^3
STEL: 6 mg/m^3

TRIPOLI, SILICA, CRYSTALLINE. [1317-95-9]
ACGIH—TLV
Use respirable mass quartz formula

TRIPROPYLALUMINIUM. See Alkylaluminums.

TRIS(AZIRIDINYL)-p-BENZOQUINONE. [68-76-8]
IARC—carcinogenic in animals

TRIS(1-AZIRIDINYL) PHOSPHINE SULFIDE. [52-24-4]
OSHA Candidate List of Potential Occupational Carcinogens (EPA Carcinogen Assessment Group List), 45 FR 53672, August 12, 1980
IARC—carcinogenic in animals
NCI—carcinogenic in animals

2,4,6-TRIS(1-AZIRIDINYL)-s-TRIAZINE. [51-18-3]
IARC—carcinogenic in animals

TRIS(2-CHLOROETHYL) PHOSPHITE. [140-08-9]
TSCA 4(e), ITC
Ninth Report of the TSCA Interagency Test-

ing Committee to the administrator, Environmental Protection Agency, October 1981; 47 FR 5456, February 5, 1982

TSCA 8(d) Unpublished Health and Safety Studies Reporting, 40 CFR 716, 47 FR 38800, September 2, 1982; 48 FR 13178, March 30, 1983

1,2,3-TRIS(CHLOROMETHOXY)PROPANE.
[38571-73-2]

IARC—carcinogenic in animals

TRIS(2,3-DIBROMOPROPYL) PHOSPHATE. [126-72-7]

OSHA Candidate List of Potential Occupational Carcinogens (EPA Carcinogen Assessment Group List), 45 FR 53672, August 12, 1980

NCI—carcinogenic in animals

*RCRA 40 CFR 261

EPA hazardous waste no. U235

TRIS(2-ETHYLHEXYL) TRIMELLITATE.
[3319-31-1]
54624, December 3, 1982

TRYPAN BLUE. [72-57-1]

OSHA Candidate List of Potential Occupational Carcinogens (EPA Carcinogen Assessment Group List), 45 FR 53672, August 12, 1980

IARC—carcinogenic in animals

*RCRA 40 CFR 261

EPA hazardous waste no. U236

TRYSBEN. [50-31-7]

40 CFR 162.11, FIFRA-RPAR, Final Action, Voluntary Cancellation
Current Status: 43 FR 5782 (2/9/78)
Criteria Possibly Met or Exceeded: Oncogenicity

TUNGSTEN AND CEMENTED TUNGSTEN CARBIDE. [7440-33-7]

NIOSH Criteria Document (Pub. No. 77-227), NTIS Stock No. PB 275594, September 26, 1977

The term "cemented tungsten carbide" or "hard metal" refers to a mixture of tungsten carbide (WC), cobalt, and sometimes other metals and metal oxides or carbides. The tungsten carbide content of hard metal is generally 80% or more, and the cobalt content is generally less than 10%, but it may be as high as 25%. When the cobalt content exceeds 2%, its contribution to the potential health hazard is judged to exceed that of tungsten carbide and all other components, and the recommended standard for such mixtures is based on the current U.S. federal standard for occupational exposure to cobalt, 0.1 mg/m^3. If a future NIOSH recommendation for an occupational exposure limit for cobalt differs from the U.S. federal standard for cobalt, this new recommendation should be considered to replace the current recommendation for an occupational exposure limit for dusts of cemented tungsten carbide containing more than 2% cobalt. If nickel is used as a binder rather than cobalt and nickel content of the mixture exceeds 0.3%, then the NIOSH recommended occupational exposure limit for nickel of 15 μg/m^3 shall apply. The recommended limits for airborne tungsten are different for "insoluble tungsten," "soluble tungsten," and "cemented tungsten carbide" because of their differing potential biologic effects and inherent toxicities. These criteria and the recommended standards shall apply to places of employment involved in the manufacture, use, storage, or handling of any of the defined materials.

The "action level" is defined as one-half the appropriate recommended time-weighted average (TWA) environmental limit. "Occupational exposure to cemented tungsten carbide, elemental tungsten, or its compounds" is defined as exposure at a concentration greater than the appropriate action level, that is, except for the case of cemented tungsten carbide products in which nickel is the binder, when the NIOSH definition of occupational exposure to nickel shall apply. Exposures to airborne tungsten concentrations equal to or less than one-half of the workplace environmental limits, as determined in accordance with Section 8, will require adherence to Sections 3(a), 4(b), 4(c), 5, 7, and 8. The term "insoluble tungsten," used throughout this document, refers to elemental tungsten and all insoluble tungsten compounds.

For the purpose of compliance with the recommended standards, insoluble tungsten compounds include all those for which water solubility is listed as "insoluble" or less than 0.01 g/100 cc. Soluble tungsten compounds are those listed as "very soluble," "soluble," "slightly soluble," equal to or greater than 0.01 g/100 cc or "decomposes." Those compounds for which no solubility information is listed should be considered soluble unless it can be demonstrated that they are insoluble in water. As a rough guide for control of airborne concentrations of tungsten compounds, the standard for insoluble tungsten should apply whenever processes within a building or independent structure involve only those stages which produce insoluble tungsten compounds from other insoluble tungsten compounds, for example, starting with ammonium-p-tungstate through tungsten carbide. Whenever soluble tungsten compounds are starting, intermediate, or final products, the standard for soluble compounds should apply throughout the building, unless it can be demonstrated through differential sampling and analysis that both limits are being met as appropriate.

1. Occupational exposure to insoluble tungsten shall be controlled so that employees are not exposed to insoluble tungsten at a concentration greater than 5 mg, measured as tungsten, per cubic meter of air, determined as a TWA concentration for up to a 10-hour work shift in a 40-hour workweek.

2. Occupational exposure to soluble tungsten shall be controlled so that employees are not exposed to soluble tungsten at a concentration greater than 1 mg, measured as tungsten, per cubic meter of air, determined as a TWA concentration for up to a 10-hour work shift in a 40-hour workweek.

3. Occupational exposure to dust of cemented tungsten carbide that contains more than 2% cobalt shall be controlled so that employees are not exposed to a concentration greater than 0.1 mg, measured as cobalt, per cubic meter of air, determined as a TWA concentration for up to a 10-hour work shift in a 40-hour workweek.

4. Occupational exposure to dust of cemented tungsten carbide that contains more than 0.3% nickel shall be controlled so that employees are not exposed at a concentration greater than 15 μg of nickel per cubic meter of air determined as a TWA concentration, for up to a 10-hour work shift in a 40-hour workweek as specified in NIOSH's Criteria for a Recommended Standard for Occupational Exposure to Inorganic Nickel.

The standard is designed to protect the health and provide for the safety of employees for up to a 10-hour work shift, 40-hour workweek, over a working lifetime.

TUNGSTEN, as W, INSOLUBLE COMPOUNDS. [7440-33-7]

ACGIH-TLV
TWA: 5 mg/m^3
STEL: 10 mg/m^3

TUNGSTEN, as W, SOLUBLE COMPOUNDS. [7440-33-7]

ACGIH-TLV
TWA: 1 mg/m^3
STEL: 3 mg/m^3

TURPENTINE. [8006-64-2]

*OSHA 29 CFR 1910.1000 Table Z-1
PEL(TWA): 100 ppm, 560 mg/m^3

ACGIH-TLV
TWA: 100 ppm, 560 mg/m^3
STEL' 150 ppm, 840 mg/m^3

NFPA—hazardous chemical (flammable chemical)
health—1, flammability—3, reactivity—0

U

UDMH. *See* Unsymmetrical dimethylhydrazine. [57-14-7]

ULTRAVIOLET RADIATION.

NIOSH Criteria Document (Pub. No. 73-11009), NTIS Stock No. PB 21468, December 20, 1972

Ultraviolet radiation (ultraviolet energy) is defined as that portion of the electromagnetic spectrum described by wavelengths from 200 to 400 nm. Adherence to the recommended standards will, it is believed, prevent occupational injury from ultraviolet radiation, that is, will prevent adverse acute and chronic cutaneous and ocular changes precipitated or aggravated by occupational exposure to ultraviolet radiation.

Sufficient technology exists to prevent adverse effects on workers, but technology to measure ultraviolet energy for compliance with the recommended standard is not now adequate, so work practices are recommended for control of exposure in cases where sufficient measurement or emission data are not available.

These criteria and the recommended standard will be reviewed and revised when relevant information warrants.

Section 1—Exposure Standards

a. For the ultraviolet spectral region of 315 to 400 nm, total irradiance incident on unprotected skin or eyes, based on either measurement data or on output data, shall not exceed 1.0 mW/cm² for periods greater than 1000 seconds, and for exposure times of 1000 seconds or less the total radiant energy shall not exceed 1000 mW · sec/cm² (1.0 J/cm²).

b. For the ultraviolet spectral region of 200 to 315 nm, total irradiance incident on unprotected skin or eyes, based on either measurement data or on output data, shall not exceed levels described below.

1. If the ultraviolet energy is from a narrow-band or monochromatic source, permissible dose levels for a daily 8-hour period can be read directly from Figure 1, or, for selected wavelengths, from Table 6.

2. If the ultraviolet energy is from a broad-band source, the effective erradiance (I_{eff}) relative to a 270-nm monochromatic source shall be calculated from the formula below. From I_{eff}, the permissible exposure time in seconds for unprotected skin or eyes shall be computed by dividing 0.003 J/cm², the permissible dose of 270-nm radiation, by I_{eff} in watts per square centimeter.

$$I_{eff} = \Sigma \, I_\lambda S_\lambda \Delta_\lambda$$

where I_{eff} = effective irradiance relative to a monochromatic source at 270 nm
I_λ = spectral irradiance (W/cm²/nm)
S_λ = relative spectral effectiveness (unitless); see Table 6 for values of S_λ at different wavelengths
Δ_λ = band width (nm)

Table 7 lists permissible exposure times corresponding to selected values of I_{eff} in microwatts per square centimeter.

If radiation intensity from a point source is known at some distance from the worker, for example, from measurement at another point or from output data at a known distance from the ultraviolet source, attenuation of radiation from that point to the worker can be calculated from the principle that radiation decreases with the square of the distance it must travel. For example, an object 3 feet away from a radiation source receives 1/9 the energy of an object 1 foot

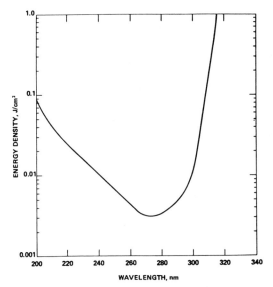

Figure 1. Recommended ultraviolet radiation exposure standard. Adapted from a figure developed and published by the American Conference of Governmental Industrial Hygienists in "Threshold Limit Values for Chemical Substances and Physical Agents in the Workroom Environment with Intended Changes for 1972."

Table 6. Total Permissible 8-Hour Doses and Relative Spectral Effectiveness of Some Selected Monochromatic Wavelengths.[a]

Wavelength, nm	Permissible 8-hour Dose, mJ/cm^2	Relative Spectral Effectiveness, S_λ
200	100.0	0.03
210	40.0	0.075
220	25.0	0.12
230	16.0	0.19
240	10.0	0.30
250	7.0	0.43
254	6.0	0.50
260	4.6	0.65
270	3.0	1.00
280	3.4	0.88
290	4.7	0.64
300	10.0	0.30
305	50.0	0.06
310	200.0	0.015
315	1000.0	0.003

[a]Adapted from a table developed and published by the American Conference of Government Industrial Hygienists in "Threshold Limit Values for Chemical Substances and Physical Agents in the Workroom Environment with Intended Changes for 1972."

away. This assumption is conservative in some instances, since ultraviolet radiation, especially at very low wavelengths, may be absorbed by some components of the atmosphere. Where information on atmospheric absorption of ultraviolet radiation is known, further correction may be applied. The calculation of intensity of radiation at any given point by use of the inverse square formula explained above does not take into consideration reflected energy.

The recommended standard is not proposed for application as a standard to lasers. It should be recognized that significant non-occupational exposure to ultraviolet radiation can occur from exposure to sunlight, particularly during the summer months.

Table 7. Maximum Permissible Exposure Times for Selected Values of I_{eff}.[a]

Duration of Exposure per Day	Effective Irradiance, I_{eff}, $\mu W/cm^2$
8 hr	0.1
4 hr	0.2
2 hr	0.4
1 hr	0.8
30 min	1.7
15 min	3.3
10 min	4.0
5 min	10.0
1 min	50.0
30 sec	100.0

[a]Adapted from a table developed and published by the American Conference of Government Industrial Hygienitst in "Threshold Limit Values for Chemical Substances and Physical Agents in the Workroom Environment with Intended Changes for 1972."

1-UNDECANETHIOL.
See Thiols, NIOSH Criteria Document.

UNSLAKED LIME. See Calcium oxide.

UNSYMMETRICAL DIMETHYL-HYDRA-ZINE (UDMH). [57-14-7]

NFPA—hazardous chemical
health—3, flammability—3, reactivity—1

URACIL, 5-[BIS(2-CHLOROMETHYL)AMINO]-.

*RCRA 40 CFR 261
EPA hazardous waste no. U237

URACIL MUSTARD. [66-75-1]
OSHA Candidate List of Potential Occupational Carcinogens (EPA Carcinogen Assessment Group List), 45 FR 53672, August 12, 1980

IARC—carcinogenic in animals

*RCRA 40 CFR 261
EPA hazardous waste no. U237

URACIL, 2-THIO-. Syn: 2-Mercapto-4-hydroxypyrimidine. [141-90-2]
OSHA Candidate List of Potential Occupational Carcinogens, 45 FR 53672, August 12, 1980

URANIUM (INSOLUBLE COMPOUNDS). [7440-61-1]
*OSHA 29 CFR 1910.1000 Table Z-1
PEL(TWA): 0.25 mg/m^3

URANIUM (SOLUBLE COMPOUNDS). [7440-61-1]
*OSHA 29 CFR 1910.1000 Table Z-1
PEL(TWA): 0.05 mg/m^3

URANIUM (NATURAL), SOLUBLE and INSOLUBLE COMPOUNDS, as U. [7440-61-1]
ACGIH—TLV
TWA: 0.2 mg/m^3
STEL: 0.6 mg/m^3

URANIUM NITRATE. *See* Uranyl nitrate. [10102-06-4, 36478-76-9]

URANYL ACETATE. [541-09-3]
*CWA 311(b)(2)(A), 40 CFR 116, 117
Discharge RQ = 5000 pounds (2270 kilograms)

URANYL NITRATE.
[10102-06-4, 36478-76-9]

NFPA—hazardous chemical
when nonfire: health—0, flammability—0, reactivity—0, oxy—oxidizing chemical
when fire: health—1, flammability—0, reactivity—0, oxy—oxidizing chemical

*CWA 311 (b)(2)(A), 40 CFR 116, 117
Discharge RQ = 5000 pounds (2270 kilograms)

UREA, 3-(p-CHLOROPHENYL)-1,1-DIMETHYL-. *See* Monuron. [150-68-5]

UREA, 1,3-DIETHYL-2-THIO-. Syn: *N,N'*Diethylthiourea. [105-55-5]
OSHA Candidate List of Potential Occupational Carcinogens, 45 FR 53672, August 12, 1980

NCI—carcinogenic in animals

UREA, 1,1,3,3-TETRAMETHYL-2-THIO-. Syn: TMTU. [2782-91-4]
OSHA Candidate List of Potential Occupational Carcinogens, 45 FR 53672, August 12, 1980

UREA, 1,1,3-TRIMETHYL-2-THIO-.
Syn: *N,N,N'*-Trimethylthiourea.
[2489-77-2]
OSHA Candidate List of Potential Occupational Carcinogens, 45 FR 53672, August 12, 1980

NCI—carcinogenic in animals

URETHANE. [51-79-6]
OSHA Candidate List of Potential Occupational Carcinogens (EPA Carcinogen Assessment Group List), 45 FR 53672, August 12, 1980

IARC—carcinogenic in animals
TSCA-CHIP available

V

VALERALDEHYDE. [110-62-3]

ACGIH-TLV
TWA: 50 ppm, 175 mg/m^3

VANADIC ACID, AMMONIUM SALT.
[7803-55-6]

*RCRA 40 CFR 261
EPA hazardous waste no. P119

VANADIUM. [7440-62-2]

NIOSH Criteria Document (Pub. No. 77-222), NTIS# PB81 225658, August 22, 1977
"Vanadium" includes vanadium compounds, metallic vanadium, and vanadium carbide. "Vanadium compounds" include all chemically combined forms of vanadium, but not alloys, intermetallics, or vanadium carbide. "Metallic vanadium" includes the element alone or in alloys or intermetallics, such as ferrovanadium and vanadium-aluminum. "Occupational exposure to vanadium" is defined as exposure to airborne vanadium above the action level. An "action level" is defined as equal to the environmental limit for that form of vanadium. Exposures to vanadium at lower concentrations will require adherence to Sections 2(a,e), 3(b), 5(a,c), and 8(a).

Occupational exposure to metallic vanadium and vanadium carbide shall be controlled so that employees are not exposed at a concentration greater than 1.0 mg of vanadium per cubic meter of air measured as a time-weighted average (TWA) concentration for up to a 10-hour workday, 40-hour workweek, over a working lifetime.

Occupational exposure to vanadium compounds shall be controlled so that employees are not exposed at a concentration greater than 0.05 mg of vanadium per cubic meter of air (0.05 mg V/m^3) measured as a ceiling concentration during any 15-minute sampling period.

The standard is designed to protect the health and provide for the safety of employees for up to a 10-hour work shift, 40-hour workweek, over a working lifetime.

VANADIUM(V) OXIDE. [1314-62-1]

*RCRA 40 CFR 261
EPA hazardous waste no. P120 (hazardous substance)

VANADIUM PENTOXIDE. [1314-62-1]

*OSHA 29 CFR 1910.1000 Table Z-1
PEL(TWA): dust 0.5 mg/m^3, fume 0.1 mg/m^3

*RCRA 40 CFR 261
EPA hazardous waste no. P120 (hazardous substance)

*CWA 311(b)(2)(A), 40 CFR 116, 117
Discharge RQ = 1000 pounds (454 kilograms)

VANADIUM, as V$_2$O$_5$, RESPIRABLE DUST AND FUME. [1314-62-1]

ACGIH-TLV (1982 Addition)
TWA: 0.05 mg/m^3

VANADIUM TETRACHLORIDE.
[7632-51-1]

NFPA—hazardous chemical
health—3, flammability—0, reactivity—2,
W—water may be hazardous in fire fighting

VANADYL SULFATE. [27774-13-6]

*CWA 311(b)(2)(A), 40 CFR 116, 117
Discharge RQ = 1000 pounds (454 kilograms)

VAT YELLOW 4. [128-66-5]

NCI—carcinogenic in animals

VEGETABLE OIL (HYDROGENATED).

NFPA—hazardous chemical (flammable chemical)
health—0, flammability—1, reactivity—0

VEGETABLE OIL MISTS.

ACGIH-TLV
TWA: nuisance particulate—30 mppcf or 10 mg/m^3 of total dust <1% quartz, or 5 mg/m^3 respirable dust

VERMICULITE. [1318-00-9]

TSCA-CHIP available

VERNDATE.

FIFRA—Data Call In, 45 FR 75488, November 14, 1980
Status: letter issued November 25, 1981

VINETHANE. See Divinyl ether. [109-93-3]

VINYL ACETATE.

[108-05-4]

ACGIH-TLV
TWA: 10 ppm, 30 mg/m^3
STEL: 20 ppm, 60 mg/m^3

NIOSH Criteria Document (Pub. No. 78-205), NTIS PB80 176993, September 7, 1978
Synonyms for vinyl acetate include: acetic acid, vinyl ester; ethenyl ester; vinyl A monomer; ethenyl ethanoate; and VyAc. "Occupational exposure to vinyl acetate" is defined as exposure to airborne vinyl acetate at concentrations above one-half the recommended ceiling limit. Exposure to airborne vinyl acetate at concentrations at or below one-half the recommended ceiling limit will require adherence to Sections 1(b), 2(a,c,d), 3, 4, 5, 6, 7, and 8(a,c).

Exposure to vinyl acetate in the workplace shall be controlled so that employees are not exposed at concentrations greater than 15 mg/m^3 of air, or 4 parts per million parts of air, measured as a ceiling concentration in samples collected during any 15-minute period.

The recommended standard is designed to protect the health and provide for the safety of employees for up to a 10-hour work shift, 40-hour workweek, over a working lifetime.

NFPA—hazardous chemical
health—2, flammability—3, reactivity—2

*CWA 311(b)(2)(A), 40 CFR 116, 117
Discharge RQ = 1000 pounds (454 kilograms)

VINYL BENZENE. See Styrene, monomer. [100-42-5]

VINYL BROMIDE. See Ethylene, bromo-. [593-60-2]

VINYL CHLORIDE. [75-01-4]

*OSHA Standard 29 CFR 1910.1017
Cancer-Suspect Agent
Permissible exposure limit: (1) No employee may be exposed to vinyl chloride at concentrations greater than 1 ppm averaged over any 8-hour period, (2) No employee may be exposed to vinyl chloride at concentrations greater than 5 ppm averaged over any period not exceeding 15 minutes. (3) No employee may be exposed to vinyl chloride by direct contact with liquid vinyl chloride.

"Action level" means a concentration of vinyl chloride of 0.5 ppm averaged over an 8-hour workday.

The standard applies to the manufacture, reaction, packaging, repackaging, storage, handling, or use of vinyl chloride or polyvinylchloride, but does not apply to the handling or use of fabricated products made of polyvinylchloride.

ACGIH-TLV
TWA: 5 ppm, 10 mg/m^3
Human carcinogen—recognized to have carcinogenic or cocarcinogenic potential

NIOSH Criteria Document (no NIOSH publication number is assigned), available only from NIOSH, March 11, 1974
NIOSH, with the assistance of expert consultants from both industry and organized labor, began development of a recommended occupational health standard. These and other activities were discussed in more detail at the OSHA Informal Fact-Finding Hearing on Possible Hazards of Vinyl Chloride

Manufacture and Use on February 15, 1974. It was also during this hearing that Professor Cesare Maltoni of Bologna, Italy, presented the preliminary results of his research which showed induction of angiosarcoma of the liver and other organs, as well as the production of other cancers in rats exposed to vinyl chloride.

Although vinyl chloride must be considered as a carcinogenic agent, the immediate problem appeared to be concentrated in polymerization facilities. Consequently, the NIOSH recommended standard only applies to such operations. This is not to say, however, that appropriate standards should not be developed for other exposures to the basic chemical. NIOSH is implementing further evaluation of the data, coupled with field observations, to determine exposure potentials in pre- and post-polymerization operations.

Based on theoretical considerations, there is probably no threshold for carcinogenesis although it is possible that with very low concentrations, the latency period might be extended beyond the life expectancy. In view of these considerations and NIOSH's inability to describe a safe exposure level as required in Section 20(a)(3) of the Occupational Safety and Health Act, the concept of a threshold limit for vinyl chloride gas in the atmosphere was rejected.

Consequently, the recommendations are such that where any employee is exposed to measurable concentrations of vinyl chloride, as determined by the recommended sampling and analytical method, he or she shall wear an air-supplied respirator. This recommendation is based on some preliminary information that the standard chemical cartridge respirators are inefficient in protecting against vinyl chloride. NIOSH is implementing a study to evaluate the degree of protection afforded by different types of respirators using vinyl chloride as the test gas. As information becomes available, it will be forwarded to OSHA as recommendations for alternative respirator usage. The employer is also required to develop a control plan to reduce airborne concentrations of vinyl chloride to levels not detectable by the recommended method.

Regulated areas shall be established where synthetic resins containing vinyl chloride are manufactured. These regulated areas shall include but are not limited to vinyl chloride loading or unloading operations, storage, and transfer facilities; synthetic resin polymerization processes and operations; and resin handling, compounding, packaging, and storage areas. Access shall be restricted to authorized employees only.

OSHA Candidate List of Potential Occupational Carcinogens (EPA Carcinogen Assessment Group List), 45 FR 53672, August 12, 1980

IARC—carcinogenic in animals

NFPA—hazardous chemical
health—2, flammability—4, reactivity—1

CWA 304(a)(1), 45 FR 79318, November 28, 1980, Water Quality Criteria

Freshwater Aquatic Life

No freshwater organisms have been tested with vinyl chloride and no statement can be made concerning acute or chronic toxicity.

Saltwater Aquatic Life

No saltwater organisms have been tested with vinyl chloride and no statement can be made concerning acute or chronic toxicity.

Human Health

For the maximum protection of human health from the potential carcinogenic effects due to exposure to vinyl chloride through ingestion of contaminated water and contaminated aquatic organisms, the ambient water concentration should be zero based on the nonthreshold assumption for this chemical. However, zero level may not be attainable at the present time. Therefore, the levels that may result in incremental increase of cancer risk over the lifetime are estimated at 10^{-5}, 10^{-6}, and 10^{-7}. The corresponding criteria are 20 μg/liter, 2.0 μg/liter and 0.2 μg/liter, respectively. If the above estimates are made for consumption of aquatic organisms only, excluding consumption of water, the levels are 5246 µg/liter, 525 µg/liter, and 52.5 µg/liter, respectively. Other concentrations representing different risk levels may be calculated by use of the guidelines. The risk estimate range is presented for information purposes and does not represent an agency judgment on an "acceptable" risk level.

CWA 307(a)(1), Priority Pollutant

*CAA 112, 40 CFR 61

*RCRA 40 CFR 261

EPA hazardous waste no. U043

VINYL CYANIDE. See Acrylonitrile. [107-13-1]

VINYL-1-CYCLOHEXENE. [100-40-3]

TSCA-CHIP available

VINYL CYCLOHEXENE DIOXIDE. [106-87-6]

ACGIH-TLV

TWA: 10 ppm, 60 mg/m^3

Industrial substance suspect of carcinogenic potential for humans—which is suspect of inducing cancer, based on either (1) limited epidemiologic evidence, exclusive of clinical reports of single cases, or (2) demonstration of carcinogenesis in one or more animal species by appropriate methods. Worker exposure by all routes should be carefully controlled to levels consistent with the animal and human experience data.

VINYL ETHER. See Divinyl ether. [109-93-3]

VINYL FLUORIDE. See Fluoroalkenes. [75-02-5]

TSCA 4(e), ITC

Seventh Report of the TSCA Interagency Testing Committee to the administrator, Environmental Protection Agency, October 1980; 45 FR 78432, November 25, 1980

*TSCA 8(d) Unpublished Health and Safety Studies Reporting, 40 CFR 716, 47 FR 38800, September 2, 1982; 48 FR 13178, March 30, 1983

*TSCA 8(d) Unpublished Health and Safety Studies Reporting, 40 CFR 716, 47 FR 387, September 2, 1982

TSCA-CHIP available

VINYL HALIDES.

NIOSH Current Intelligence Bulletin No. 28—cancer-related bulletin

VINYLIDENE BROMIDE. [593-92-0]

TSCA-CHIP available

VINYLIDENE CHLORIDE. See 1,1-Dichloroethylene. [75-35-4]

VINYLIDINE FLUORIDE (VDF). See Fluoroalkenes. [75-38-7]

TSCA 4(e), ITC

Seventh Report of the TSCA Interagency Testing Committee to the administrator, Environmental Protection Agency, October 1980; 45 FR 78432, November 25, 1980

*TSCA 8(d) Unpublished Health and Safety Studies Reporting, 40 CFR 716, 47 FR 38800, September 2, 1982; 48 FR 13178, March 30, 1983

TSCA-CHIP available

VINYL METHYL KETONE. See Methyl vinyl ketone. [78-94-4]

VINYLSTYRENE. See Divinylbenzene. [108-57-6]

VINYL TOLUENE. [25013-15-4]

*OSHA 29 CFR 1910.1000 Table Z-1

PEL(TWA): 100 ppm, 480 mg/m^3

ACGIH-TLV

TWA: 50 ppm, 240 mg/m^3

STEL: 100 ppm, 485 mg/m^3

NFPA—hazardous chemical

health—2, flammability—2, reactivity—1

VINYL TRICHLORIDE. See 1,1,2-Trichloroethane. [79-00-5]

VM & P NAPHTHA. [8030-30-6]

ACGIH-TLV

TWA: 300 ppm, 1350 mg/m^3

STEL: 400 ppm, 1800 mg/m^3

W

WARFARIN. [81-81-2]

*OSHA 29 CFR 1910.1000 Table Z-1
PEL(TWA): 0.1 mg/m^3
ACGIH—TLV
TWA: 0.1 mg/m^3
STEL: 0.3 mg/m^3
FIFRA—Registration Standard—Issued August 1981 (rodent toxicant and anticoagulant)
*RCRA 40 CFR 261
EPA hazardous waste no. P001

WASTE ANESTHETIC GASES AND VAPORS.

NIOSH Criteria Document (Pub. No. 77-140), NTIS Stock No. PB 274238, May 4, 1977

"Waste anesthetic gases" are defined as inhalation anesthetic gases and vapors that are released into work areas associated with, and adjacent to, the administration of a gas or volatile liquid used for anesthetic purposes. Currently, the six most commonly used anesthetic agents are nitrous oxide, halothane, enflurane, methoxyflurane, diethyl ether, and cyclopropane.

Occupational exposure to waste anesthetic gases includes exposure to any inhalation anesthetic agents that escape into locations associated with, and adjacent to, anesthetic procedures. Such locations shall include, but shall not be limited to, operating rooms, delivery rooms, labor rooms, recovery rooms, and dental operatories.

1. Occupational exposure to halogenated anesthetic agents shall be controlled so that no worker is exposed at concentrations greater than 2 ppm of any halogenated anesthetic agent, based on the weight of the agent collected from a 45-liter air sample by charcoal adsorption over a sampling period not to exceed 1 hour. Agents that shall be controlled, with their respective weights corresponding to 2 ppm, are as follows: chloroform, 9.76 mg/m^3; trichloroethylene, 10.75 mg/m^3; halothane, 16.15 mg/m^3; methoxyflurane, 13.5 mg/m^3; enflurane, 15.1 mg/m^3; fluroxene, 10.31 mg/m^3. When such agents are used in combination with nitrous oxide, levels of the halogenated agent well below 2 ppm are achievable. In most situations, control of nitrous oxide to a time-weighted average (TWA) concentration of 25 ppm during the anesthetic administration period will result in levels of approximately 0.5 ppm of the halogenated agent.

2. Occupational exposure to nitrous oxide, when used as the sole anesthetic agent, shall be controlled so that no worker is exposed at TWA concentrations greater than 25 ppm during anesthetic administration. Available data indicate that with current control technology, exposure levels of 50 ppm and less for nitrous oxide are attainable in dental offices.

The standard is designed to protect the health and safety of workers during their working lifetime in locations where exposures to waste anesthetic gases and vapors occur.

WELDING FUMES (not otherwise classified).

ACGIH—TLV
TWA: 5 mg/m^3
Total particulate, TLV 5 mg/m^3
Most welding, even with primitive ventila-

tion, does not produce exposures inside the welding helmet above 5 mg/m^3. That which does should be controlled.
STEL
Total particulate, TLV 5 mg/m^3
Most welding, even with primitive ventilation, does not produce exposures inside the welding helmet above 5 mg/m^3. That which does should be controlled.

WOOD DUST (certain hard woods as beech and oak).

ACGIH-TLV
TWA: 1 mg/m^3

WOOD DUST, soft wood.

ACGIH-TLV
TWA: 5 mg/m^3
STEL: 10 mg/m^3

X

XYLENE (XYLOL). [1330-20-7, 95-47-6, 106-42-3, 108-38-3]

*OSHA 29 CFR 1910.1000 Table Z-1
PEL(TWA): 100 ppm, 435 mg/m^3

NIOSH Criteria Document (Pub. No. 75-168), NTIS Stock No. PB 246702, May 23, 1975

Regardless of the source of raw materials from which produced, "xylene" (also known as "xylol" or "dimethylbenzene") refers to any one of or combination of the isomers of xylene: *ortho-*, *meta-*, or *para*-dimethylbenzene. "Exposure to xylene" is defined as exposure above half the recommended time-weighted average environmental limit.

Occupational exposure to xylene shall be controlled so that workers are not exposed to xylene at a concentration greater than 100 parts per million parts of air by volume (approximately 434 mg/m^3 of air) determined as a time-weighted average (TWA) exposure for up to a 10-hour workday, 40-hour workweek, with a ceiling concentration of 200 parts per million parts of air by volume (approximately 868 mg/m^3 of air) as determined by a sampling period of 10 minutes.

The standard is designed to protect the health and safety of workers over a working lifetime.

*RCRA 40 CFR 261
EPA hazardous waste no. U239 (hazardous substance, ignitable waste)

XYLENE (*o-*, *m-*, *p*-ISOMERS).
[1330-20-7, 95-47-6, 106-42-3, 108-38-3]

ACGIH-TLV

Skin

TWA: 100 ppm, 435 mg/m^3
STEL: 150 ppm, 655 mg/m^3

NFPA—hazardous chemical
health—2, flammability—3, reactivity—0

*CWA 311(b)(2)(A), 40 CFR 116, 117
Discharge RQ = 1000 pounds (454 kilograms)

*TSCA 8(d) Unpublished Health and Safety Studies Reporting, 40 CFR 716, 47 FR 387, September 2, 1982

CAA 112
Chemical proposed to be assessed as a Hazardous Air Pollutant by a House of Representatives bill to amend the CAA

***m*-XYLENE.** [108-38-3]

TSCA 4(e), ITC
Initial Report to the administrator, Environmental Protection Agency, TSCA Interagency Testing Committee, October 1, 1977; 42 FR 55026, October 12, 1977; responded to by the EPA administrator 43 FR 50134, October 26, 1978; 47 FR 56392, December 16, 1982

***o*-XYLENE.** [95-47-6]

TSCA 4(e), ITC
Initial Report to the administrator, Environmental Protection Agency, TSCA Interagency Testing Committee, October 1, 1977; 42 FR 55026, October 12, 1977; responded to by the EPA administrator 43 FR 50134, October 26, 1978; 47 FR 56392, December 16, 1982

***p*-XYLENE.** [106-42-3]

TSCA 4(e), ITC
Initial Report to the administrator, Environmental Protection Agency, TSCA Interagency Testing Committee, October 1, 1977; 42 FR 55026, October 12, 1977; responded to by the EPA administrator 43 FR 50134,

300 m-XYLENE α,α'-DIAMINE

October 26, 1978; 47 FR 56392, December 16, 1982

m-XYLENE α,α'-DIAMINE. [1477-55-0]
ACGIH-TLV
TWA: ceiling limit, 0.1 mg/m^3

XYLENOL. [1300-71-6]
*CWA 311(b)(2)(A), 40 CFR 116, 117
Discharge RQ = 1000 pounds (454 kilograms)

XYLIDINE. [1300-73-8]
*OSHA 29 CFR 1910.1000 Table Z-1
Skin
PEL(TWA): 5 ppm, 25 mg/m^3
Skin
TWA: 2 ppm, 10 mg/m^3
NFPA—hazardous chemical
health—3, flammability—1, reactivity—0
RCRA 40 CFR 261
EPA hazardous waste No. U093

2,6-XYLIDINE. [87-62-7]
NCI—carcinogenic in animals

XYLOL. *See* Xylene. [1330-20-7, 95-47-6, 106-42-3, 108-38-3]

Y

YELLOW OB. [131-79-3]
IARC—carcinogenic in animals

YOHIMBIN-16-CARBOXYLIC ACID, 11,17-DIMETHOXY-18-[(3,4,5-TRIMETHOXY-BENZOYL)OXY]-, METHYL ESTER. [50-55-5]
*RCRA 40 CFR 261
EPA hazardous waste no. U200

YTTRIUM. [7440-65-5]
*OSHA 29 CFR 1910.1000 Table Z-1
PEL(TWA): 1 mg/m^3
ACGIH-TLV
TWA: 1 mg/m^3
STEL: 3 mg/m^3

Z

ZEARALENONE. [17924-92-4]

NCI—carcinogenic in animals

ZINC. [7440-66-6]

CWA 304(a)(1), 45 FR 79318, November 28, 1980, Water Quality Criteria

Freshwater Aquatic Life

For total recoverable zinc the criterion to protect freshwater aquatic life as derived using the guidelines is 47 µg/liter as a 24-hour average and the concentration (in micrograms per liter) should not exceed the numerical value given by $e^{(0.83[\ln(\text{hardness})] + 1.95)}$ at any time. For example, at hardnesses of 50, 100, and 200 mg/liter $CaCO_3$ the concentration of total recoverable zinc should not exceed 180, 320, and 570 µg/liter at any time.

Saltwater Aquatic Life

For total recoverable zinc the criterion to protect saltwater aquatic life as derived using the guidelines is 58 µg/liter as a 24-hour average and the concentration should not exceed 170 µg/liter at any time.

Human Health

Sufficient data are not available for zinc to derive a level that would protect against the potential toxicity of this compound. Using available organoleptic data, for controlling undesirable taste and odor quality of ambient water, the estimated level is 5 mg/liter. It should be recognized that organoleptic data as a basis for establishing a water quality criterion have limitations and have no demonstrated relationship to potential adverse human health effects.

CWA 307(a)(1), Priority Pollutant

*National Secondary Drinking Water Regulations, 40 CFR 143; 44 FR 42198, July 19, 1979, effective January 19, 1981
Maximum contaminant level—5 mg/liter

ZINC (POWDER OR DUST).

NFPA—hazardous chemical
health—0, flammability—1, reactivity—1

ZINC ACETATE. [55-73-46]

*CWA 311(b)(2)(A), 40 CFR 116, 117
Discharge RQ = 1000 pounds (454 kilograms)

ZINC AMMONIUM CHLORIDE.
[14639-97-5, 14639-98-6, 52628-25-8]

*CWA 311(b)(2)(A), 40 CFR 116, 117
Discharge RQ = 5000 pounds (2270 kilograms)

ZINC BICHROMATE. See Dichromates.

ZINC BORATE. [1332-07-6]

*CWA 311(b)(2)(A), 40 CFR 116, 117
Discharge RQ = 1000 pounds (454 kilograms)

ZINC BROMIDE. [7699-45-8]

*CWA 311(b)(2)(A), 40 CFR 116, 117
Discharge RQ = 5000 pounds (2270 kilograms)

ZINC CARBONATE. [3486-35-9]

*CWA 311(b)(2)(A), 40 CFR 116, 117
Discharge RQ = 1000 pounds (454 kilograms)

ZINC CHLORATE. [10361-95-2]

NFPA—hazardous chemical
when nonfire: health—1, flammability—0, reactivity—2, oxy—oxidizing chemical
when fire: health—2, flammability—0, reactivity—2, oxy—oxidizing chemical

ZINC CHLORIDE. [7646-85-7]
*CWA 311(b)(2)(A), 40 CFR 116, 117
Discharge RQ = 5000 pounds (2270 kilograms)

ZINC CHLORIDE FUME. [7646-85-7]
*OSHA 29 CFR 1910.1000 Table Z-1
PEL(TWA): 1 mg/m^3
ACGIH-TLV
TWA: 1 mg/m^3
STEL: 2 mg/m^3

ZINC CHROMATE, as Cr. [13530-65-9]
ACGIH-TLV
TWA: 0.05 mg/m^3
Industrial substance suspect of carcinogenic potential for humans—which is suspect of inducing cancer based on either (1) limited epidemiologic evidence, exclusive of clinical reports of single cases, or (2) demonstration of carcinogenesis in one or more animal species by appropriate methods. Worker exposure by all routes should be carefully controlled to levels consistent with the animal and human experience data.

ZINC CHROMATE HYDROXIDE.
[15930-94-6]
IARC—carcinogenic in animals

ZINC CYANIDE. [557-21-1]
*CWA 311(b)(2)(A), 40 CFR 116, 117
Discharge RQ = 10 pounds (4.54 kilograms)
*RCRA 40 CFR 261
EPA hazardous waste no. P121 (hazardous substance)

ZINC DICHROMATE. See Dichromates.

ZINC DIETHYL. See Diethylzinc.
[557-20-0]

ZINC ETHYL. See Diethylzinc.

ZINC FLUORIDE. [7783-49-5]
*CWA 311(b)(2)(A), 40 CFR 116, 117
Discharge RQ = 1000 pounds (454 kilograms)

ZINC FORMATE. [557-41-5]
*CWA 311(b)(2)(A), 40 CFR 116, 117
Discharge RQ = 1000 pounds (454 kilograms)

ZINC HYDROSULFITE. [7779-86-4]
*CWA 311(b)(2)(A), 40 CFR 116, 117
Discharge RQ = 1000 pounds (454 kilograms)

ZINC NITRATE. [7779-88-6]
*CWA 311(b)(2)(A), 40 CFR 116, 117
Discharge RQ = 5000 pounds (2270 kilograms)

ZINC OXIDE. [1314-13-2]
NIOSH Criteria Document (Pub. No. 76-104), NTIS Stock No. PB 246693, October 10, 1975
"Exposure to zinc oxide" is defined as exposure above half the recommended time-weighted average environmental limit.

Occupational exposure to zinc oxide shall be controlled so that workers are not exposed to zinc oxide at an environmental concentration greater than 5 mg ZnO/m^3 determined as a time-weighted average (TWA) exposure for up to a 10-hour workday, 40-hour workweek, with a ceiling of 15 mg ZnO/m^3 as determined by a sampling time of 15 minutes. The standard is designed to protect the health and safety of workers for up to a 10-hour workday, 40-hour workweek, over a working lifetime.

ZINC OXIDE, DUST. [1314-13-2]
ACGIH-TLV
TWA: nuisance particulate—30 mppcf or 10 mg/m^3 of total dust <1% quartz, or 5 mg/m^3 respirable dust

ZINC OXIDE FUME. [1314-13-2]
*OSHA 29 CFR 1910.1000 Table Z-1
PEL(TWA): 5 mg/m^3
ACGIH-TLV
TWA: 5 mg/m^3
STEL: 10 mg/m^3

ZINC PHENOLSULFONATE. [127-82-2]
*CWA 311(b)(2)(A), 40 CFR 116, 117
Discharge RQ = 5000 pounds (2270 kilograms)

ZINC PHOSPHATE.
FIFRA—Registration Standard Development Data Evaluation and Development of Regulatory Position (rodent toxicant and anticoagulant)

ZINC PHOSPHIDE. [1314-84-7]
*CWA 311(b)(2)(A), 40 CFR 116, 117
Discharge RQ = 1000 pounds (454 kilograms)
*RCRA 40 CFR 261
EPA hazardous waste no. P122 (hazardous substance reactive waste, toxic waste)

ZINC SILICOFLUORIDE. [16871-71-9]
*CWA 311(b)(2)(A), 40 CFR 116, 117
Discharge RQ = 5000 pounds (2270 kilograms)

ZINC STEARATE. [557-05-1]
ACGIH-TLV
TWA: nuisance particulate—30 mppcf or 10 mg/m^3 of total dust < 1% quartz, or 5 mg/m^3 respirable dust
STEL: 20 mg/m^3

ZINC SULFATE. [7733-02-0]
*CWA 311(b)(2)(A), 40 CFR 116, 117
Discharge RQ = 1000 pounds (454 kilograms)

ZIRAM (ZINC DIEMTHYLDITHIOCARBAMATE). [137-30-4]
NCI—carcinogenic in animals

ZIRCONIUM (POWDER or SPONGE). [7740-67-7]
NFPA—hazardous chemical
health—1, flammability—4, reactivity—1

ZIRCONIUM CHLORIDE. See Zirconium tetrachloride. [10026-11-6]

ZIRCONIUM COMPOUNDS. [7440-67-2]
*OSHA 29 CFR 1910.1000 Table Z-1
PEL(TWA): 5 mg/m^3 (as Zr)
ACGIH-TLV
TWA: 5 mg/m^3
STEL: 10 mg/m^3

ZIRCONIUM NITRATE. [13746-89-9]
*CWA 311(b)(2)(A), 40 CFR 116, 117
Discharge RQ = 5000 pounds (2270 kilograms)

ZIRCONIUM POTASSIUM FLUORIDE. [16923-95-8]
*CWA 311(b)(2)(A), 40 CFR 116, 117
Discharge RQ = 5000 pounds (2270 kilograms)

ZIRCONIUM SULFATE. [14644-61-2]
*CWA 311(b)(2)(A), 40 CFR 116, 117
Discharge RQ = 5000 pounds (2270 kilograms)

ZIRCONIUM TETRACHLORIDE. [10026-11-6]
NFPA—hazardous chemical
health—3, flammability—0, reactivity—1
*CWA 311(b)(2)(A), 40 CFR 116, 117
Discharge RQ = 5000 pounds (2270 kilograms)
40 CFR 162.11, FIFRA-RPAR completed
Current Status: PD 1 published, 41 FR 52792, December 1, 1976: NTIS# PB80 216823.
PD 2 published, 45 FR 13189, February 28, 1980: NTIS# PB80 216831, returned to registration imposing certain restrictions on use. Criteria Possibly Met or Exceeded: Acute Toxicity to Mammalian and Avian Species, Reduction in Endangered and Nontarget Species, and Acute Toxicity without Antidote